PPINCOTT'S QUICK REFERENCES

T3-BPK-311

ick Reference to

Fluid
Balance

rma Milligan Metheny, R.N., M.S.N., Ph.D.
Associate Professor of Nursing
Graduate Medical–Surgical Nursing Department
Saint Louis University
St. Louis, Missouri

J. B. Lippincott Company Philadelphia
London   Mexico City   New York   St. Louis   São Paulo   Sydney

Sponsoring Editor: Paul R. Hill
Manuscript Editor: Lee Henderson
Indexer: Julie Schwager
Art Director: Tracy Baldwin/Earl Gerhart
Designer: Carol C. Bleistine
Cover Illustration: John Nearing Design
Production Supervisor: N. Carol Kerr
Production Assistant: S. M. Gassaway
Compositor: International Computaprint Corporation
Printer/Binder: R. R. Donnelley & Sons Company

6 5 4

Library of Congress Cataloging in Publication Data

Metheny, Norma Milligan.
  Quick reference to fluid balance.
  (Lippincott's quick references)
  Bibliography: p.
  Includes index.
1. Body fluid disorders — Handbooks, manuals, etc.
I. Title. [DNLM: 1. Body fluids — Nursing texts.
2. Water–electrolyte balance — Nursing texts. 3. Water–
electrolyte imbalance — Nursing texts. WY 150 M592q]
RC630.M47   1984   616   83-7910
ISBN 0-397-54448-0

The author and publisher have exerted every effort to ensure that drug selection and dosage set forth in this text are in accord with current recommendations and practice at the time of publication. However, in view of ongoing research, changes in government regulations, and the constant flow of information relating to drug therapy and drug reactions, the reader is urged to check the package insert for each drug for any change in indications and dosage and for added warnings and precautions. This is particularly important when the recommended agent is a new or infrequently employed drug.

preface

The purpose of *Quick Reference to Fluid Balance* is to provide nurses and nursing students with a practical and quick reference to body-fluid disturbances. Emphasis throughout the text is on nursing assessment of patients with fluid and electrolyte disturbances and on interventions to prevent or alleviate these disturbances. The format of the book is such that information is easy to find and comprehend.

The book has been organized into three major sections in order to facilitate understanding of the nurse's role in caring for patients with fluid and electrolyte problems.

The first section on basic concepts has chapters on physiologic regulatory mechanisms, nursing assessment, evaluation of laboratory data, and principles of parenteral therapy.

The second section furnishes an easy-to-follow overview of fluid and electrolyte disturbances. Discussions of normal functions of major electrolytes are followed by lists of common causes of their imbalances, along with nursing responsibilities in their management.

The third section deals with selected treatments and clinical situations commonly associated with fluid and electrolyte disturbances. For example, chapters are included on total parenteral nutrition, tube feedings, and blood transfusions. Other chapters deal with fluid and electrolyte problems in surgical patients and in those with head injuries, heart and renal failure, burns, bowel obstruction, loss of gastrointestinal fluids, cirrhosis, pancreatitis, heat disorders, toxemia of pregnancy, diabetic ketoacidosis, nonketotic hyperosmolar coma, and salicylate intoxication. Finally, a glossary of commonly used fluid balance terms is included for quick reference.

Norma Milligan Metheny, R.N., M.S.N., Ph.D.

acknowledgments

I am most grateful to Paul Hill, Sponsoring Editor, and Lee Henderson, Manuscript Editor, of the J. B. Lippincott Company for their valuable advice and continuing support throughout the preparation of the book.

I also wish to thank the following nurses for their highly valued suggestions during the preparation of the manuscript:

Patricia Eisenberg, R.N., M.S.N.
Medical–Surgical Clinical Specialist
The Jewish Hospital of St. Louis
St. Louis, Missouri

Darnell Roth, R.N.
Intravenous Therapy Consultant
Alexian Brothers Hospital
St. Louis, Missouri

Martha Spies, R.N., M.S.N.
Assistant Professor of Nursing
Saint Louis University
St. Louis, Missouri

Finally, I wish to thank Helen Wells and Kathy Imbs for their patience and skill in typing the manuscript.

contents

section one

Basic Concepts   1

1 **Regulation of Body Fluids**   3

Body Fluids   3
   Amount and Composition   3
   Location   4
   Normal Routes of Fluid Loss   5
   Electrolytes   7
   Homeostatic Mechanisms   8
   Changes in Water and Electrolyte Homeostasis in
   the Aged   13
   Differences in Water and Electrolyte Homeostasis
   in Infants, Children, and Adults   16
   References   18

2 **Nursing Assessment of Fluid Balance Status**   20
   Patient History   20
   Evaluation of Laboratory Data   28
   Clinical Assessment   28
   References   45

3 **Evaluation of Laboratory Data**   46
   References   56
   Bibliography   56

4 **Principles of Parenteral Fluid Therapy**   57
   Parenteral Fluids   57
   Total Parenteral Nutrition   72

Complications of Intravenous Fluid
Administration   72

References   77

section two

Overview of Fluid and Electrolyte Problems   79

5 Fluid Volume Imbalances   81

Fluid Volume Deficit   81

Fluid Volume Excess   84

Reference   85

6 Third-Space Shifting of Body Fluids   86

Clinical Situations Associated With Third-Space
Fluid Shifting   86

Phases of Third-Space Shifting of Body Fluids   87

References   89

7 Sodium Imbalances   90

Facts About Sodium   90

Sodium Deficit (Hyponatremia)   90

Sodium Excess (Hypernatremia)   100

References   102

8 Potassium Imbalances   103

Facts About Potassium   103

Potassium Deficit (Hypokalemia)   106

Potassium Excess (Hyperkalemia)   111

References   115

9 Calcium Imbalances   116

Facts About Calcium   116

Calcium Deficit (Hypocalcemia)   117

Calcium Excess (Hypercalcemia)   122

References   125

10 Magnesium Imbalances   127

Facts About Magnesium   127

Magnesium Excess (Hypermagnesemia)   130

References   132

11 **Phosphorus, Lithium, and Zinc** 133
Phosphorus Imbalances 133
Lithium 136
Zinc Imbalances 138

12 **Acid–Base Imbalances** 141
Regulation of Acid–Base Balance 141
Acid–Base Imbalances 145
Obtaining Samples for Arterial Blood
Gases 159
References 162

section three _____

Selected Clinical Situations and Treatments Associated With Fluid and Electrolyte Problems 163

13 Total Parenteral Nutrition 165
Composition of TPN Solutions 165
Preparation and Storage of TPN Solutions 166
Metabolic Complications of TPN 167
Infectious Complications of TPN 172
References 173

14 Tube Feedings 174
Complications of Tube Feedings 174
General Facts About Administration of Tube
Feedings 180
References 185

15 Postoperative Period 186
The Body's Response to Surgical Trauma 186
Prevention of Postoperative Complications by
Preoperative Assessment and Preparation 189
Postoperative Gains and Losses of Water,
Electrolytes, and Other Nutrients 193
Postoperative Problems in Water and Electrolyte
Balance 196
Shock 201
References 208

16 Blood Administration   209

Complications of Blood Transfusion   211

References   220

17 Loss of Gastrointestinal Fluids   221

Character of Gastrointestinal Secretions   221

Vomiting and Gastric Suction   222

Diarrhea, Intestinal Suction, and Ileostomy   226

Fistulas   228

Prolonged Use of Laxatives and Enemas   229

Imbalances Associated With Preparations for Diagnostic Studies of the Colon   230

18 Intestinal Obstruction   231

Simple Mechanical Obstruction   231

Fluid and Electrolyte Imbalances   232

Fluid Replacement Therapy   233

References   236

19 Cirrhosis of the Liver and Acute Pancreatitis   237

Cirrhosis of the Liver   237

Pathophysiology   237

Clinical Manifestations   238

Associated Water and Electrolyte Disturbances   238

Medical Therapy   242

Acute Pancreatitis   247

Causes and Pathologic Changes   247

Clinical Signs   247

Associated Fluid and Electrolyte Disturbances   248

Medical Management   249

References   252

20 Diabetic Ketoacidosis and Nonketotic Hyperosmolar Coma   253

Diabetic Ketoacidosis   253

Nonketotic Hyperosmolar Coma   271

References   276

21 Head Injuries   **278**

Overview of Fluid Balance Problems   **278**
Hyponatremia Due to SIADH   279
Hypernatremia Associated With Central Diabetes Insipidus   281
Other Causes of Hypernatremia   284
Nonketotic Hyperosmolar Coma   286
Cerebral Edema and Elevated Intracranial Pressure   286
References   293

22 Renal Failure   **295**
Acute Renal Failure   295
Chronic Renal Failure   307
References   313

23 Congestive Heart Failure   **315**
Left-Sided Heart Failure   315
Right-Sided Heart Failure   316
Adaptive Mechanisms in Decreased Left Ventricular Function   316
Nursing Assessment of Patients With Heart Failure   318
Treatment of Congestive Heart Failure: Nursing Implications   322
Pulmonary Edema   328
References   330

24 Burns   **331**
Evaluation of Burn Severity   331
Water and Electrolyte Changes in Burns   333
Physiologic Basis for Treatment and Nursing Care During the Fluid Accumulation Phase   335
Physiologic Basis for Treatment and Nursing Care During the Fluid Remobilization Phase   350
References   351

25 Heat Disorders   **353**
Heat Cramps   354
Heat Exhaustion   354
Heatstroke   355

Summary of Measures to Prevent Heat
Disorders   361
References   363

26 Toxemia of Pregnancy   365
Pathophysiology   366
Clinical Manifestations   367
Treatment and Nursing Interventions   368
References   371

27 Salicylate Intoxication   373
Pathophysiology   373
Clinical Manifestations   374
Treatment   375
References   376

Glossary   377

Index   381

section one

Basic Concepts

chapter 1

Regulation of Body Fluids

It is important to understand certain fundamental facts about body fluids, their location, electrolyte content, and how they are regulated. Comprehension of these facts facilitates understanding of nursing interventions in the care of patients with fluid and electrolyte problems.

Body Fluids

Amount and Composition

- The major portion of the human body is composed of fluid (water and electrolytes). Total body fluid content varies with fat content of the body, sex, and age.
- The amount of fluid in adults can range from approximately 45% to 60% of body weight, depending on the amount of body fat. The reason for this variance becomes clear when one considers that fat cells contain little water and that lean tissue is rich in water content. Thus, an obese person has substantially less body fluid than does a lean, muscular person of the same weight.
- Women have a lower percentage of body fluid than men. For example, the average fluid content of young adult men is 60% of body weight, but only 52% in young adult women. The difference is largely related to differences in body fat content (women having proportionately more fat and less muscle mass than men).
- The premature infant's body is approximately 90% fluid by weight, as compared to 70% to 80% in the full-term newborn.

- In addition to having proportionately more body fluid than the adult, the infant has proportionately more fluid in the extracellular compartment. (For example, more than half of the newborn's body fluid is in the extracellular compartment, as compared to a third or less in the adult.)
- As the infant becomes older, his total body fluid percentage decreases; the decrease is particularly rapid during the first 6 months.
- After the first year the total body fluid is about 64% (34% in the cellular compartment and 30% in the extracellular compartment).
- By the end of the second year, the total body fluid approaches the adult percentage of approximately 60% (36% in the cellular compartment and 24% in the extracellular compartment).
- At puberty the adult body fluid composition is attained (40% in the cellular compartment and 20% in the extracellular compartment). For the first time there is a sex differentiation.
- After 40 years of age, mean values for total body fluid in percentage of body weight decrease for both men and women; however, the sex differentiation remains.
- After 60 years of age the percentage may decrease to approximately 52% in males and 46% in females (even less in obese persons).
- The reduction in body fluid is explained by the fact that there is a decrease in lean body mass and a relative increase in body fat with aging. (Recall that body fluid decreases as fat content increases.)
- Variations in total body fluid with age are listed in Table 1-1.

Location

Body fluid is divided into two major compartments:
- Intracellular (fluid in the cells)
- Extracellular (fluid outside the cells), further subdivided into

Table 1-1
Approximate Values of Total Body Fluid as a Percentage of Body Weight in Relation to Age and Sex

Age	Total Body Fluid (% body weight)
Full-term newborn	70%–80%
1 year	64%
Puberty to 39 years	Men: 60%
	Women: 52%
40 to 60 years	Men: 55%
	Women: 47%
More than 60 years	Men: 52%
	Women: 46%

1. Intravascular fluid (plasma)
2. Interstitial fluid (fluid lying between the cells)

- Approximately 30% to 40% of the body weight in adults is made up of intracellular fluid, the largest portion of which is in the skeletal muscle mass.
- Extracellular fluid represents 20% of the body weight in adults and is divided between the interstitial fluid (15% of body weight) and the plasma (5% of body weight). The plasma volume averages 3 liters in the average adult.
- Special fluids that are also considered part of the extracellular fluid include cerebrospinal fluid, ocular fluid, gastrointestinal secretions, fluid in lymph, and fluid in the potential spaces of the body.
- The amount of fluid in a typical 70-kg (154-lb) young adult male is depicted in Fig. 1-1.
- As children grow to puberty there is a continuous decrease in the proportion of total body fluid (particularly in the extracellular compartment).

Note: The extracellular fluid of the newborn may constitute 42% of his body weight, as compared to only 20% in the adult. Because extracellular fluid is more readily lost from the body than is cellular fluid, *the infant is more vulnerable to fluid volume deficit than the adult*. This explains why a profound fluid volume deficit can be incurred in a relatively short time by an infant with diarrhea.

Normal Routes of Fluid Loss

Organs of water loss include the kidneys, skin, lungs, and gastrointestinal tract.

Kidneys

The usual urine volume in the adult is between 1 and 2 liters each day.

Skin

Visible water and electrolyte loss through the skin occurs by sweating (*sensible perspiration*). *Sweat* is a hypotonic fluid containing several solutes; the chief ones are sodium chloride and potassium. Actual sweat losses vary according to environmental temperature (can vary from 0 ml to 1000 ml or more every hour). Significant sweat losses occur if the patient's body temperature exceeds 101° F (38.3° C) or if room temperature exceeds 90° F (32.2° C). Continuous water loss by evaporation (approximately 600 ml/day) occurs through the skin as *insensible perspiration*; it is a nonvisible form of water loss. The presence of fever greatly increases in-

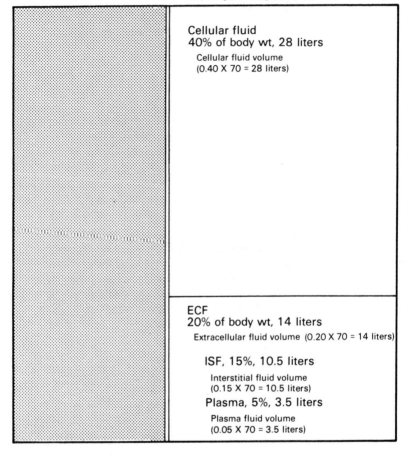

| SOLIDS | FLUID |
| 40% by weight | 60% by weight (42 liters) |

Total body fluid volume in liters equals 60% of the
body weight in kilograms (60% X 70 = 42 liters)

Cellular fluid
40% of body wt, 28 liters
Cellular fluid volume
(0.40 X 70 = 28 liters)

ECF
20% of body wt, 14 liters
Extracellular fluid volume (0.20 X 70 = 14 liters)

ISF, 15%, 10.5 liters
Interstitial fluid volume
(0.15 X 70 = 10.5 liters)
Plasma, 5%, 3.5 liters
Plasma fluid volume
(0.05 X 70 = 3.5 liters)

Figure 1-1.   Fluid compartments and the amount of fluid in a 70-kg (154-lb) young adult male.

sensible water loss through both the lungs and the skin. Loss of the natural skin barrier in major burns also increases water loss (as much as 3 to 5 liters each day) by this route.

Lungs

The lungs normally eliminate water vapor (insensible loss) at the rate of approximately 300 ml to 400 ml every day. The loss is much greater with increased respiratory rate or depth, or both.

Gastrointestinal Tract

The usual loss through the gastrointestinal tract is only about 100 ml to 200 ml every day even though approximately 8 liters of fluid circulate through the gastrointestinal system every 24 hours. (The bulk of the fluid is reabsorbed in the small intestine.) Large losses can occur because of diarrhea or intestinal fistulas.

Electrolytes

- The electrolyte content of body fluids varies within the major compartments. Table 1-2 lists the electrolytes in plasma; Table 1-3 lists the electrolytes in the intracellular fluid.
- Because special techniques are required to measure the concentration of electrolytes in the intracellular fluid, it is customary to measure the electrolytes in the extracellular fluid, namely plasma. Plasma electrolyte concentrations are used in assessing and managing patients with electrolyte imbalances.
- The major electrolytes in the extracellular fluid are sodium (Na^+) (cation) and chloride (Cl^-) (anion). Note in plasma (see Table 1-2) the great preponderance of sodium ions (142 mEq/liter) compared to other cations (positively charged ions). About 95% of the extracellular fluid osmolality (concentration) is determined by the sodium concentration. Sodium is of primary importance in regulating body fluid volume; it has long been noted that retention of sodium is as-

Table 1–2
Plasma Electrolytes

Electrolytes	mEq/liter
Cations	
Sodium (Na^+)	142
Potassium (K^+)	5
Calcium (Ca^{2+})	5
Magnesium (Mg^{2+})	2
Total cations	154
Anions	
Chloride (Cl^-)	103
Bicarbonate (HCO_3^-)	26
Phosphate (HPO_4^{2-})	2
Sulfate (SO_4^{2-})	1
Organic acids	5
Proteinate	17
Total anions	154

Table 1–3
**Approximation of Major Electrolyte
Content in Cellular Fluid**

Electrolytes	mEq/liter
Cations	
Potassium (K$^+$)	150
Magnesium (Mg^{2+})	40
Sodium (Na$^+$)	10
Total cations	200
Anions	
Phosphates / Sulfates	150
Bicarbonate (HCO$_3^-$)	10
Proteinate	40
Total anions	200

sociated with fluid retention. Whenever excessive quantities of sodium are lost, body fluid volume tends to decrease.

- The electrolyte content of interstitial fluid is not measured in clinical situations; it is essentially the same as that of plasma except that it contains much less proteinate. (Recall that plasma protein is necessary to maintain oncotic pressure and keep the intravascular fluid inside the blood vessels.)

- The major electrolytes in the intracellular fluid are potassium (K$^+$) (cation) and phosphate (HPO$_4^{2-}$) (anion). Since the extracellular fluid can tolerate only small potassium levels (approximately 5 mEq/liter), release of large intracellular potassium stores by cellular trauma can be extremely dangerous.

- The body expends a great deal of energy maintaining the extracellular preponderance of sodium and the intracellular preponderance of potassium; it does so by means of cell membrane pumps, which exchange sodium and potassium ions.

Homeostatic Mechanisms

- The body has a remarkable system that maintains the normal state of fluid and electrolyte balance (Fig. 1-2). The actions of major organs involved in homeostasis are as follows.

Kidneys

- The kidneys are vital to the regulation of fluid and electrolyte balance. They normally filter 170 liters of plasma every day in the adult, while excreting only 1.5 liters of urine.

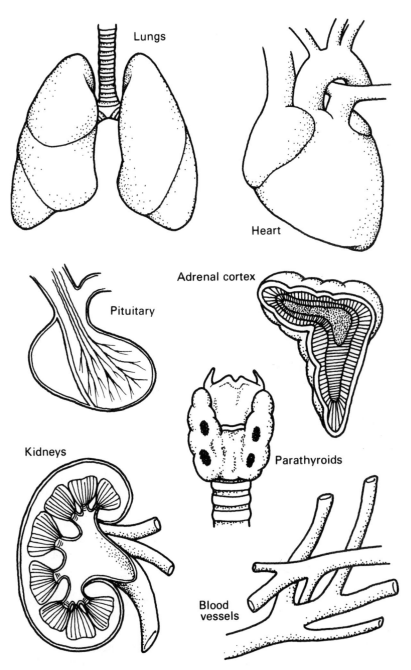

Figure 1–2. Homeostatic mechanisms.

- They act both autonomously and in response to blood-borne messengers, such as aldosterone and antidiuretic hormone (ADH).
- Major functions of the kidneys in fluid balance homeostasis include
  1. Regulation of extracellular fluid volume and osmolality by selective retention and excretion of body fluids.
  2. Regulation of electrolyte levels in the extracellular fluid by selective retention of needed substances and excretion of unneeded substances.
  3. Regulation of *p*H of extracellular fluid by excretion or retention of hydrogen ions.
  4. Excretion of metabolic wastes (primarily acids) and toxic substances.
- Renal failure results in multiple fluid and electrolyte problems (see Chap. 22).

Heart and Blood Vessels

- Plasma must reach the kidneys in sufficient volume to permit regulation of water and electrolytes. The pumping action of the heart provides circulation of blood through the kidneys under sufficient pressure for urine to form; renal perfusion makes renal function possible.
- There are stretch receptors in the atria and blood vessels that react to hypovolemia by stimulating fluid retention.

Lungs

- The lungs are also vital in maintaining homeostasis. Alveolar ventilation is responsible for the daily elimination of about 13,000 mEq of hydrogen ions (H^+), as opposed to only 40 mEq to 80 mEq excreted daily by the kidneys. Under the control of the medulla, the lungs act promptly to correct metabolic acid-base disturbances.
- The lungs regulate H^+ concentration (*p*H) by controlling the level of carbon dioxide (CO_2) in the extracellular fluid as follows:
  1. Metabolic alkalosis causes compensatory hypoventilation, resulting in CO_2 retention (and thus increased acidity of the extracellular fluid).
  2. Metabolic acidosis causes compensatory hyperventilation, resulting in CO_2 excretion (and thus decreased acidity of the extracellular fluid).
- Pulmonary dysfunction can produce a rapid change (matter of seconds) in the plasma H^+ concentration (acid–base balance). Hypoventilation causes respiratory acidosis; hyperventilation causes respiratory alkalosis. When the lungs are at fault, the kidneys must

compensate for the *p*H disturbances. Acid–base regulation is discussed in depth in Chapter 12.

- The lungs also remove approximately 300 ml of water daily through exhalation (insensible water loss) in the normal adult. Abnormal conditions such as hyperpnea or continuous coughing increase this loss; mechanical ventilation with excessive moisture decreases it.

Pituitary Gland

The hypothalamus manufactures a substance known as *antidiuretic hormone* (ADH), which is stored in the posterior pituitary gland and released as needed. Because ADH makes the body retain water, it is sometimes referred to as the "water-conserving" hormone.

Functions of ADH include

- Maintenance of osmotic pressure of the cells by controlling renal water retention or excretion.
    1. When osmotic pressure of the extracellular fluid (ECF) is greater than that of the cells (as in hypernatremia [excess sodium] or hyperglycemia), ADH secretion is increased, causing renal retention of water.
    2. When osmotic pressure of the ECF is less than that of the cells (as in hyponatremia [sodium deficit]), ADH secretion is decreased, causing renal excretion of water.
- Control of blood volume (less influential than aldosterone).
    1. An increased secretion of ADH occurs when blood volume is decreased, resulting in water conservation.
    2. A decreased secretion of ADH occurs when blood volume is increased, resulting in water loss.

Excessive secretion of ADH is discussed in Chapter 7; diabetes insipidus (decreased production of ADH) is discussed in Chapter 21.

Adrenal Glands

- The adrenal glands are made up of two major sections, the adrenal cortex and the adrenal medulla. It is the adrenal cortex that is most important in fluid and electrolyte homeostasis.
- The primary adrenocortical hormone in the influence of fluid balance is *aldosterone*, a mineralocorticoid secreted by the zona glomerulosa (outer zone) of the adrenal cortex.
- Aldosterone is instrumental in sodium and potassium balance in the following way:
    1. An increased secretion of aldosterone causes sodium retention (and thus water retention) and potassium loss.
    2. Conversely, a decreased secretion of aldosterone causes sodium and water loss and potassium retention.

- A drop in the extracellular sodium level triggers aldosterone secretion. (For example, a drop in sodium from 145 mEq/liter down to 135 mEq/liter may double aldosterone secretion.) A *rise* in the extracellular sodium level causes decreased aldosterone secretion.
- A rise in the extracellular potassium level triggers aldosterone secretion. (For example, an elevation from the normal of 5 mEq/liter to 8 mEq/liter can cause as much as a tenfold increase in aldosterone secretion.) Conversely, hypokalemia is associated with reduced aldosterone secretion.
- Aldosterone helps regulate blood volume by regulating sodium retention as follows:
  1. An increased secretion of aldosterone occurs when the blood volume is decreased (as in sodium depletion or hemorrhage).
  2. Conversely, a decreased secretion of aldosterone occurs when the blood volume is increased.
- Cortisol, another adrenocortical hormone, has only a fraction of the mineralocorticoid potency of aldosterone. However, when secreted in large quantities, it can produce sodium and fluid retention, and potassium deficit.

Parathyroid Glands

Most persons have four parathyroid glands; some have less and some have as many as seven. These pea-sized glands, embedded in the corners of the thyroid gland, regulate calcium (Ca^{2+}) and phosphate (HPO_4^{2-}) balance by means of *parathyroid hormone* (PTH). PTH influences bone resorption, calcium absorption from the intestines, and calcium reabsorption from the renal tubules.

The following facts are important to consider:

- Increased secretion of PTH causes
  1. Elevated serum calcium concentration.
  2. Lowered serum phosphate concentration.
- Conversely, decreased secretion of PTH causes
  1. Lowered serum calcium concentration.
  2. Elevated serum phosphate concentration.
- Hypocalcemia stimulates (within minutes) the release of PTH, which, in turn, elevates the serum calcium level (approximately a 1% decrease in calcium can give as much as a 100% increase in PTH release).
- Hypercalcemia depresses the release of ADH.
- A high serum phosphate concentration causes a secondary depression of serum calcium; as a result, PTH release is stimulated (a condition often seen in renal failure).

- It should be remembered that there is often a reciprocal relationship between serum calcium and phosphate levels; an elevation of one usually causes depression of the other (however, this is not always true).
- Another hormone that bears consideration in the regulation of calcium is *calcitonin*. It is secreted by the thyroid gland; for this reason it is also referred to as *thyrocalcitonin*. The action of calcitonin on calcium is opposite that of PTH; in other words, it reduces serum calcium concentration.

  An increase in serum calcium concentration causes an increased secretion of calcitonin. (For example, a 10% increase in serum calcium can cause an immediate twofold increase in the rate of calcitonin secretion.) The pharmacologic use of calcitonin in the treatment of hypercalcemia is discussed in Chapter 9.

Changes in Water and Electrolyte Homeostasis in the Aged

Homeostasis may be defined as the body's ability to restore equilibrium under stress. The aged are particularly vulnerable to fluid and electrolyte imbalances when under stress because their major homeostatic mechanisms have undergone numerous physiologic changes.

Renal Changes

- A 50% reduction in kidney function may occur between early maturity and old age. With advanced age, there is a reduction in blood flow to the kidneys, a decrease in the number of nephrons, a decline in the glomerular filtration rate (GFR), and diminished renal concentrating ability. Renal blood flow is reduced due to decreased cardiac output and increased peripheral resistance. The normal GFR of 120 ml/minute at age 40 decreases to 60 to 70 ml/minute at age 85.[1]
- One of the most common changes in renal function in old age is the inability to concentrate urine when fluid intake is reduced. A young person may be able to concentrate urine to a specific gravity (SG) of 1.035 to 1.040; persons in the seventh and eighth decades of life are often not able to concentrate urine beyond an SG of 1.022 to 1.026.[2]
- A decrease in renal concentrating ability causes a mild increase in urinary volume and can result in nocturia. Due to loss of maximum concentrating ability, it should be noted that the definition of oliguria in the elderly should encompass higher urine volumes than for young adults. For example, a urine output of 900 ml/day may be inappropriately low for many elderly persons and reflect oliguric renal

failure.[3] Because the aged patient has diminished renal concentrating ability, he needs more fluid in order to excrete a given amount of solute than a younger adult.

- It is interesting to note that the mean blood urea nitrogen (BUN) concentration at age 40 is 12.9 mg/100 ml; after age 70 it is 21.2 mg/100 ml.[4] (This reflects the reduction in GFR in the aged kidney.)
- The elderly are more vulnerable to drug toxicity due to decreased renal function.

    Potentially toxic antibiotics that are excreted virtually unchanged by the kidney (such as streptomycin, kanamycin, tobramycin, and neomycin) should be used cautiously in aged persons. Also, antacids containing calcium or magnesium should be used with extreme caution in aged patients with renal insufficiency.

Respiratory Changes

- Problems related to the respiratory mechanism in aged persons include increased rigidity of chest walls, weakening of respiratory muscles, decreased elasticity of lung tissue, defective alveolar ventilation, inadequate diffusion of respiratory gases, and interference with bronchial removal of secretions.
- The vital capacity is reduced with age, and an increase of residual air volume occurs.
- There is a decrease in the normal partial pressure of oxygen (PO_2) in the elderly; a general guideline advocated by one authority is to subtract 1 mm Hg from the minimum value of 80 mm Hg for every year over the age of 60 (up to age 90).[5]
- Recall that regulation of pH is largely controlled by the respiratory system by either elimination or retention of carbon dioxide (controlling carbonic acid content).

    In the aged, diminished respiratory function leads to less reserve in the face of major illness, surgery, burns, or trauma and thus interferes with acid–base regulation. Plasma pH tends to remain fixed on the low side of normal, largely because of decreased pulmonary and renal function.

Cardiovascular Changes

- With advanced age, blood flow through the coronary arteries is decreased and there is diminished ability of the myocardium to use oxygen. Cardiac output and stroke volume are decreased. In general, the aged person's heart is less well equipped to deal with stressful situations.
- The myocardium and circulatory system of the aged are less able to manage the hypotension associated with shock; shock becomes irreversible sooner in the aged and thus requires aggressive treatment.

- A moderate increase in blood pressure occurs in men and women up to the age of 65; systolic pressure and pulse pressure are particularly increased. Systolic hypertension develops as the aorta becomes less elastic and there is increased resistance to blood flow. Closely related is an increase in cholesterol and calcium deposits in the arteries.
- From pumping against increased resistance and from advanced age, heart contractions weaken and cause diminished cardiac output. Build up of pressure on the venous side of the circulatory system can raise capillary pressure and force fluid out into the tissues, producing edema.
- Orthostatic hypotension often develops in the aged due to increased lability of the vasopressor control mechanism; changing positions slowly helps control this problem.

Skin Changes

- The dermis loses elasticity in the aged and becomes relatively dehydrated; thus skin turgor is not a valid sign of fluid balance status in the aged.
- Atrophying sweat glands and a diminished capillary bed in the skin reduce the ability to sweat and control body temperature; because of this, heatstroke is a formidable problem in the elderly.
- Aged persons (with their atrophic dermal papillae, sweat glands, and hair follicles) will develop deep burns from the same amount of heat intensity that produces only moderate second-degree burns in middle-aged adults.

Gastrointestinal Changes

- Both volume and acidity of gastric juice gradually decrease with age due to atrophy of the gastric mucosa.
- Loss of muscle tone of the intestinal tract and a lessened sense of the need to defecate can lead to chronic constipation, a problem that often becomes uppermost in the minds of many aged persons.

Neurologic Changes

- There is a progressive loss of central nervous system cells with advancing age; the loss of neurons affects both the brain and the spinal cord.
- A decrease in the sense of smell and in the number of taste buds perhaps contributes to decreased appetite in the elderly.
- The aged patient also experiences a decreased tactile sense and increased pain tolerance.

- The thirst mechanism in the aged may be diminished and serve as a poor guide for fluid needs in ill persons. The aged patient may be too weak to verbalize his thirst or to reach for a glass of water, or he may have a clouded sensorium interfering with an appropriate response to the thirst mechanism. (Recall that the sensation of thirst is dependent on excitation of the cortical centers of consciousness.)
- The use of antianxiety agents, sedatives, or hypnotics in the elderly can lead to confusion and disorientation, causing the patient to forget to drink fluid. Because of this, hypernatremia is a common imbalance in the aged. One should also remember that elderly persons often deliberately restrict fluid intake in order to avoid embarrassing urinary incontinence.
- The autonomic nervous system is sluggish in the aged person, causing slower response to shock and to high environmental temperatures.

Endocrine Changes

- As age advances, there is a corresponding decrease in glucose tolerance until the age of 75 to 80. Thus, return to the fasting level in the glucose tolerance test is slower in the aged than in younger adults. Because of this, standards for the glucose tolerance test are adjusted for age; if "normal" standards were applied to the elderly, half of them would be classified as diabetics.
- When diabetes does occur in the aged, the most frequently encountered metabolic problem is nonketotic hyperosmolar coma (NKHC) (see Chap. 20).
- Hormonal deficiencies occurring with aging are decreased estrogen at menopause and decreased testosterone during the male climacteric. Osteoporosis in postmenopausal women is discussed in Chapter 9.

Differences in Water and Electrolyte Homeostasis in Infants, Children, and Adults

- An obvious and important difference between small children and adults is size. Yet, children are not merely miniature adults, for the child's body composition and homeostatic controls differ from those of the adult.
- It is helpful to compare the child's body composition with that of the adult and to review the salient characteristics of the child's homeostatic and metabolic functioning.

Daily Body Water Turnover in Infants and Adults

- As stated earlier, infants have proportionately more water in their extracellular compartment than adults. Because of this the infant is more vulnerable to fluid volume deficit than the adult.
- An infant may exchange half of his extracellular fluid daily, while the adult may exchange only one sixth of his in the same period. Proportionately, therefore, the infant has less reserve of body fluid than the adult.
- The daily fluid exchange is relatively greater in infants, in part because their metabolic rate is two times higher per unit of body weight than that of adults. Owing to the high metabolic rate, the infant has a large amount of metabolic wastes to excrete. Because water is needed by the kidneys in order to excrete these wastes, a large urine volume is formed each day.
- Contributing to this relatively large urine volume is the inability of the infant's immature kidney to concentrate urine efficiently.
- In addition, relatively greater fluid loss occurs through the infant's skin because of his proportionately greater body surface area. (The premature infant has approximately five times as much body surface area in relation to weight, and the newborn three times, as do the older child and adult.)
- Therefore, any condition causing a pronounced decrease in intake or an increase in the output of water and electrolytes threatens the body fluid economy of the infant.
- According to Gamble, an infant can live only 3 to 4 days without water, though an adult may live 10 days.

Renal Function

- The newborn's renal function is not yet completely developed. Thus, if infant and adult renal functions are compared on the basis of total body water, the infant's kidneys appear to become mature by the end of the first year of life. However, if body surface area is used as the criterion for comparison, the child's kidneys appear immature for the first 2 years of life.
- Since the infant's kidneys have a limited concentrating ability and require more water in order to excrete a given amount of solute, he has difficulty in conserving body water when it is needed. Also, he may be unable to excrete an excess fluid volume.

Body Surface Area

- The infant's relatively greater body surface area is present until the child is 2 to 3 years of age.

- The skin represents an important route of fluid loss, especially in illness. Since the gastrointestinal membranes are essentially an extension of the body surface area, their area is also relatively greater in the young infant than in the older child and adult. Hence, relatively greater losses occur from the gastrointestinal tract in the sick infant than in the older child and adult.
- In comparing fluid losses in infants to those in adults, one might regard the infant as a "smaller vessel with a larger spout."

Electrolyte Concentrations and Metabolic Acid Formation

- Serum electrolyte concentrations do not vary strikingly between infants, small children, and adults.
- The serum sodium concentration changes little from birth to adulthood.
- The serum potassium concentration is higher in the first few months of life than at any other time, as is the serum chloride concentration.
- Magnesium and calcium are both low in the first 24 hours after birth.
- The serum phosphorus level is higher in infants and children than in adults.
- Inability of the premature infant to regulate its calcium ion concentration can bring on hypocalcemic tetany.
- Newborn and premature infants have less homeostatic buffering capacity than do older children. Newborns and premature infants have a tendency toward metabolic acidosis with pH averages slightly lower (7.30 to 7.38) than normal.[6] The mild metabolic acidosis is thought to be related to high metabolic acid production and to renal immaturity.
- The premature infant is even more acidotic than the newborn. Because cow's milk has a higher phosphate and sulfate content than breast milk, newborns fed cow's milk have a lower pH than do breast-fed babies.

References

1. Papper S: The effects of age in reducing renal function. Geriatrics, May 1973, p 84
2. Cole W: Medical differences between the young and the aged. J Am Geriatr Soc, Aug 1970, p 592
3. Mitchell J: Chronic renal failure in the elderly: What to look for when it starts. Geriatrics, Nov 1980, p 28

4. Papper, Geriatrics, p 87
5. Shapiro B, Harrison R, Walton J: Clinical Application of Blood Gases, 3rd ed, p 126. Chicago, Year Book Medical Publishers, 1982
6. Maxwell M, Kleeman C (eds): Clinical Disorders of Fluid and Electrolyte Metabolism, 3rd ed, p 1570. New York, McGraw-Hill, 1980

chapter 2

Nursing Assessment of Fluid Balance Status

- Nurses caring for patients with actual or potential fluid and electrolyte imbalances must assume a primary monitoring role. Patient evaluation requires a review of the *patient's history* and *laboratory data*, as well as careful *clinical assessment*.
- One must have a knowledge of the conditions that may give rise to water and electrolyte abnormalities. The suspicion that these abnormalities exist is strengthened by the presence of clinical features known to occur with them and often by the results of laboratory tests and electrocardiographic examination.
- Changes in the patient developing a fluid and electrolyte imbalance are often subtle at first, requiring the nurse to be suspicious of expected imbalances and to be alert for their occurrence.
- The first area of consideration is the patient's history.

Patient History

Initial questions to consider include the following:
- Is there a disease or injury state present that can disrupt fluid and electrolyte balance?
- Is the patient receiving any medication, parenteral fluid, or other treatment that can disrupt fluid and electrolyte balance?
- Is there any abnormal loss of body fluids that can disrupt fluid and electrolyte balance?
- How does the total intake compare with the total output?

Some Disease or Injury States Commonly Associated With Imbalances

The nurse should be aware that certain disease or injury states put the patient at risk for specific fluid and electrolyte imbalances. Some of these states are listed in Table 2-1 with commonly associated imbalances.

Treatment-Induced Causes of Fluid and Electrolyte Problems

- Unfortunately, there are a number of imbalances *created* by medical therapy. The nurse should be aware of expected disturbances and should be on the alert for their appearance. Many times the imbalance can be *prevented* by early nursing and medical interventions.
- Some medical therapies that can produce electrolyte disturbances are listed in Table 2-2.

Imbalances Associated With Loss of Body Fluids

Types of imbalances that accompany the loss of a specific body fluid vary with the content of the lost fluid.

Gastric Juice

The usual daily volume of gastric juice is approximately 2500 ml, and the pH is usually 1 to 3 (very acidic). Gastric juice contains H^+, Na^+, K^+, and Mg^{2+}.

Imbalances that may result from severe vomiting or prolonged suction include

- Metabolic alkalosis (due to loss of H^+ and Cl^-).
- Potassium deficit (due to direct loss and, more importantly, to metabolic alkalosis).
- Deficit of ionized calcium (recall that calcium ionization is diminished when the pH is too alkaline).
- Magnesium deficit (although rare, this imbalance can occur with the loss of gastric juice, particularly when prolonged nasogastric suction is used and magnesium-free parenteral fluids are administered).
- Sodium deficit (due to actual loss).

Note: Patients sometimes vomit a mixture of gastric and duodenal fluids. Vomiting of alkaline duodenal fluids in addition to gastric juice can cause the plasma pH to remain essentially normal or even cause metabolic acidosis if the loss of alkaline fluid is greater than the loss of gastric fluid.

(*Text continues on p. 27.*)

Table 2–1
Disease or Injury States Commonly Associated With Imbalances

Condition	Imbalances Likely to Occur	Reason
Uncontrolled severe diabetes mellitus (ketoacidosis)	Metabolic acidosis	Increased utilization of fat for energy needs causes accumulation of ketone bodies (acids)
	Fluid volume deficit	Hyperglycemia causes osmotic diuresis
Uncontrolled severe diabetes mellitus (nonketotic hyperosmolar coma)	Fluid volume deficit	Hyperglycemia causes profound osmotic diuresis
Adrenal insufficiency	Na^+ deficit	Inadequate amounts of adrenal cortical hormones produced, causing the body to retain K^+ and release too much Na^+
	K^+ excess	
Hyperaldosteronism	Fluid volume excess	Excessive secretion of aldosterone by adrenal cortex causes the body to retain Na^+ and water, and to lose K^+
	K^+ deficit	
Hyperparathyroidism	Ca^{2+} excess	Increased secretion of parathyroid hormone causes Ca^{2+} to be released from bone matrix into the plasma
Hypoparathyroidism	Ca^{2+} deficit	Decreased bone resorption causes depressed plasma Ca^{2+} level
Acute pancreatitis	Ca^{2+} deficit	Fixation of Ca^{2+} by fatty acids liberated from necrotic mesenteric fat deposits? Altered parathyroid hormone (PTH) secretion?
Perforated viscus and chemical peritonitis	Plasma-to-interstitial fluid shift	Loss of fluid into peritoneal cavity as a result of inflammatory process
	Ca^{2+} deficit	Fixation of calcium by fatty acids liberated from necrotic mesenteric fat deposits
Hysteria (psychogenic hyperventilation)	Respiratory alkalosis	Overbreathing, resulting in excessive loss of CO_2
Oxygen-lack with hyperpnea	Respiratory alkalosis	Overbreathing, resulting in excessive loss of CO_2
Tracheobronchitis	Na^+ excess	Excessive water vapor loss caused by excessive coughing
Emphysema	Respiratory acidosis	Inability to exchange CO_2

Disease or Injury States Commonly Associated With Imbalances (continued)

Condition	Imbalances Likely to Occur	Reason
Congestive heart failure	Fluid volume excess	Increased retention of Na^+ and water by kidneys caused by: • Decreased renal perfusion secondary to failure of pumping action of heart • Increased aldosterone secretion
End-stage renal failure	Metabolic acidosis K^+ excess Fluid volume excess	Inability of the kidneys to excrete acid metabolites, K^+, and fluid adequately
Chronic alcoholism	Mg^{2+} deficit	Alcohol produces magnesium diuresis; also, dietary intake of magnesium is often low
Oat-cell lung tumor	Na^+ deficit	Tumor produces an antidiuretic hormonelike substance, causing water retention
Bony metastases	Ca^{2+} excess	Bony destruction
Tumors secreting ectopic PTH-like substances (such as epidermoid lung tumor, renal adenocarcinoma, and oral cavity squamous tumors)	Ca^{2+} excess	PTH-like substance promotes an elevated serum Ca^{2+} level (PTH is an abbreviation for parathyroid hormone)
Breast malignant tumors	Ca^{2+} excess	Increased osteolytic activity of tumor
Massive crushing injuries	K^+ excess Plasma-to-interstitial fluid shift	Due to release of K^+ from injured cells Due to leakage of plasma through damaged capillaries at the injury site
Head injuries	Na^+ deficit	If inappropriate secretion of ADH occurs (see Chap. 7) If diabetes insipidus occurs
	Na^+ excess	
Burns (early)	K^+ excess	Due to cellular destruction and release of intracellular K^+
	Hypovolemia	Due to shift of fluid from plasma to the interstitial space (plasma leaks out through the damaged capillaries at the burn site)
Burns (after 3rd day)	K^+ deficit	Due to shift of K^+ back into cells
	Hypervolemia	Due to shift of fluid from interstitial space (occurs as the capillaries heal; it is more apt to be severe if the patient has been overloaded with intravenous fluids)

Table 2–2
Treatment-Induced (Iatrogenic) Causes of Fluid and Electrolyte Problems

Medical Therapy	Imbalances Likely to Occur	Reason
Potassium-depleting diuretics (thiazides, furosemide, ethacrynic acid, mercurials)	K^+ deficit Metabolic alkalosis	These agents promote renal loss of K^+; hypokalemia leads to metabolic alkalosis
Potassium-sparing diuretics (spironolactone, triamterene, amiloride)	K^+ excess	These agents, in varying ways, promote K^+ retention
Corticosteroids (in large doses)	Fluid volume excess K^+ deficit Metabolic alkalosis	These agents promote Na^+ and water retention, and K^+ loss Associated with hypokalemia
Aspirin poisoning	Respiratory alkalosis	Can occur early due to overstimulation of the respiratory center by toxic salicylate level
	Metabolic acidosis	Can occur later due to inadequate utilization of carbohydrates and, thus, increased metabolism of fats with ketone formation
Vitamin D overdose	Ca^{2+} excess	Due to increased bone resorption and intestinal absorption of Ca^{2+}
Cyclophosphamide (Cytoxan) or vincristine (Oncovin) in large doses	Na^+ deficit	These anticancer drugs favor water retention; they are most likely to produce hyponatremia in patients with tumors that secrete an ADH-like substance (such as oat-cell lung tumor)
Thiazides	Ca^{2+} excess (mild)	Due to decreased urinary Ca^{2+} excretion (furosemide does just the opposite—it increases urinary Ca^{2+} excretion)
Pitocin	Na^+ deficit	This agent has water-retentive qualities, similar to ADH; it is most likely to cause hyponatremia when administered in electrolyte-free parenteral fluids, such as D_5W

Treatment-Induced (Iatrogenic) Causes of Fluid and Electrolyte Problems (continued)

Medical Therapy	Imbalances Likely to Occur	Reason
Laxatives	K^+ deficit	Due to excessive loss of intestinal K^+
Chlorpropamide (Diabinese)	Na^+ deficit	Apparently, in a small percentage of patients, this drug causes increased ADH release in addition to potentiating its effect
Morphine, meperidine, or barbiturates	Respiratory acidosis	These agents, in excessive doses, depress respirations and thus decrease CO_2 elimination
Mafenide (Sulfamylon)	Metabolic acidosis	This agent can cause excessive loss of HCO_3^-
Excessive administration of isotonic saline (0.9% NaCl)	Fluid volume excess Metabolic acidosis	Due to excessive Na^+ content Due to excessive Cl^- content, causes compensatory decrease in HCO_3^-
Excessive or too rapid administration of K^+-containing solutions	K^+ excess	Amount or rate exceeds kidney's ability to rid plasma of excessive K^+ It must be remembered that symptoms of hyperkalemia can be induced by rapid elevation of serum K^+ even when the actual concentration is not extremely high
Excessive administration of electrolyte-free solutions (such as D5W), particularly during the early postoperative period	Na^+ deficit	Due to actual dilution of plasma Na^+ level Recall that water is abnormally retained during periods of stress due to increased ADH secretion
Excessive administration of sodium bicarbonate ($NaHCO_3$)	Fluid volume excess	Due to sodium and water retention (can be dangerous in patients with cardiac or renal dysfunction) Due to HCO_3^- content of fluid
Prolonged administration of calcium-free and magnesium-free solutions (longer than 2 wk)	Metabolic alkalosis Ca^{2+} deficit Mg^{2+} deficit	Due to plasma dilution
Hypodermoclysis with electrolyte-free solution (usually D_5W)	Na^+ deficit	Na^+ is drawn from the plasma and tissue fluid into the infusion site to render the solution isotonic for absorption; it becomes trapped in the edematous site

(continued)

Table 2-2
Treatment-Induced (Iatrogenic) Causes of Fluid and Electrolyte Problems (continued)

Medical Therapy	Imbalances Likely to Occur	Reason
Excessive administration of citrated blood, particularly to patients with liver disease	Ca^{2+} deficit	Citrate ions combine with calcium ions in bloodstream
Excessive administration of stored blood to oliguric patients	K^+ excess	Stored blood has an elevated K^+ level due to leakage of K^+ from anoxic cells (see Chap. 16)
High-solute tube feedings, with inadequate water supplements	Na^+ excess	Excessive water loss occurs from the extracellular fluid to allow for renal elimination of solutes (see Chap. 14)
Prolonged immobilization	Ca^{2+} excess	Absence of weight-bearing causes resorption of Ca^{2+} from the bone matrix into the extracellular fluid
Gastric suction plus drinking plain water	Metabolic alkalosis K^+ deficit Na^+ deficit	Due to "electrolyte-washout." Loss of H^+ and Cl^- causes alkalosis; hypokalemia is partially due to loss of K^+ but mainly is associated with the alkalosis; Na^+ is lost in the drainage
	Mg^{2+} deficit	Mg^{2+} deficit can occur when gastric suction is prolonged
Tight abdominal binders and dressings, or chest restraints	Respiratory acidosis	Due to decreased respiratory excursions, resulting in CO_2 retention

Intestinal Juice

The daily volume of intestinal juice is about 3000 ml, and its pH is alkaline (7.8 to 8.0). It contains substantial amounts of K^+, HCO_3^-, and Na^+. Diarrhea, intestinal suction, or fistulas can result in

- Fluid volume deficit (fluid losses can be great)
- Metabolic acidosis (due to loss of HCO_3^-)
- Potassium deficit (due to loss of K^+, which is plentiful in intestinal juice)
- Sodium deficit (due to loss of Na^+)

Bile

The normal daily secretion of bile is approximately 1500 ml, and the pH is alkaline. It contains substantial amounts of Na^+ and HCO_3^-. Abnormal losses of bile can occur from fistulas or from T-tube drainage following gallbladder surgery.

Imbalances that can occur when bile is lost include

- Sodium deficit
- Metabolic acidosis
- Fluid volume deficit

Pancreatic Juice

The normal daily secretion of pancreatic juice is approximately 1000 ml, and the pH is 8 (alkaline). It should be remembered that pancreatic juice is an integral part of other intestinal secretions.

Imbalances that can result from pancreatic fistulas include

- Metabolic acidosis (due to loss of HCO_3^-)
- Sodium deficit
- Calcium deficit
- Fluid volume deficit

Sweat

Sweat is a hypotonic fluid containing primarily water, sodium, chloride, and potassium. The sodium concentration in sweat averages between 50 and 80 mEq/liter, considerably less than the sodium content in plasma. Sweat can vary in volume from 0 to 1000 ml/hr, or more. Excessive perspiration can cause

- Sodium deficit in persons not accustomed (acclimated) to high temperatures, particularly if large quantities of plain water are ingested. (Persons accustomed to heat are able to conserve sodium more effectively, thus, less sodium is lost in their sweat).
- Potassium deficit (most likely to occur in persons acclimated to heat, since these persons tend to lose more potassium in sweat due to the effects of aldosterone).

- Sodium excess can occur in the heavily perspiring person who has an inadequate water supply. (Heat disorders are discussed in Chap. 25.)

Intake and Output

One should be alert for major discrepancies between fluid intake and output.

- Is the patient losing substantially more fluid than he is taking in?
- Or, is he retaining fluids abnormally?

Answers to these questions are often crucial. See the section on intake and output under Clinical Assessment.

Evaluation of Laboratory Data

The reader is referred to Chapter 3 for a review of common laboratory tests used to evaluate fluid balance status.

Clinical Assessment

- After the history described above has been reviewed, the nurse should be able to formulate potential nursing diagnoses (*i.e.*, identify likely imbalances).
- At this point a thorough nursing assessment is indicated. However, it is important to note that nursing assessment is not a "one-time" procedure; instead, assessment must be done at regular intervals to detect changes. It is the nursing staff that is charged with the 24-hour care of hospitalized patients; because of this the nurse has an excellent opportunity to monitor changes.
- Parameters to consider in clinical assessment include

    Comparison of total intake and output measurements

    Urine volume and concentration

    Skin turgor

    Tongue turgor

    Degree of moisture in oral cavity

    Thirst

    Appearance of skin, and skin temperature

    Edema

    Body weight

    Body temperature

    Pulse

    Respiration

Blood pressure
Neck veins
Hand veins
Central venous pressure
Neuromuscular irritability
Other signs

- Nursing observations must be interpreted, using one's knowledge of the patient's history and pathophysiologic condition, to determine if the information gained from the assessment suggests normal or abnormal signs or measurements, and to what degree. Then, the appropriate nursing action(s) must be determined. One must also know when medical intervention is required; a close working relationship with the physician is necessary in order to provide optimal patient care.

Comparison of Intake and Output

- Many serious fluid balance problems can be averted by maintaining a careful vigil on patient intake and output (I & O), keeping accurate records. (Totals for several consecutive days should be compared.)

  If the total intake is substantially less than the total output, it is obvious that the patient is in danger of fluid volume deficit. On the other hand, if the total intake is substantially more than the output, the patient is in danger of fluid volume excess (or, in the case of inappropriate secretion of antidiuretic hormone (ADH), water excess).

- To better understand abnormal states, it is helpful to review the 24-hour average intake and output in a normal adult. (Table 2-3.)
- It is important to initiate intake and output records for any patient

Table 2–3
**Average Intake and Output
in an Adult for a 24-Hour Period**

Intake		Output	
Oral liquids	1300 ml	Urine	1500 ml
Water in food	1000 ml	Stool	200 ml
Water produced by metabolism	300 ml	*Insensible*	
		Lungs	300 ml
Total	2600 ml	Skin	600 ml
		Total	2600 ml

with a real or potential water and electrolyte problem; do not wait for an order from the physician.

1. Intake should include all fluids taken in by the gastrointestinal route (including foods that are liquid at room temperature, such as gelatin). Of course, fluids gained by the parenteral route must also be included.

2. Output should include urine, vomitus, diarrhea, drainage from fistulas, and drainage from suction apparatus. Perspiration should be noted and its amount estimated. The presence of prolonged hyperventilation should also be noted, since it is an important route of water-vapor loss. Drainage from lesions (such as from large decubitus ulcers) should be noted and estimated.

3. The I & O record should include the time of day and the type of fluid gained and lost. This information is necessary in planning therapy.

Summary of Common Sources of Error in I & O Measurement

Failure to communicate to all members of the staff which patients require I & O measurements (may result in discarded body fluids or intakes not recorded, making the entire record inaccurate)

Failure to explain the need for recording all oral fluids to the patient and his family

Failure to instruct the ambulatory patient to use a bedpan, urinal, or other collecting device for voiding and to save the specimen until it can be measured and recorded

Failure to measure fluids that can be directly measured because it takes less time to guess at their amounts

Failure to designate the specific volume of glasses, cups, bowls, and other fluid containers used in the hospital (Each person may ascribe a different volume to the same glass of water.)

Well-meaning intentions to record a drink of water or an emptied urinal at a later, more convenient time are often forgotten.

Failure to estimate "uncaught" vomitus or incontinent urine; frequently are recorded as lost specimens with no mention as to the approximate volume.

Failure to measure sips of water. A sip of water, or mouth rinsing, if repeated often, might add up to as much as 1000 ml per day!

Failure to consider that parenteral fluid bottles are overfilled—a liter bottle may actually contain 1100 ml; a 500-ml bottle may contain as much as 600 ml!

Failure to record the amount of solution used to irrigate tubes and the amount of fluid withdrawn during the irrigation

Failure to estimate fluid lost as wound exudate; drainage from a large decubitus can be extensive.

Failure to estimate fluid as perspiration (a necessary linen change may represent as much as a liter of perspiration)

Urine Volume and Concentration

- As a general rule of thumb, the normal urinary output is about 1 ml per kilogram of body weight per hour.[1]
- Note (see Table 2-3) that the usual urine volume in adults is approximately 1500 ml/24 hr (ranges from 1000 to 2000 ml/24 hr). This is equivalent to approximately 40 to 80 ml/hr. Urine volume in children is less, dependent on age and weight.
- During periods of stress, the 24-hour urine volume in the adult may diminish to 750 to 1000 ml/24 hr (or 30 to 50 ml/hr). Urine volume is somewhat less during periods of stress because of increased aldosterone and ADH secretion.
- A low urine volume suggests fluid volume deficit, and a high urine volume suggests fluid volume excess.
- The reader is encouraged to review the normal values for urinary specific gravity and osmolality (see Table 3-1).
- A number of factors can alter urinary output, such as
  1. Amount of fluid intake
  2. Losses from skin, lungs, and gastrointestinal tract
  3. Amount of waste products for excretion (urine volume is increased in conditions with high solute loads, such as diabetes mellitus, high-protein tube feedings, thyrotoxicosis, and fever)
  4. Renal concentrating ability (a person with a urinary specific gravity (SG) of 1.030 requires approximately 15 ml of water to excrete 1 Gm of solute; a person with an SG of 1.010 owing to renal disease needs at least 40 ml of water in order to excrete 1 Gm of solute)
  5. Blood volume (hypovolemia causes decreased renal perfusion and, thus, oliguria; hypervolemia causes increased urinary volume if the kidneys are functioning normally)
  6. Hormonal influences (primarily aldosterone and ADH)

Summary of Important Nursing Considerations Related to Urine Output

Maintain I & O records on all patients with real or potential fluid balance problems. Measure all fluid gains and losses according to routes.

Be alert for fluid intake greatly exceeding fluid output, or fluid output greatly exceeding fluid intake. Total the I & O records for several consecutive days to get a clearer understanding of fluid balance status.

Be aware that the usual urine output in adults is 1 to 2 liters/day (or approximately 750 to 1200 ml/day during periods of stress).

Use a device calibrated for small volumes of urine when hourly urine volumes must be measured (Fig. 2-1.).

Be aware that the usual urine output in adults is 40 to 80 ml/hr (or 30 to 50 ml/hr during periods of stress).

(continued)

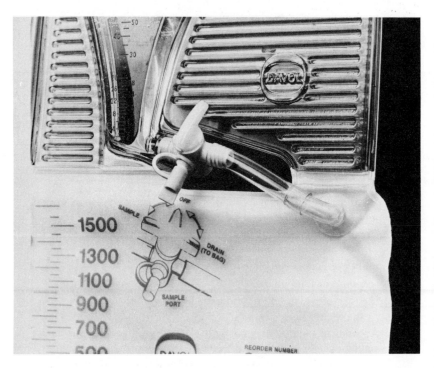

Figure 2–1. Davol #3590 Uri-Meter drainage bag, a collecting device that can be used for precise measurement of small volumes of urine. After hourly collection, the urine can be diverted into the large drainage bag. (Courtesy of Davol, Inc., Cranston, RI)

Be aware that patients taking in high-solute loads (as in high-protein tube feedings) need extra water to aid in solute elimination.

Be aware that individuals with diminished renal concentrating ability (such as the aged) need more fluid in order to excrete solutes than do those with normal renal function.

Be aware that a low urine volume with a high SG indicates fluid volume deficit.

Be aware that a low urine volume with a low SG is indicative of renal disease.

Evaluate I & O levels and urinary SG in relation to other clinical signs.

Be aware of common sources of errors in I & O measurements. (See Summary after Comparison of Intake and Output section.)

Skin Turgor

- In a normal person, pinched skin will immediately fall back to its normal position when released. This elastic property, referred to as turgor, is partially dependent on interstitial fluid volume.[2]

- In an individual with a fluid volume deficit, the skin may remain slightly elevated for many seconds.

- Tissue turgor is best measured by pinching the skin over the sternum or forehead. Some prefer to test skin turgor in children over the abdominal area and on the medial aspects of the thighs.
- Skin turgor in children begins to diminish after 3% to 5% of the body weight is lost.
- Severe malnutrition, particularly in infants, can cause depressed skin turgor even in the absence of fluid depletion.
- Obese infants with fluid volume deficit may have skin turgor that is deceptively normal.
- Infants with hypernatremia may have firm skin that feels thick.
- Reduced skin turgor is common in older patients (those more than 55 to 60 years of age) due to a primary decrease in skin elasticity.[3] Skin turgor in these patients is probably most valid on the skin overlying the sternum or on the inner aspects of the thighs.

Tongue Turgor

- In a normal person the tongue has one longitudinal furrow. In the person with fluid volume deficit there are additional longitudinal furrows, and the tongue is smaller (due to fluid loss).
- Tongue turgor is not affected appreciably by age and thus is a useful assessment for all age groups.
- Sodium excess causes the tongue to appear red and swollen.

Moisture in Oral Cavity

- A dry mouth may be due to fluid volume deficit or to mouth breathing. When in doubt, the nurse should run a finger along the oral cavity and feel the membrane where the cheek and gum meet; dryness in this area indicates a true fluid volume deficit. (This area will be moist if the problem is due to mouth breathing).
- Dry, sticky mucous membranes are noted in sodium excess (the oral cavity feels like "flypaper").

Thirst

- Thirst is a subjective sensory symptom and has been defined as an awareness of the desire to drink.
- The sense of thirst is so protective (maintaining the serum sodium level in normal persons) that hypernatremia never occurs unless thirst is impaired or rendered ineffective because the patient is unconscious or is denied access to water.
- Thirst is prominent in patients who have increased water losses (as in hyperglycemia, high fever, or diarrhea).
- Unfortunately, patients with altered states of consciousness do not

experience normal thirst; also, many patients are debilitated and unable to respond to thirst. It is no wonder that hypernatremia is a common imbalance on a neurologic unit.

- An infant can be tested for thirst by offering water, although the presence of nausea may mask the symptom.

Tearing and Salivation

The absence of tearing and salivation in a child is a sign of fluid volume deficit; it becomes obvious with a fluid loss of 5% of the total body weight.

Appearance of Skin and Skin Temperature

- Metabolic acidosis can cause warm, flushed skin (due to peripheral vasodilatation).
- Severe fluid volume deficit causes the skin to be pale and cool (due to peripheral vasoconstriction, which occurs to compensate for hypovolemia).
- The condition of the skin in heat disorders is discussed in Chapter 25.

Facial Appearance

- An individual with a severe fluid volume deficit has a pinched, drawn facial expression (Fig. 2-2).
- A fluid volume deficit of 10% of body weight causes decreased intraocular pressure, and thus the eyes appear sunken and feel soft to the touch.

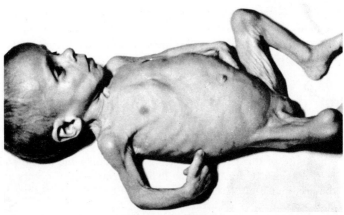

Figure 2-2. Severely dehydrated infant. (Waechter E, Blake F: Nursing Care of Children, 9th ed. Philadelphia, JB Lippincott, 1976)

Edema

- Edema is defined as the presence of excessive interstitial fluid (the fluid between the cells).
- It is the result of an increase in the total body sodium content. (Barely perceptible pitting edema usually implies an excess of at least 400 mEq of sodium.)
- There is no peripheral edema with only water retention (as occurs in excessive secretion of ADH).
- Clinically, edema is not usually apparent in the adult until the retention of 5 to 10 pounds of excess fluid.
- Pitting edema is a phenomenon manifested by a small depression that remains after one's finger is pressed over an edematous area and then removed (Fig. 2-3). Gradually, within 5 to 30 seconds after the pressure is removed, the "pit" disappears. Usually, pitting edema is not evident until at least a 10% increase in weight has occurred.
- Edema can be generalized or dependent. (*Dependent edema* refers to the flow of excess fluid by gravity to the most dependent portions of the body—the feet and ankles, if standing; the back and buttocks, if lying down.)
- Description of edema by appearance is somewhat subjective. For example, it is sometimes indicated using plus signs to represent amount, ranging from +1 to +4 (+1 indicating barely perceptible edema, +2 and +3 in between, and +4 indicating severe edema). Measurement of an extremity or body part with a millimeter tape, in the same area each day, is the most exact method.

Body Weight

- Daily weighing of patients with potential or actual fluid balance problems is of great clinical value because
    1. Accurate body weight measurements are much easier to obtain than accurate intake-output measurements.
    2. Rapid variations in weight closely reflect changes in body fluid volume.
- The use of body weight as an accurate index of fluid balance is based on the assumption that the patient's dry weight remains relatively stable. (Even under starvation conditions, an individual loses no more than $1/3$ lb to $1/2$ lb of dry weight/day.)
- A rapid loss of body weight will occur when the total fluid intake is less than the total fluid output.
    1. Rapid loss of 2% total body weight (TBW) indicates mild fluid volume deficit.

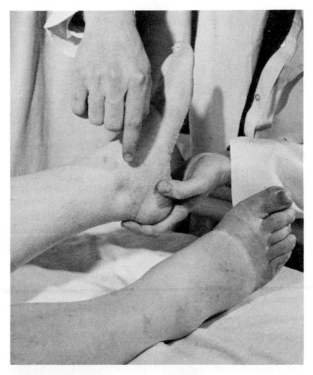

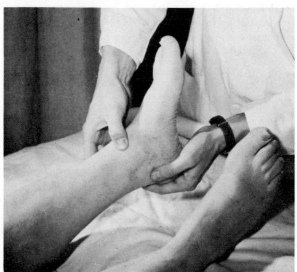

Figure 2–3. (*Top*) Pitting edema of feet and lower legs. (*Bottom*) The same patient has been relieved by treatment. (Courtesy of CIBA Pharmaceutical Co, Summit, NJ)

2. Rapid loss of 5% TBW indicates moderate fluid volume deficit.
3. Rapid loss of 8% or more of TBW indicates severe fluid volume deficit.

- A rapid gain of body weight will occur when the total fluid intake is greater than the total fluid output.
  1. Rapid gain of 2% TBW indicates mild fluid volume excess.
  2. Rapid gain of 5% TBW indicates moderate fluid volume excess.
  3. Rapid gain of 8% or more of TBW indicates severe fluid volume excess.
- A rapid gain or loss of 1 kg (2.2 lb) of body weight is approximately equivalent to the gain or loss of 1 liter of fluid (or, expressed another way, a gain or loss of 500 ml of fluid is equivalent to a gain or loss of 1 lb).
- It should be remembered that a patient may have a severe fluid volume deficit even though body weight is essentially unchanged when there is a "third-space" loss of body fluid. (See Chap. 6.)
- The following practices should be followed in weighing patients:
  1. Use the same scale each time. (Unfortunately, there is significant variation among scales.)
  2. Measure weight in the morning before breakfast and after voiding.
  3. Be sure the patient is wearing the same, or similar, clothing each time (the clothing should be dry).

Body Temperature

- Changes in body temperature as a result of fluid and electrolyte imbalances may include
  1. Elevation of body temperature in hypernatremia (dehydration) probably related to lack of available fluid for sweating. Also, dehydration almost certainly has a direct effect on the hypothalamus.
  2. Decrease in body temperature in fluid volume deficit, when uncomplicated by infection (probably as a result of decreased metabolism).

- May be 97° F (36.1° C) to 99° F (37.2° C) rectally in moderate fluid volume deficit
- May be 95° F (35° C) to 98° F (36.7° C) rectally in severe fluid volume deficit
- An elevated body temperature can cause fluid balance problems if extra fluids are not supplied as indicated.

A temperature elevation between 101° F (38.3° C) and 103° F (39.4° C) increases the 24-hour fluid requirement by at least 500 ml, and a temperature above 103° F increases it by at least 1000 ml.

Pulse

- Increased pulse rate can be the result of fluid volume deficit, magnesium deficit, severe potassium deficit, or sodium excess.
- Decreased pulse rate can occur in severe potassium excess or magnesium excess.
- A weak, low-volume pulse is associated with fluid volume deficit.
- A bounding full-volume pulse is associated with fluid volume excess.
- Irregular pulse rates occur with potassium imbalances and magnesium deficit.

Respirations

- Deep, rapid respirations may be a compensatory mechanism for metabolic acidosis or may be a primary disorder causing respiratory alkalosis.
- Slow, shallow respirations may be a compensatory mechanism for metabolic alkalosis or may be a primary disorder causing respiratory acidosis.
- Weakness or paralysis of respiratory muscles is likely in severe hypo- or hyperkalemia and in severe magnesium excess (the respiratory center may be paralyzed at a serum magnesium level of 10 to 15 mEq/liter).
- Moist rales, in the absence of cardiopulmonary disease, indicate fluid volume excess.

Blood Pressure

- A fall in systolic pressure exceeding 10 mm Hg from the lying to the standing or the sitting position (postural hypotension) usually indicates fluid volume deficit.
- Hypotension may occur with magnesium excess (first occurring at a level of 3 to 5 mEq/liter).
- Hypertension can occur with magnesium deficit and with fluid volume excess.

Neck Veins

- The jugular veins provide a built-in manometer for following changes in central venous pressure (CVP)—no invasive maneuvers are required, and the procedure can be highly reliable when done correctly.

- Changes in fluid volume are reflected by changes in neck vein filling, provided that the patient is not in heart failure.
- Normally, with the patient supine, the external jugular veins fill to the anterior border of the sternocleidomastoid muscle. Flat neck veins in the supine position indicate a decreased plasma volume.
- Normally, with the patient positioned sitting at a 45-degree angle, the venous distentions should not extend higher than 2 cm above the sternal angle. Elevated venous pressure is indicated by neck veins distended from the top portion of the sternum to the angle of the jaw (Fig. 2-4).
- To estimate cervical venous pressure, the nurse should

    1. Position the patient in a semi-Fowler's position (head of bed elevated to a 30- to 45-degree angle), keeping the neck straight.
    2. Remove any of the patient's clothing that could constrict the neck or upper chest.
    3. Provide adequate lighting in order to visualize effectively the external jugular veins on each side of the neck.
    4. Measure the level to which the veins are distended on the neck or above the level of the manubrium.

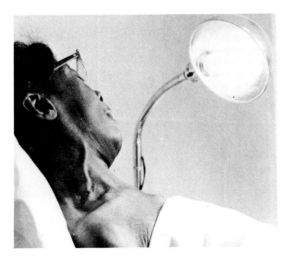

Figure 2–4.   When lying in a semireclined position, this patient has distended neck veins, which indicates that the heart is incapable of receiving and pumping adequately all of the incoming venous blood. (Reproduced with permission. ©American Heart Association)

Hand Veins

- Observation of hand veins can be helpful in evaluating the patient's plasma volume.
- Usually, elevation of the hands causes the hand veins to empty in 3 to 5 seconds; placing the hands in a dependent position causes the veins to fill in 3 to 5 seconds (Figs. 2-5 and 2-6).
- A decreased plasma volume causes the hand veins to take longer than 3 to 5 seconds to fill when the hands are in a dependent position.
- An increased plasma volume causes the hand veins to take longer than 3 to 5 seconds to empty when the hands are elevated. When this is the case, the peripheral veins are engorged and clearly visible.

Central Venous Pressure

- Central venous pressure (CVP) refers to pressure in the right atrium or vena cava and provides information about the following parameters:
    1. Blood volume

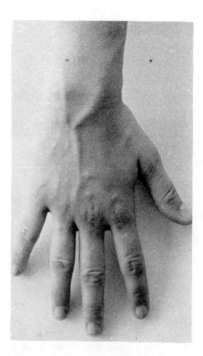

Figure 2–5.   Appearance of hand veins when the hand is held in a dependent position.

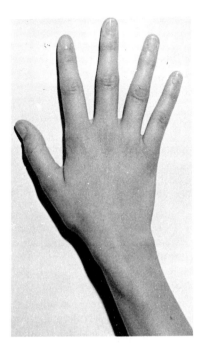

Figure 2–6. Appearance of hand veins when the hand is held in an elevated position.

2. Effectiveness of the heart's pumping action
3. Vascular tone

- Pressure in the right atrium is usually 0 cm to 4 cm of water; pressure in the vena cava is approximately 4 cm to 11 cm of water.
- A low CVP may indicate
    1. Decreased blood volume or
    2. Drug-induced vasodilatation (causing pooling of blood in peripheral veins)
- A high CVP may indicate
    1. Increased blood volume or
    2. Heart failure or
    3. Vasoconstriction (causing the vascular bed to become smaller)
- More important than absolute values are the upward or downward trends; these trends are determined by taking frequent readings (often every 30 to 60 minutes).
- It is always important to evaluate CVP in reference to other available clinical data such as
    1. Blood pressure

2. Pulse
3. Respirations
4. Breath and heart sounds
5. Fluid intake
6. Urinary output

Example: A rise in CVP paralleling that of systolic BP is an indication of adequate fluid volume replacement. A low CVP persisting after fluid volume replacement may be a sign of continued occult bleeding.

- In those patients with normal cardiac function and relatively normal pulmonary function the CVP remains an acceptable guide to blood volume.
- Patients with acute cardiopulmonary decompensation require more extensive hemodynamic monitoring with a device that reflects pressures in both sides of the heart.
- Sometimes the rate of an infusion is titrated according to the patient's CVP; when this is necessary, the physician should designate the desired limits so that the nurse can adjust the flow rate accordingly. For example, in acute hypercalcemia, the physician may order isotonic saline at a rate of 250 ml/hr, provided that the CVP does not exceed 10 cm.

Nursing Considerations in Measurement of Central Venous Pressure

- Measure CVP with the patient flat in bed, if possible. If not, make the readings with the patient in the same position each time. Indicate the position on the chart when recording the pressure. Place the zero point of the manometer at the level of the patient's right atrium (Fig. 2-7).
- Be aware that fluid should fluctuate 3 cm to 5 cm in the manometer with respirations when the catheter is properly positioned in the vena cava.
- Take the following steps to read CVP:
  1. Turn the stopcock so that the solution will flow from the container to the manometer, allowing it to fill and reach a level of 30 cm (Fig. 2-8, System 2).
  2. Then turn the stopcock to direct manometer flow to the patient (Fig. 2-8, System 3). The fluid level should drop, reaching a reading level in about 15 seconds.
  3. Record the reading at the upper level of the respiratory fluctuation of fluid in the manometer (fluid falls slightly on inspiration and rises slightly on expiration).
  4. Turn the stopcock to allow resumption of the infusion to keep the catheter patent and to supply needed fluids (Fig. 2-8, System 1).

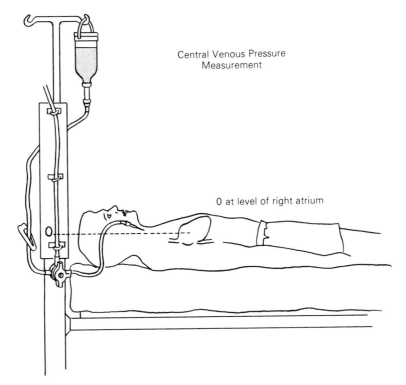

Central Venous Pressure
Measurement

0 at level of right atrium

Figure 2–7. Central venous pressure measurement with zero-point of manometer at level of right atrium.

- Use meticulous aseptic technique for dressing changes, catheter care, and tubing changes.
- Check frequently for signs of redness, swelling, and purulent drainage at the injection site.
- Secure connections to prevent the occurrence of air embolism. Remember that air emboli are more likely to occur when a catheter is placed in the central veins where pressure is low. (Air embolism is discussed in Chap. 4.)

Neuromuscular Irritability

- It is sometimes necessary to assess patients for increased or decreased neuromuscular irritability, particularly when imbalances in calcium, magnesium, and sodium are suspected.

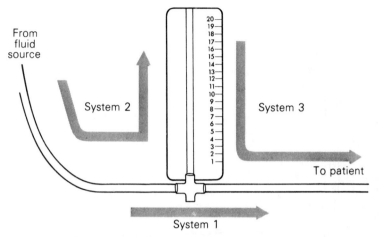

System 1 allows flow from the container to the patient (routine infusion).

System 2 allows flow from the container to the manometer (allows manometer to fill).

System 3 allows flow from the manometer to the patient (allows reading of CVP).

Figure 2–8.   Fluid flow systems in central venous pressure measurement. (Hudak CM, Lohr T, Gallo B: Critical Care Nursing, 3rd ed, p 151. Philadelphia, JB Lippincott, 1982)

- The nurse may, as necessary, check for Chvostek's sign and Trousseau's sign; also, deep tendon reflexes can be tested to monitor neuromuscular irritability.
- To test Chvostek's sign, the facial nerve should be percussed about 2 cm anterior to the earlobe. A positive response shows a unilateral twitching of the facial muscles, including the eyelid and lips. Chvostek's sign is indicative of hypocalcemia or hypomagnesemia.

Assessment of Deep Tendon Reflexes

GENERAL INFORMATION

- The common deep tendon reflexes include the biceps, triceps, brachioradialis, patellar and Achilles reflexes.
- A deep tendon reflex is elicited by briskly tapping a partially stretched tendon with a rubber percussion hammer, preferably over the tendon insertion of the muscle. The broad head of the hammer is used to stroke easily accessible tendons (*e.g.*, the Achilles) and the pointed end for less accessible tendons (*e.g.*, the biceps). The response in the prospective muscle is a sudden contraction. The muscle being tested should be slightly stretched (by the position of the limb), and the patient should be relaxed. With too little or too much muscle stretch the reflex cannot be elicited.
- The reflexes are usually graded on a 0 to +4 scale

0 = no response
1+ = somewhat diminished, but present
2+ = normal
3+ = brisker than average and possibly but not necessarily indicative of disease
4+ = hyperactive

- Deep tendon reflexes may be hyperactive in the presence of hypocalcemia, hypomagnesemia, hypernatremia, and alkalosis. They may be hypoactive in the presence of hypercalcemia, hypermagnesemia, hyponatremia, hypokalemia, and acidosis.

- Of course, many factors other than electrolyte disturbances can produce abnormalities in deep tendon reflexes. As with most other signs, deep tendon reflexes should be evaluated in light of other clinical signs, patient history, and laboratory data.

Other Signs

Other areas to consider in clinical assessment include changes in

- Behavior
- Sensation
- Fatigue level

Since these changes are often vague, they are best evaluated in context with specific imbalances. (See Chaps. 5 through 12.)

References

1. Pestana C: Fluids and Electrolytes in the Surgical Patient, 2nd ed, p 43. Baltimore, Williams & Wilkins, 1981
2. Rose B: Clinical Physiology of Acid–Base and Electrolyte Disorders, p 233. New York, McGraw-Hill, 1977
3. Ibid

chapter 3

Evaluation of Laboratory Data

- Data from laboratory tests provide the nurse with valuable information about the patient's fluid and electrolyte status.
- It is important to know the normal values for common laboratory tests to facilitate quick interpretation of test results and foster effective communication with the physician and other members of the health care team.
- Table 3-1 lists laboratory tests on blood and urine that are commonly used to evaluate fluid and electrolyte status. Also included in the table are normal values and significance of variations in these tests. Of course, normal values for laboratory tests vary slightly from institution to institution. Therefore, the values in Table 3-1 may be slightly different from those in the reader's clinical setting. (Always evaluate test results in light of the normal values listed by the laboratory performing the test.)
- Keep in mind that laboratory data can be misleading at times and must always be evaluated with the patient's clinical status and history in mind.
- Laboratory findings are also discussed in the chapters dealing with specific fluid and electrolyte imbalances (Chaps. 5 through 12) and in the chapters dealing with clinical situations commonly causing alterations in fluid and electrolyte balance (Chaps. 13 through 27).

Table 3–1
Laboratory Tests Used to Evaluate Fluid and Electrolyte Status

Laboratory Test	Normal Values	Significance
Blood Chemistry and Electrolytes		

Blood urea nitrogen (BUN)	10–20 mg/100 ml	• Elevated BUN can be due to
		  1. Reduced renal blood flow secondary to fluid volume deficit (causing reduced urea clearance)
		  2. Excessive protein intake (causing increased urea production)
		  3. Increased catabolism due to trauma, starvation, bleeding into the intestines, or catabolic drugs (causing increased urea production)
		• Low BUN (i.e., 6–8 mg/100 ml) is often associated with overhydration; may also be associated with low protein intake

Serum creatinine	0.7–1.5 mg/100 ml	• Creatinine is a more specific and sensitive indicator of renal disease than is the BUN, since nonrenal causes of elevation are few • Elevated in all diseases of kidneys in which 50% or more of the nephrons are destroyed • Elevated creatinine level may occur in severe fluid volume depletion, which results in a reduction in GFR (prerenal failure) • May be slightly over normal in patients with large muscle mass or acromegaly • A low creatinine is not clinically significant

(continued)

Table 3–1
Laboratory Tests Used to Evaluate Fluid and Electrolyte Status (continued)

Laboratory Test	Normal Values	Significance
Blood Chemistry and Electrolytes *(continued)*		
BUN:Creatinine ratio	10:1	• When the ratio increases in favor of the BUN (i.e., ratio > 10:1), conditions such as hypovolemia, low perfusion pressures to the kidney, or increased protein metabolism may be present • When the ratio is < 10:1, conditions such as low protein intake, hepatic insufficiency, or repeated dialysis may be present • When both the BUN and creatinine levels rise, maintaining the 10:1 ratio, the problem is likely intrinsic renal disease (although it may also be seen when fluid volume depletion results in reduction in the GFR)
Hematocrit	Male 44%–52% Female 39%–47%	• Increased in fluid volume deficit (relative loss of fluid makes blood cells appear more concentrated) • Decreased in fluid volume excess (gain in intravascular fluid dilutes blood cell concentration in plasma) • Decreased 4 hr to 6 hr after hemorrhage as interstitial fluid moves into the bloodstream to build up diminished plasma volume
Serum osmolality*	280–295 mOsm/kg Can be measured by lab or can be roughly calculated by the following formula: $$Osm = 2(Na^+ + K^+) + \dfrac{G}{18} + \dfrac{BUN}{2.8}$$	• Determined mainly by serum Na^+ concentration (Recall that Na^+ makes up 90% of the osmotic pressure generated by plasma) • Increased in dehydration (hypernatremia) • Decreased in overhydration (hyponatremia) • Increased in hyperglycemia and in presence of an elevated BUN

Serum glucose	70–110 mg/100 ml	• Markedly elevated glucose level in bloodstream causes osmotic diuresis and fluid volume deficit (see Chap. 20)
Serum electrolytes Sodium	135–145 mEq/liter	• < 135 mEq/liter indicates hyponatremia • > 145 mEq/liter indicates hypernatremia • See Chapter 7
Potassium	3.5–5.5 mEq/liter	• < 3.5 mEq/liter indicates hypokalemia • > 5.5 mEq/liter indicates hyperkalemia
Calcium	Total 8.5–10.5 mg/100 ml or 4.0–5.5 mEq/liter Ionized About 50% of total value	• Calcium exists in plasma in an ionized fraction and a protein-bound fraction; total Ca^{2+} refers to the sum of these fractions • Total serum Ca^{2+} values must be evaluated in light of the serum albumin concentration because each fall (or rise) of serum albumin level (beyond the normal range of 4–5 Gm/100 ml) is associated with a fall (or rise) of serum Ca^{2+} concentration of approximately 0.8 mg/100 ml • Some laboratories are able to measure the ionized serum Ca^{2+} fraction directly • Ionized Ca^{2+} level is decreased in alkalosis and increased in acidosis
Magnesium	1.5–2.5 mEq/liter or 1.8–3.0 mg/100 ml	• < 1.5 mEq/liter (or 1.8 mg/dl) indicates hypomagnesemia • > 2.5 mEq/liter (or 3.0 mg/dl) indicates hypermagnesemia • See Chapter 10

(continued)

*The terms *osmolality* and *osmolarity* are frequently confused. Osmolality refers to the number of osmoles per kg of water; as a result, the total volume will be 1 liter of water plus the small volume occupied by the solutes (measured in mOsm/kg water).[1] Osmolarity refers to the number of osmoles per liter of solution; as a result, the volume of water is less than 1 liter by the amount equal to the solute volume (measured in mOsm/liter).[2] The difference is negligible in practice, since solute concentrations in the body fluids are low.

Table 3–1
Laboratory Tests Used to Evaluate Fluid and Electrolyte Status (continued)

Laboratory Test	Normal Values	Significance
Blood Chemistry and Electrolytes *(continued)*		
Chloride	100–106 mEq/liter	• < 100 mEq/liter indicates hypochloremia (commonly associated with hypokalemia and metabolic alkalosis) • > 106 mEq/liter indicates hyperchloremia (may be associated with excessive administration of isotonic saline or ammonium chloride)
Carbon dioxide (CO_2) content	24–30 mM/liter	• Measures total bicarbonate (HCO_3^-) and carbonic acid (H_2CO_3) in plasma • In the absence of COPD, an elevated value indicates metabolic alkalosis • Below normal in metabolic acidosis (however, if arterial *p*H reveals alkalosis, a low CO_2 Content indicates respiratory alkalosis) • Obviously, acid-base balance in acutely ill patients is best determined by arterial blood gas studies (particularly when respiratory problems are present) (see Chap. 12)
Phosphorus	Adults 3.0–4.5 mg/100 ml (1.8–2.6 mEq/liter) Children 4.0–7.0 mg/100 ml (2.3–4.1 mEq/liter)	• Less than normal indicates hypophosphatemia • Greater than normal indicates hyperphosphatemia • See Chapter 11
Zinc	77–137 μg/100 ml (by atomic absorption)	• See Chapter 11

Lithium		• 0.8 mEq/liter (therapeutic level 8 to 12 hr after administration) • Toxic level > 1.6 mEq/liter • Severe intoxication at 3–6 mEq/liter • See Chapter 11
Serum proteins Total Albumin Globulin	6.0–8.0 Gm/100 ml 3.5–5.5 Gm/100 ml 1.5–3.0 Gm/100 ml	• Decreased serum protein level causes reduced osmotic pull in intravascular space, allowing fluid to shift to the interstitial space (edema) • Important to know albumin level when evaluating serum Ca^{2+} levels
Lactate (arterial blood)	1.5 mEq/liter (approximately 10 mg/100 ml)	• Abnormal when > 2 mEq/liter (although there is no change in plasma *p*H when no higher than 2 to 3 mEq/liter) • Usually > 7 mEq/liter in lactic acidosis (may be as high as 30 mEq/liter)
Serum ketones		• Often > 50 mg/100 ml in diabetic ketoacidosis • Usually < 20 mg/100 ml in salicylate intoxication
Serum salicylates		• Therapeutic range is 20–25 mg/100 ml • Toxic range: > 30 mg/100 ml • Toxic salicylate level can induce both respiratory alkalosis (due to stimulation of medulla) and metabolic acidosis (due to disrupted carbohydrate metabolism and ketone formation) • See Chapter 27
Serum alcohol (ethanol)		• High serum alcohol levels lead to serum hyperosmolality, since alcohol depresses antidiuretic hormone secretion • Marked intoxication at 0.3%–0.4% • Alcoholic stupor at 0.4%–0.5% • Alcoholic coma > 0.5%

(continued)

Table 3–1
Laboratory Tests Used to Evaluate Fluid and Electrolyte Status (continued)

Laboratory Test	Normal Values	Significance
Blood Chemistry and Electrolytes (*continued*)		
Anion gap (AG)	12–15 mEq/liter AG = $Na^+ - (Cl^- + HCO_3^-)$	• Useful in ascertaining cause of metabolic acidosis • A level greater than 15 mEq/liter indicates the presence of excessive organic acids (as in diabetic ketoacidosis, lactic acidosis, uremic renal failure, and salicylate intoxication) • A normal anion gap acidosis may be due to diarrhea, ureteroenterostomies, excessive chloride administration, and distal tubular acidosis • An anion gap less than 9 mEq/liter is extremely rare and probably represents a lab error
Urine Chemistry and Electrolytes		
Urine electrolytes* Sodium	80–180 mEq/24 hr (varies with Na^+ intake)	• Urine Na^+ <10 mEq/liter 1. Hyponatremia due to Na^+ loss, as from GI tract, in burns, and late in use of diuretics 2. Oliguria due to hypovolemia • Urine Na^+ >20 mEq/liter (in presence of hypovolemia) 1. Adrenal insufficiency 2. Osmotic diuretics 3. Diuretics (early) 4. Salt-wasting renal disease • Urine Na^+ >20 mEq/liter (in presence of normal or slightly increased circulating blood volume) 1. Excessive ADH hormone secretion 2. Psychogenic polydipsia

Potassium	40–80 mEq/24 hr (varies with dietary intake) Normal ratio of urinary $Na^+:K^+$ is 2:1	• Increased in hyperaldosteronism (the normal $Na^+:K^+$ ratio of 2:1 may be reversed in this situation) • Decreased in adrenal insufficiency ($Na^+:K^-$ ratio may be 10:1 in this situation)

Chloride	110–250 mEq/24 hr	• Helpful in differentiating between types of metabolic alkalosis
		  1. <10 mEq/liter when metabolic alkalosis is due to vomiting, gastric suction, or diuretic use (late)
		  2. Usually >20 mEq/liter when metabolic alkalosis is due to hyperaldosteronism or profound K^+ depletion

Calcium	100–250 mg/24 hr (if on average diet) Varies with dietary intake	• Increased in hyperparathyroidism, osteolytic bone disease, idiopathic hypercalciuria, and immobilization (especially in children) • Decreased in hypoparathyroidism and vitamin D deficiency
Urine osmolality	Typical urine is 500–800 mOsm/liter (extreme range is 50–1400 mOsm/liter) Usually about 1½ to 3 times greater than serum osmolality	• Elevated in fluid volume deficit (healthy kidneys conserve needed fluid, causing urine to be more concentrated) • Decreased in fluid volume excess (healthy kidneys excrete unneeded fluid, causing urine to be dilute) • Simultaneous measurement of serum and urine osmolality is a more accurate way to measure renal concentrating ability than is urinary SG
Urinary specific gravity (SG)	1.003–1.030 (most random samples have an SG of 1.012 to 1.025)	• Elevated in fluid volume deficit • Decreased in fluid volume excess • SG persistently below 1.015 is a sign of significant renal disease

(continued)

*May be of limited value because of (1) the wide range of normal values (due to wide range of dietary intake of water and electrolytes) or (2) recent administration of diuretics. The measurement of urinary electrolytes without knowledge of the dietary intake is obviously of limited value.

Table 3–1
Laboratory Tests Used to Evaluate Fluid and Electrolyte Status (continued)

Laboratory Test	Normal Values	Significance
Urine Chemistry and Electrolytes *(continued)*		
		• As stated above, urine osmolality is a better way to measure renal function than is urinary SG Urinary SG depends on *weight* of the dissolved particles in urine, not on their *concentration* (as in osmolality). Therefore, heavy molecules, such as glucose and albumin, will elevate the SG out of proportion to the osmolality. Thus it is more accurate to measure urine osmolality than SG in patients with glycosuria or proteinuria
Urinary *p*H	4.5–8.0 (pooled daily output averages around 6.0; most random samples are <6.6)	• Usually increased (more alkaline) in alkalotic states; however, urine may be acidic when alkalosis is accompanied by severe hypokalemia
		• Increased (more alkaline) in urea-splitting urinary infections (*E. coli, Proteus,* and *Pseudomonas*)
		• Increased with use of alkalinizing agents, such as 1. Sodium bicarbonate 2. Potassium citrate
		• Decreased (more acidic) in acidotic states
		• Decreased with use of acidifying agents, such as 1. Ascorbic acid 2. Ammonium chloride 3. Sodium acid phosphate 4. Methenamine mandalate
Arterial Blood Gases		
*p*H	7.35–7.45	• <7.35 indicates acidosis • >7.45 indicates alkalosis

PaCO$_2$	38–42 mm Hg	• <38 mm Hg can indicate primary carbonic acid deficit (respiratory alkalosis) or can be a compensatory reaction to metabolic acidosis • >42 mm Hg can indicate primary carbonic acid excess (respiratory acidosis) or can be a compensatory reaction to metabolic alkalosis
PaO$_2$	80–100 mm Hg	• Lower than normal PaO$_2$ predisposes to anaerobic metabolism and accumulation of lactic acid (metabolic acidosis)
Bicarbonate	22–26 mEq/liter	• <22 mEq/liter can indicate primary bicarbonate deficit (metabolic acidosis) or can be a compensatory reaction to respiratory alkalosis • >26 mEq/liter can indicate primary bicarbonate excess (metabolic alkalosis) or can be a compensatory reaction to respiratory acidosis
Base excess	−2 to +2	• Expresses the *nonrespiratory* component of acid–base balance • < −2 in metabolic acidosis • > +2 in metabolic alkalosis

References

1. Rose B: Clinical Physiology of Acid-Base and Electrolyte Disorders, p 14. New York, McGraw-Hill, 1977
2. Ibid

Bibliography

Goldberger E: A Primer of Water, Electrolyte and Acid-Base Syndromes, 6th ed. Philadelphia, Lea & Febiger, 1980

Kempe C, Silver H, O'Brien D: Pediatric Diagnosis and Treatment. 4th ed. Los Altos, CA, Lange Medical Publications, 1976

Rose B: Clinical Physiology of Acid-Base and Electrolyte Disorders. New York, McGraw-Hill, 1977

Tilkian S, Conover M, Tilkian A: Clinical Implications of Laboratory Tests. St Louis, CV Mosby, 1979

Wallach J: Interpretation of Diagnostic Tests, 3rd ed. Boston, Little, Brown & Co. 1978

chapter 4

Principles of Parenteral Fluid Therapy

The nurse has a primary role in parenteral fluid therapy, for it is the nurse who starts parenteral fluids and monitors their administration. One must be knowledgeable about the contents of parenteral fluids, their purposes, and the contraindications and complications associated with their use, because they are medications. Also, it is important to know how to assess the patient's response to parenteral therapy and prevent complications when possible.

Parenteral Fluids

Before discussing the types of parenteral fluids available and the factors the nurse should consider in their administration, it is helpful to review how the physician calculates fluid orders.

Factors Affecting Fluid Orders

Factors the physician considers in prescribing parenteral fluids include
- Renal function
- Daily maintenance requirements
- Clinical status (including existing fluid and electrolyte imbalances)
- Avoiding the creation of new disturbances as a result of parenteral fluid therapy

57

Renal Function

- Before treatment is begun for the patient with a real or potential fluid balance problem, renal status must be evaluated. There is a danger of causing other imbalances if the kidneys are not functioning adequately. For example, administration of potassium-containing fluids to a patient with inadequate renal function can induce serious hyperkalemia.

- If the patient with a severe fluid volume deficit is oliguric, it is necessary to determine if the depressed renal function is the result of the fluid volume deficit (prerenal azotemia), or, more seriously, acute tubular necrosis secondary to fluid volume deficit. To differentiate between these conditions, a fluid load is infused in a short period of time and the patient is observed closely. If urine output improves, the problem was simple fluid volume deficit; if urine flow is not re-established, the problem is likely renal insufficiency. Description of a fluid challenge test to differentiate simple fluid volume deficit from a deficit complicated by acute tubular necrosis is described in Chapter 5. It is important to differentiate between these two conditions because *prompt treatment of prerenal azotemia can prevent acute tubular necrosis.* (Prompt reporting of low urine output to the physician is thus imperative.)

- Treatment of a patient with adequate renal function is simplified greatly as long as sufficient water and electrolytes are provided, because the kidneys can select needed substances and keep the patient in fluid and electrolyte balance.

Maintenance and Replacement Needs

- Various methods for calculating daily fluid requirements are available.

- One method operates on the assumption that approximately 30 ml of fluid are needed per kilogram of body weight (or 15 ml/lb) for maintenance needs.

Example: A 70-kg adult would require approximately 2100 ml daily. This amount is for maintenance only; the presence of abnormal losses increases the need for fluids. This method does not take into consideration that fluid needs vary with age (children needing more fluid per kilogram of body weight than adults).

- A more precise method for calculating fluid needs utilizes body surface area and thus is more valid across age groups. The assumption with this method is that approximately 1500 ml of fluid are required per M^2 of body surface area for maintenance needs. (Table 4-1 lists body surface areas for various body weights.) Extra fluid is needed to manage existing fluid volume deficits.

Table 4–1
**Conversion of Body Weight
to Body Surface Area (BSA)***

| Weight | | Approximate BSA |
kg	lb	in Square Meters (m²)
3	6.6	0.20
6	13.2	0.30
10	22.0	0.45
20	44.0	0.80
30	66.0	1.00
40	88.0	1.30
50	110.0	1.50
65	143.0	1.70
70	154.0	1.76
85	187.0	2.00

*Figures are approximate and apply only to individuals of average body build.

1. A moderate fluid volume deficit increases the 24-hour fluid needs to approximately 2400 ml/M^2 of body surface area (includes maintenance and volume replacement).
2. A severe fluid volume deficit increases the 24-hour fluid needs to approximately 3000 ml/M^2 of body surface area (includes maintenance and volume replacement).

- High environmental humidity (as in an incubator) minimizes insensible water loss and, thus, decreases the total amount of fluid to be given.
- Fever increases fluid maintenance requirements by approximately 15% for each 1° C rise in body temperature.
- Maintenance requirements for sodium in the adult are usually met if 100 mEq/day are provided; the daily maintenance allowance for potassium is approximately 40 mEq to 60 mEq.
- Magnesium, calcium, and phosphorus supplements are needed if parenteral therapy is necessary for more than 1 or 2 weeks.
- Approximately 100 Gm to 150 Gm of carbohydrate are needed daily to minimize protein catabolism and prevent ketosis of starvation. A liter of a 5% dextrose solution has 50 Gm of dextrose; a liter of a 10% dextrose solution has 100 Gm of dextrose.
- Abnormal losses of body fluids are treated according to the type of fluid lost. For example, there are commercial fluids available for replacement of gastric and intestinal fluid losses.
- Usually abnormally lost fluids are replaced in the volume they are lost. However, a urine output exceeding 80 ml/hr should *not* be replaced on a volume for volume basis since it is an indication of fluid

overload; indeed, fluids should be limited to decrease the output to a more appropriate level (such as 40 to 60 ml/hr). In prolonged polyuric states, it is necessary to measure urine electrolytes to determine replacement needs for sodium and potassium.

Clinical Status

- Fluid orders should be readjusted frequently as indicated by changes in the patient's clinical status.
- Factors included in clinical assessment include
    1. Comparison of intake-output measurements (used frequently to measure the effectiveness of treatment)
    2. Daily body weights
    3. Skin and tongue turgor
    4. Vital signs
    5. Central venous pressure
    6. Urinary specific gravity
    7. Laboratory values
- The nurse needs to be constantly alert for significant changes in these indices of physiologic functioning and quickly share the findings with the physician so that appropriate adjustments in fluid therapy can be made.
- Head-injured patients are usually given slightly less fluid than normal to lessen the danger of cerebral edema. Authors vary as to recommended restrictions; however, most agree that the amount should be based on the patient's plasma osmolality and electrolyte status. Many authorities favor a volume of 1200 to 1800 ml/day in head-injured adults. (Fluid therapy in head-injured patients is discussed further in Chap. 21.)
- Surgical patients should not receive excessive volumes of fluid (particularly dextrose in water), because of excessive secretion of ADH in the first 24 to 48 hours postoperatively. (Fluid therapy in postoperative patients is discussed further in Chap. 15.)

Water and Electrolyte Solutions

- Table 4-2 lists some common water and electrolyte solutions with comments about their use.
- Electrolyte solutions are considered *isotonic* if the total electrolyte content (anions plus cations) approximates 310 mEq/liter. They are considered *hypotonic* if the total electrolyte content is less than 250 mEq/liter and *hypertonic* if the total electrolyte content exceeds 375 mEq/liter.
- In addition to commercially prepared solutions ready for use, pharmaceutical companies prepare additives so that the physician can tailor a solution specifically to the patient's needs.

(*Text continues on p. 64.*)

Table 4–2
Contents of Selected Water-and-Electrolyte Solutions With Comments About Their Use

Solution	Comments
0.9% NaCl (isotonic saline) Na$^+$ 154 mEq/liter Cl$^-$ 154 mEq/liter	An isotonic solution that expands plasma volume—used in hypovolemic states Supplies an excess of Na$^+$ and Cl$^-$—can cause fluid volume excess and hyperchloremic acidosis if used in excessive volumes (particularly in patients with compromised renal function) Sometimes used to correct mild metabolic alkalosis Sometimes used to correct mild Na$^+$ deficit Not desirable as a routine maintenance solution since it provides only Na$^+$ and Cl$^-$ (and these are provided in excessive amounts)
0.45% NaCl (half-strength saline) Na$^+$ 77 mEq/liter Cl$^-$ 77 mEq/liter	A hypotonic solution that provides Na$^+$, Cl$^-$, and free water Free water is desirable to aid the kidneys in elimination of solute Na$^+$ and Cl$^-$ are provided, allowing the kidneys to select needed amounts of these electrolytes Lacking in other electrolytes needed for daily replacement
0.3% NaCl (third-strength saline) Na$^+$ 51 mEq/liter Cl$^-$ 51 mEq/liter	A hypotonic solution that provides Na$^+$, Cl$^-$, and free water Often used to treat hypernatremia (because this solution contains a small amount of Na$^+$, it dilutes the plasma sodium while not allowing it to drop too rapidly)
3% NaCl Na$^+$ 513 mEq/liter Cl$^-$ 513 mEq/liter **5% NaCl** Na$^+$ 855 mEq/liter Cl$^-$ 855 mEq/liter	Grossly hypertonic solutions used *only* to treat severe hyponatremia Must be administered *cautiously* in small volumes over a long period of time (such as 200–300 ml over a 4-hr period) Not more than 400 ml of a 5% NaCl solution should be infused in 1 day Patients receiving these fluids require close clinical observation—excessive administration can cause serious fluid volume excess with pulmonary edema due to the high Na$^+$ content Administer in volume-controlled container with microdrip setup or with a volumetric pump

(continued)

Table 4–2
**Contents of Selected Water-and-Electrolyte
Solutions With Comments About Their Use** (continued)

Solution	Comments
Lactated Ringer's solution (Hartmann's solution) Na^+ 130 mEq/liter K^+ 4 mEq/liter Ca^{2+} 3 mEq/liter Cl^- 109 mEq/liter Lactate (metabolized to bicarbonate) 28 mEq/liter	An isotonic solution that contains multiple electrolytes in roughly the same concentration as found in plasma (note that this solution is lacking in Mg^{2+}) Used in the treatment of hypovolemia, burns, and fluid lost as bile or diarrhea Useful in treating *mild* metabolic acidosis Does not supply free water for renal excretory purposes; excessive *use* without provision for free water (as with D5W or hypotonic electrolyte solutions) can cause elevation of the serum sodium level in persons not deficient in sodium
5% dextrose in water (D_5W) No electrolytes 50 Gm of dextrose	Supplies approximately 170 cal/liter and free water to aid in renal excretion of solutes Should not be used in excessive volumes in the early postoperative period (when ADH secretion is increased due to stress reaction) Some authorities caution against the administration of electrolyte-free solutions in head-injured patients
Isotonic multiple electrolyte solutions Plasma-Lyte (Travenol) Isolyte E (McGaw) Na^+ 140 mEq/liter K^+ 10 mEq/liter Ca^{2+} 5 mEq/liter Mg^{2+} 3 mEq/liter Cl^- 103 mEq/liter HCO_3^- 55 mEq/liter	Isotonic solution with electrolyte content similar to plasma except that it has twice as much K^+ and a higher HCO_3^- content Sometimes used to replace intestinal fluid losses
Potassium chloride 0.2% in dextrose 5% K^+ 27 mEq/liter Cl^- 27 mEq/liter Glucose 50 Gm 0.3% in dextrose 5% K^+ 40 mEq/liter Cl^- 40 mEq/liter Glucose 50 Gm	Both solutions provide KCl, water, and calories Follow rules for administration of potassium solutions (listed later in this chapter)

Table 4-2
Contents of Selected Water-and-Electrolyte Solutions With Comments About Their Use (continued)

Solution	Comments
Gastric replacement solution Electrolyte No. 3 (Travenol) Isolyte G (McGaw) Ionosol G (Abbott) Na^+ 63 mEq/liter K^+ 17 mEq/liter NH_4^+ 70 mEq/liter Cl^- 150 mEq/liter	An isotonic solution used to replace gastric fluid lost in vomiting or gastric suction pH is acidic (3.3–3.7) —irritating to vein Give in large peripheral veins with good blood flow to protect venous wall from irritation Contraindicated in presence of hepatic or renal failure
Duodenal replacement solution Na^+ 138 mEq/liter K^+ 12 mEq/liter HCO_3^- 50 mEq/liter Cl^- 100 mEq/liter	Used to replace water and electrolytes lost as a result of intestinal suction, drainage, or fistulas K^+ concentration is similar to that in intestinal secretions
Sodium bicarbonate, 1.5% Na^+ 178 mEq/liter HCO_3^- 178 mEq/liter	Isotonic solution used to treat severe metabolic acidosis Due to high Na^+ content, observe for fluid overload in patients with renal or cardiac impairment Observe for signs of hypocalcemia (which may be induced with rapid alkalinization of plasma)
Sodium lactate solution, 1/6 molar Na^+ 167 mEq/liter Lactate 167 mEq/liter	Used to correct severe metabolic acidosis (lactate is metabolized to bicarbonate in 1–2 hr by the liver) Calcium salts may be needed to correct symptomatic hypocalcemia that may occur with rapid alkalinization of plasma Not used in patients with liver disease, since lactate cannot be converted to bicarbonate in such individuals; also, not used in patients with oxygen lack (unable to adequately convert lactate to bicarbonate)
Ammonium chloride, 0.9% NH_4^+ 168 mEq/liter Cl^- 168 mEq/liter	An acidifying solution used to correct severe metabolic alkalosis Due to high ammonia content, must be administered cautiously to patients with compromised hepatic function

(continued)

Table 4–2
**Contents of Selected Water-and-Electrolyte
Solutions With Comments About Their Use** (continued)

Solution	Comments
Electrolyte no. 48 Na^+ 25 mEq/liter K^+ 20 mEq/liter Mg^{2+} 3 mEq/liter Cl^- 22 mEq/liter PO_4^{2-} 3 mEq/liter HCO_3^- 23 mEq/liter (or equivalent)	A pediatric hypotonic maintenance solution Supplies numerous electrolytes plus free water Note high K^+ content
Electrolyte no. 75 Na^+ 40 mEq/liter K^+ 35 mEq/liter Cl^- 40 mEq/liter PO_4^{2-} 15 mEq/liter HCO_3^- 20 mEq/liter (or equivalent)	Hypotonic maintenance solution Supplies numerous electrolytes plus free water

Nursing Considerations in Administration of Potassium Solutions

- Because potassium is a vital electrolyte tolerated only in very limited amounts, it is important to review important points related to its safe administration. Potassium may be given IV as potassium chloride, potassium phosphate, and potassium acetate.
- Nursing considerations in potassium administration include the following:
  1. Mix small ampules containing concentrated solutions of potassium salts with at least 500 ml to 1000 ml of solution (except in the treatment of extreme hypokalemia in patients unable to tolerate a large fluid load, described below in No. 6). Potassium should *never* be directly administered in concentrated form by IV push because of the danger of cardiac arrest.
  2. Thoroughly mix KCl when it is added to an IV solution by shaking the bag or bottle to prevent "crowning" or "layering." "Crowning" of the KCl allows the administration of a large bolus of the drug at one time; phlebitis, tissue necrosis, and even cardiac arrest may occur. It is important to squeeze the medicine ports of plastic bottles while they are in the upright position and then to mix the solution thoroughly. Remember that KCl should never be added to an IV container in the hanging position!
  3. Limit the potassium concentration in 1 liter of fluid to 20 mEq

to 40 mEq, since an accidental rapid infusion rate is less dangerous when potassium content in the solution is moderate. (No more than 80 mEq of potassium should ever be added to 1 liter of fluid!)

4. Administer potassium only after adequate urine flow has been established. A decrease in urine volume to less than 20 ml/hr for 2 consecutive hours is an indication to stop potassium infusion until the situation is evaluated:

   • Urinary suppression may be due to either inadequate fluid intake or renal impairment; the rapid infusion of a hydrating solution (such as a hypotonic electrolyte solution) should cause an increase in urine output if the problem is fluid volume deficit. Once urinary output is adequate, potassium infusion may be resumed. However, failure of the hydrating solution to increase urinary output indicates renal impairment and is an indication to withhold potassium. (Recall that potassium is mainly excreted by way of the kidneys; when the kidneys are not functioning properly, a high potassium level builds up in the bloodstream.)

5. Be aware that for usual IV potassium replacement therapy, 5 to 10 mEq/hr (suitably diluted) may be given as a constant infusion. Except in extreme hypokalemia, the maximum rate should not exceed 20 mEq/hr (suitably diluted).

6. Be aware that in severe cases of hypokalemia (such as a serum $K^+ < 2.5$ mEq/liter), it may be necessary to administer a larger dose of potassium (preferably no more than 20 mEq/hr, certainly no more than 40 mEq/hr) while continuously monitoring the ECG. Serum potassium levels should be closely monitored.

   • In intensive care units when fluid volume must be severely limited, the hourly dose is sometimes diluted in as little as 100 ml of fluid and administered in a volume-controlled set-up under constant observation. Of course, grossly concentrated potassium solutions should not be given in peripheral veins because venous trauma and pain would result.

7. Be aware that potassium replacement must be accomplished slowly; a rapid rise in serum potassium predisposes to the effects of hyperkalemia, even when the serum level remains near the upper limits of normal.

8. Be aware that, according to one source, it takes approximately 40 mEq to 60 mEq of potassium to raise the serum potassium level 1.0 mEq/liter.[1] Another source states that if the serum potassium is less than 3.0 mEq/liter, an infusion of 200 mEq to 400 mEq of potassium is generally necessary in order to raise the serum potassium by 1 mEq/liter; if the serum potassium is between 3.0 and 4.5 mEq/liter, 100 mEq to 200 mEq will raise the serum potassium by 1 mEq/liter.[2]

9. Be aware that a solution containing sizable amounts of potassium (30 to 40 mEq/liter) is frequently associated with pain in the vein as it enters, especially if infused into a vein where a previous venipuncture has been performed. Slowing the administration rate usually relieves this sensation.

10. Take care to avoid accidental subcutaneous infiltration of potassium solutions, since severe tissue damage may result, particularly when the potassium content in the solution exceeds 10 mEq/liter. Should infiltration of KCl occur, the infusion should be stopped immediately.
    - One manufacturer suggests local infiltration of the affected site with 1% procaine HCl, to which hyaluronidase may be added, to reduce venospasm and dilute the potassium remaining in the tissues locally.[3] Also, local application of heat may be helpful.

Nursing Considerations in Administration of Calcium Solutions

- Calcium may be administered as calcium gluceptate, calcium gluconate, or calcium chloride. It is important that the nurse consider the following facts when administering calcium solutions intravenously:

1. Do not add calcium salts to IV solutions containing sodium bicarbonate since the precipitate calcium carbonate will form.
2. Be aware that intravenous calcium administration usually is contraindicated in digitalized patients since calcium ions exert an effect similar to that of digitalis and can cause digitalis toxicity with adverse cardiac effects.
3. Be aware that excessive or too rapid administration of calcium intravenously can cause cardiac arrest, preceded by bradycardia. Recall that hypercalcemia can cause the heart to go into spastic contraction.
4. Be aware that calcium preparations other than calcium chloride are preferred, since calcium chloride may cause venous irritation. (Calcium chloride is three times more potent than calcium gluconate.)
5. Warm the solution to body temperature before administration.
6. Be aware that necrosis and sloughing will occur if the calcium solution infiltrates into the subcutaneous tissue.

Nursing Considerations in Administration of Magnesium Solutions

- The following facts related to magnesium administration should be considered:

1. Question the use of magnesium solutions in patients with oliguria (recall that 99% of parenterally administered magnesium is excreted via the kidneys).

2. Avoid rapid administration of magnesium solutions since they may cause uncomfortable sensations of heat; more importantly, they may cause coma, respiratory depression, and cardiac arrest.

3. Observe the patient's respirations and deep tendon reflexes. A respiratory rate below 12 to 14 per minute should be reported to the physician, as should poor to absent tendon reflexes (such as the knee jerk).

4. Have calcium gluconate immediately available to counteract serious symptoms of hypermagnesemia, should they occur.

Parenteral Nutrients

- The recommended energy need of an adult at bedrest is approximately 1600 calories; this is a basal figure and does not allow for fever or other causes of increased metabolism.

- It is difficult to restore the nutritionally depleted patient by the intravenous route using "routine" fluids; however, it is possible to maintain the state of nutrition fairly well with their use for a short period of time. The advent of total parenteral nutrition (hyperalimentation) has made it possible to support life and maintain growth and development for prolonged periods. (See Chap. 13 for a discussion of total parenteral nutrition.)

Carbohydrates

Carbohydrates that can be administered and absorbed intravenously include

1. Dextrose
2. Fructose
3. Invert sugar

- Dextrose in water is commercially available in strengths of 2.5%, 5%, 10%, 20%, 25%, and 50%. The caloric content and tonicity of some carbohydrate solutions are listed in Table 4-3.

- It is impossible to meet daily caloric needs with isotonic solutions of carbohydrate solutions alone. For example: to supply 1600 calories with a 5% dextrose solution would require 9 liters, a volume exceeding the tolerance of most patients.

- Concentrated solutions, such as 20% or 50% dextrose, are useful for supplying calories. For the dextrose to be utilized, concentrated solutions must be administered slowly. When administered rapidly, such solutions act as an osmotic diuretic and pull interstitial fluid into the plasma for subsequent renal excretion.

- If a hypertonic carbohydrate solution is infused too rapidly, a transient hyperinsulin reaction may occur. (The pancreas secretes extra insulin to metabolize the infused carbohydrate.) Sudden discontinuance of the fluid may leave a temporary excess of insulin in the

Table 4–3
Parenteral Carbohydrate Solutions

Types of Solutions	Cal/liter	Tonicity
Dextrose		
2½% Dextrose	85	Hypotonic
5% Dextrose	170	Isotonic
10% Dextrose	340	Hypertonic
20% Dextrose	680	Hypertonic
50% Dextrose	1700	Hypertonic
Fructose		
10% Fructose	375	Hypertonic
Invert Sugar		
5% Invert Sugar	190	Isotonic
10% Invert Sugar	375	Hypertonic

body, causing nervousness, sweating, and weakness. It is not uncommon for small amounts of isotonic carbohydrate solution to be given after hypertonic solutions to "cover" the extra insulin and allow the return to normal secretion.

- Dextrose is thought to be the closest to ideal carbohydrate available because it is well metabolized by all tissues.
- Fructose has essentially the same caloric equivalent as glucose and may be less irritating to veins; it can be metabolized by adipose tissue independent of insulin. Fructose is contraindicated when lactic acidosis, or other forms of metabolic acidosis, are present or likely.
- Invert sugar contains equimolar quantities of glucose and fructose; its caloric equivalent is the same as that of fructose. At similar rates of administration, considerably less invert sugar than dextrose is lost in the urine.

Proteins

- Parenteral proteins are usually indicated when intravenous fluid therapy is necessary for more than 3 to 4 days (provided the patient is not eating). Protein is necessary for cellular repair, wound healing, growth, and for the synthesis of certain enzymes, hormones, and vitamins. Proteins are available commercially as either protein hydrolysates or crystalline amino acid solutions; they can be administered through peripheral or central veins.
- Nitrogen balance is the accepted measure for protein balance. A patient is in nitrogen balance if the rate of protein synthesis equals that of protein breakdown; he is in negative nitrogen balance if protein catabolism (breakdown) exceeds protein anabolism (buildup).

- Positive nitrogen balance can be maintained when protein preparations are administered intravenously with sufficient calories to prevent breakdown of the protein for energy purposes. A healthy adult requires approximately 1 Gm of protein per kg of body weight daily to replace normal protein losses; healthy growing infants and children require 1.4 Gm to 2.2 Gm of protein per kg of body weight.
- A liter of 5% amino acid solution contains 6.2 Gm of nitrogen; this is equivalent to 39 Gm of protein (the approximate daily requirement for an adult).

Nursing Considerations in Administration of Protein Solutions

- The nurse should keep the following facts in mind when administering protein solutions:

    1. Examine all protein solutions carefully before they are infused; either particulate matter or cloudiness should call for discarding the solution.
    2. Use protein solutions immediately after being opened since they are subject to spoilage.
    3. Be aware that the solution should be started slowly and the rate gradually increased (per manufacturer's recommendations). Never exceed a rate of 4 ml/kg body weight per hour.
    4. Be alert for side-effects such as vomiting, fever, vasodilatation, abdominal pain, and anaphylaxis.

Fat Emulsions

- Fats supply more than twice the calories of proteins or carbohydrates since 1 Gm of fat yields 9 calories, whereas 1 Gm of carbohydrates or protein yields only 4 calories.
- The presently available fat emulsion preparations in this country are Intralipid 10% and 20% (Cutter Laboratories), Liposyn 10% and 20% (Abbott Laboratories), and Travamulsion 10% (Travenol Laboratories).
- Ten percent fat emulsions provide 1.1 cal per ml and have an osmolarity of approximately 280 mOsm/liter; 20% fat emulsions provide 2 cal per ml and have an osmolarity of approximately 330 mOsm/liter.[4]
- Fat emulsions can be administered through peripheral or central veins. They have been in use for some time in many parts of the world; when given according to manufacturer's directions, fat emulsions are usually tolerated without incident.

Nursing Considerations in Administration of Fat Emulsions

- The nurse should keep the following facts in mind when responsible for the administration of fat emulsions:

1. Inspect the emulsion for separation or an oily appearance; if present; do not administer.
2. Do not place any additives in the fat emulsion container since this can cause instability of the emulsion.
3. Do not mix fat emulsions with electrolyte or other nutrient solutions (unless manufacturer's directions state that it is allowable to do so).
4. When it is necessary to administer a fat emulsion in conjunction with amino acids and hypertonic dextrose, it should be administered via a Y-connector or three-way stopcock near the infusion site. One manufacturer (Cutter Laboratories) states that it is possible to mix their fat emulsion (Intralipid) with selected amino acids, dextrose, and other IV nutrients to provide a single-container system for nutritional admixtures.
5. Be aware that refrigeration is no longer required for fat emulsions; however, they should be protected from excessive heat during storage. (Do not store Intralipid 10% and 20% above 25° C (77° F); do not store Liposyn 10% above 30° C (86° F).[5]
6. Be aware that the Center for Disease Control recommends that intravenous fat emulsions be used within a 12-hour period after starting the infusion.
7. Observe for the following adverse reactions that may occur with intravenous fat emulsion administration:
   - Dyspnea and cyanosis
   - Allergic reactions
   - Hyperlipemia (monitor lipid levels; lipemia should clear between daily infusions)
   - Hypercoagulability
   - Nausea and vomiting
   - Headache
   - Flushing
   - Back or chest pain
   - Dizziness
   - Leukopenia
   - Hepatomegaly and splenomegaly (monitor liver function frequently; if tests indicate liver impairment, the infusion should be discontinued)
8. Be aware that initial infusion rates should be very slow (as indicated by the manufacturer) while the patient is observed for adverse reactions. If no untoward reactions occur, the rate can be increased (as recommended by the manufacturer).

Example: In the adult, 500 ml of a 10% fat emulsion should be infused no faster than over a 4-hour period; 500 ml of a 20% fat emulsion should be administered no faster than over an 8-hour period. Note that dosages and infusion rates vary with the concentration of the fat emulsion and the patient's condition.

Alcohol Solutions

- One gram of alcohol yields 6 to 8 calories; alcohol is sometimes combined with carbohydrate to provide high-caloric intake (for example, a solution of 5% alcohol with 5% dextrose contains approximately 450 cal/liter).
- When alcohol is infused with carbohydrate, it is burned preferentially, permitting the glucose to be stored as glycogen; alcohol spares body protein by providing readily accessible calories.
- The major uses of alcohol solutions are probably in temporary situations where a euphoric effect is desirable or a high-energy solution is needed. The sedating effect of alcohol solutions is highly desirable for patients with pain.

Nursing Considerations in Administration of Alcohol Solutions

Points the nurse should keep in mind when administering alcohol solutions include

- Sedation can be achieved in the average adult without symptoms of intoxication by giving 200 ml to 300 ml of a 5% solution per hour.
- Parenterally administered alcohol causes dulling of memory, an improved sense of well-being, a loss of ability to concentrate, increased pulse and respiration, and vasodilation.
- Alcohol solutions should not be used in the presence of shock, epilepsy, severe liver disease, or in patients with coronary thrombosis.
- Alcohol inhibits the secretion of ADH, predisposing to water excretion (and, when used in excess, serum hyperosmolality).
- Tissue necrosis can occur if the alcohol solution infiltrates into the subcutaneous tissue; care should be taken to see that the needle is carefully anchored in the vein and that the site is inspected frequently to detect infiltration.

Parenteral Vitamins

- Vitamins should be administered parenterally when there is inadequate oral intake or when parenteral fluid therapy is necessary for longer than 2 or 3 days.
- Although not foods in themselves, vitamins are essential for the utilization of other nutrients.
- The need for vitamins is increased during periods of stress such as acute illness, infection, surgery, burns, injury, and convalescence.
- The vitamins most frequently needed in parenteral therapy are vitamin C and several of the B complex vitamins, all of which are water soluble and stored by the body only in small amounts.

- Vitamin deficiency has been observed after only 1 week of parenteral therapy of dextrose and water alone.
- If intravenous therapy is prolonged, fat-soluble vitamins A and D should be supplied weekly.

Total Parenteral Nutrition

Total parenteral nutrition (TPN; hyperalimentation) is discussed in Chapter 13.

Complications of Intravenous Fluid Administration

Pyrogenic Reactions

- The presence of pyrogenic substances in either the infusion solution or the administration setup can induce a febrile reaction. (Pyrogens are foreign proteins capable of producing fever.)
- Severity of symptoms depends upon the number of pyrogens infused, and the patient's susceptibility. (Patients with fever or liver disease are more susceptible than others.)
- *Nursing actions to prevent and detect pyrogenic reactions*
  1. Handle infusion containers and administration sets aseptically since contaminants can enter after they are opened—use particular care when adding medications to the solution.
  2. Prior to initiating an infusion, squeeze plastic containers to detect leaks and inspect glass bottles against light for cracks or bright reflections that penetrate the glass wall of the bottle; if either are present, the solution is no longer sterile and must be discarded.
  3. Indicate on the bottle the date and time the seal was broken (parenteral solutions should be discarded after 24 hr).
  4. Use 500-ml bottles for "keep-open" infusions to assure change of containers within a safe period of time. (Recall that the lowest volume that can be given to keep the vein open is desired; with a microdrip set, as little as 350 ml can be given in 24 hr.)
  5. Change administration sets at least every 24 to 48 hours.
  6. Discard any normally clear solution that appears cloudy or has particulate matter in it.
  7. Change administration sets after the administration of blood or other protein-containing solutions because these substances become a growth medium for bacteria.
  8. Be aware that symptoms of a pyrogen reaction include

- An abrupt temperature elevation accompanied by severe chills (the reaction usually begins about 30 minutes after the start of the infusion).
- Headaches, general malaise, nausea and vomiting.
9. Stop the infusion at once if symptoms (listed above) occur. Check the vital signs, and notify the physician. Save the solution, administration set, and the venipuncture device (aseptically capped) so that they can be cultured if necessary.

Infiltration

- *Infiltration* is defined as the entry of fluid into the subcutaneous tissue when the needle inadvertently leaves the vein.
- Fluids particularly irritating to the subcutaneous tissue include potassium solutions, hypertonic solutions, and those with a pH varying greatly from that of the body (such as ammonium chloride, 1/6 molar sodium lactate, or gastric replacement solution).
- *Nursing actions to prevent and detect infiltration*
    1. Select a site for venipuncture that is not over an area of flexion (such as the wrist or elbow). If necessary to use these sites, apply an armboard.
    2. Anchor venipuncture devices securely with tape to keep them from accidentally dislodging.
    3. Educate patients to protect the extremity receiving the infusion.
    4. Avoid rough handling of the extremity receiving the infusion and of the equipment during transportation of patient to other departments.
    5. Observe for infiltration at frequent intervals to allow early detection (before significant tissue trauma results).
       Signs of infiltration include
       - Edema at the injection site.
       - Cooler temperature of skin around the injection site than elsewhere (because the IV fluid is cooler than the body).
       - Significant decrease in infusion rate, or a complete stop of the infusion.
       - Failure to get a blood return into the tubing when the bottle is lowered below the level of the needle. (This method is not always foolproof, however, since the needle lumen may be partially in the vein with its tip in the subcutaneous tissue.)
    6. Test for infiltration, when in doubt, by applying a tourniquet (or applying pressure with the fingers) to restrict venous flow proximal to the injection site; if the solution continues to flow, regardless of the venous obstruction, the needle is obviously not in the vein.
    7. Discontinue the infusion immediately when infiltration is noted.

Thrombophlebitis

- *Thrombophlebitis* is a condition with clot formation in an inflamed vein.
- Although some degree of venous irritation accompanies all intravenous infusions, it is usually of significance only in infusions kept going in the same site for more than 12 hours and when particularly irritating fluids are administered (such as alcohol, vesicant agents, and hypertonic solutions).
- *Nursing actions to prevent and detect thrombophlebitis*
    1. Change intravenous devices as indicated. Recall that it is recommended that a peripheral venipuncture device not be left in place longer than 72 hours, and preferably no longer than 48 hours.
    2. Infuse irritating solutions in large veins. Recall that large veins have a higher blood flow and thus can quickly dilute irritants.
    3. Dilute irritating additives with manufacturer's guidelines in mind. (Recall that the more dilute the solution, the less likely the irritant will cause venous trauma.)
    4. Stabilize venipuncture devices with proper taping to avoid irritating the vein.
    5. Apply an armboard when the venipuncture is near an area of flexion to avoid irritation of the vein by the device.
    6. Observe the venipuncture site for signs of thrombophlebitis.
        - Pain along the course of the vein
        - Redness and edema at the injection site (a red streak may form above the site)
        - Vein feels hard, warm, and cordlike to touch
        - Flow rate sluggish due to venospasm
        - If severe, systemic reaction to the infection will occur (tachycardia, fever, and malaise)
    7. Stop the infusion when thrombophlebitis is detected; restart the infusion in another site.
    8. Apply cold compresses to the injured site; later, warm compresses may be used to relieve discomfort and promote healing.

Circulatory Overload

- Circulatory overload refers to an excessive fluid volume in the bloodstream (often caused by excessive parenteral fluid administration).
- *Nursing actions related to circulatory overload*
    1. Be alert for symptoms of circulatory overload, particularly in patients with cardiac decompensation.
       Symptoms include

- Rise in blood pressure and central venous pressure
- Venous distension (engorged peripheral veins)
- Wide variance between fluid intake and output (intake greatly exceeds output)
- Coughing, shortness of breath, and rales
- Pulmonary edema with severe dyspnea and cyanosis

2. Be aware that many physicians favor monitoring CVP on all patients with a history of congestive heart failure or in whom rates of fluid administration will exceed 500 ml/hr.

3. Slow the infusion to a keep-open rate if symptoms of overload occur; elevate the patient to a sitting position to facilitate breathing. Notify the physician.

Air Embolism

- The danger of air embolism is present in all intravenous infusions, even though it does not occur frequently.
- Cannulation of central veins (as for hyperalimentation or CVP measurement) is far more likely to be associated with air embolism than is cannulation of peripheral veins.
- The exact quantity of air that can lead to death in humans is unknown, but appears to be related to the *rate* of entry, as follows:

1. A fatal episode has been reported following the sudden administration of 100 ml of air intravenously.[6]

2. One source states that normal adults can tolerate as much as 200 ml of air intravenously; but in seriously ill patients, as little as 10 ml may be fatal.[7]

3. Another source extrapolated data from animal studies and concluded that the average lethal dose in humans would be between 70 and 150 ml/sec.[8]

Nursing Actions to Prevent Air Embolism

GENERAL CONSIDERATIONS

- Tightly secure all connections in the administration setup to prevent air from entering the system. If a stopcock is part of the IV setup, the outlets not in use should be completely shut off. (Avoid overtightening connections so that a forceps is needed to loosen them, which can cause cracking of the tubing or device and "create" an entry site for air.)
- Inspect plastic bags and tubing for cracks or other defects that may allow the entry of air.
- Discontinue an infusion before the container and tubing are completely empty.
- Be aware that the danger of air embolism is greater when vented glass containers are used than when plastic containers are used.

(continued)

- Be aware that the potential for air embolism exists when fluid from two vented containers run simultaneously through the same needle (Fig. 4-1). Completely clamp off the first bottle to empty; otherwise, air will be drawn into the vein from the empty vented container.
- Read and follow manufacturer's directions for safe use of infusion pumps since some types can pump air into the vein if the infusion bottle is allowed to empty.
- Be alert for signs of air embolism.
  1. Dyspnea and cyanosis
  2. Hypotension
  3. Weak, rapid pulse
  4. Loud continuous churning sound over the precordium (not always present)
  5. Loss of consciousness
- Clamp the administration tubing immediately if the above symptoms occur; some authorities recommend turning the patient on the left side with the head down and the lower extremities elevated (left lateral and modified Trendelenburg position). This allows air to rise into

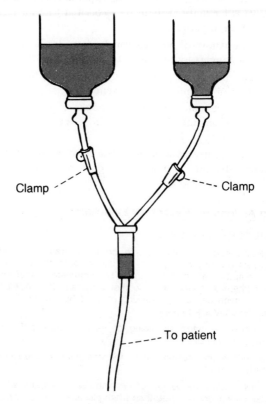

Clamp

Clamp

To patient

Figure 4–1. Bottles connected in parallel (Y-type) set-up.

the right atrium and away from the pulmonary outflow tract. Oxygen is administered by mask to achieve a high oxygen concentration. The physician may attempt to aspirate air from the right ventricle with an intracardiac needle.

Central Veins

- Position the patient flat in bed and have him perform the Valsalva maneuver during the time (seconds) that the tubing is disconnected from the catheter for tubing change.

  1. The Valsalva maneuver is accomplished by telling the patient to "take a deep breath and hold it while bearing down" (forced expiration against a closed glottis).

  2. If the patient cannot do the Valsalva maneuver or has a tracheal tube, he may be ventilated with an Ambu bag, changing the tubing during a prolonged inspiratory phase. (The catheter hub is opened to air only after the inspiratory phase has been held for 3 sec to 5 sec, and before it is released.)[9] Two nurses are required for this procedure.

  3. Recall that there is a low pressure in the central veins that can pull air in during tubing changes or when connections in the administration apparatus are not air tight. This low pressure is particularly pronounced in the hypovolemic patient; hypovolemia generates an increased "sucking force" in the central veins. Valsalva's maneuver increases the intrathoracic pressure as well as the mean pressure in the central veins, decreasing the danger of air being sucked in.

  4. Be sure that the connection between the catheter hub and the tubing is secure. Adequate taping of the tubing and catheter lessens the possibility of accidental dislodgment. This is particularly important when the patient is elevated from a flat to a sitting or standing position since the danger of air embolism is greater when the patient is upright.

  5. *Occlude the catheter hub immediately* if accidental separation of the tubing occurs.

- Apply an occlusive dressing to the site immediately after removal of the central venous catheter, since the catheter tract can be a source of air embolism.

Peripheral Veins

- Do not elevate the extremity receiving the infusion above the level of the heart since this results in venous collapse and negative venous pressure. Negative pressure in the vein receiving the infusion can draw air in if there are any defects in the apparatus or if the solution flask empties.

- Keep the clamp that regulates the flow rate below the level of the heart. If the clamp is above the level of the heart and is adjusted so that it only partially occludes the tubing, any existing defect between the clamp and the level of the heart will allow air to enter the system.[10] On the other hand, if the clamp is placed below the level of the heart, a defect between the drip chamber and the clamp will cause leakage of fluid rather than entry of air (because of the greater pressure in the vein than in the tubing).

References

1. Wilson R (ed): Principles and Techniques of Critical Care, Sect K, p 24. Kalamazoo, Upjohn, 1976

2. Maxwell M, Kleeman C (eds): Clinical Disorders of Fluid and Electrolyte Metabolism, 3rd ed, p 472. New York, McGraw-Hill, 1980

3. Invenex Laboratories, Gibco Division, The Dexter Corporation: Potassium Chloride Injection, USP (Product insert). Chagrin Falls, OH, Nov., 1981

4. Plumer A: Principles and Practice of Intravenous Therapy, 3rd ed, p 224. Boston, Little, Brown & Co, 1982

5. Ibid, p 225

6. Yeakel A: Lethal air embolism from plastic blood storage containers. JAMA 204:267, 1969

7. Freitag J, Miller L (eds): Manual of Medical Therapeutics, 23rd ed, p 293. Boston, Little, Brown & Co, 1980

8. Ordway C: Air embolus via CVP catheter without positive pressure. Ann Surg 179:479–481, 1974

9. Colley R, Wilson J: Meeting patients nutritional needs with hyperalimentation. Nurs 79:83, 1979

10. Gottlieb J, Ericsson J, Sweet R: Venous air embolism—A review. Anesth Analg 44, No. 6:776, 1965

section two

Overview of Fluid and Electrolyte Problems

chapter 5

Fluid Volume Imbalances

Fluid Volume Deficit

Fluid volume deficit (FVD) is caused by a loss of both water and electrolytes in approximately the same proportions as they exist in normal extracellular fluid. It should not be confused with the term *dehydration,* which refers to a loss of water alone primarily (leaving the patient with a sodium excess). Fluid volume deficit is a common imbalance that may occur alone or in combination with other imbalances.

Causes

Common causes of FVD include
- Loss of water and electrolytes, as in
    1. Vomiting
    2. Diarrhea
    3. Fistulas
    4. Polyuria
    5. Gastrointestinal suction
- Trapped fluids (third-space effect), as in bowel obstruction, burns, peritonitis, and crushing injuries. (See Chap. 6.)
- Decreased intake of water and electrolytes.
- Most often, it is due to a combination of fluid loss and decreased intake.

Clinical Signs

Clinical signs of FVD may include
- Weight loss over a short period of time

1. 2% loss of body weight, mild FVD
2. 5% loss of body weight, moderate FVD
3. 8% or greater loss of body weight, severe FVD

- Slow filling of hand veins. (See Chap. 2.)
- Soft, small tongue with longitudinal furrows (normally, the tongue has only one furrow in the midline).
- Dry mucous membranes.
- Urine output <30 ml/hr in the adult.
- Drop in systolic pressure by 10 mm Hg (or more) when changed from a lying to a sitting or standing position (orthostatic hypotension).
- Weak, rapid pulse.
- Flat neck veins in supine position. (See assessment of neck veins in Chap. 2.)
- Central venous pressure (CVP) below normal (<4 cm water, assuming the normal pressure in the vena cava is 4–11 cm water).
- Decreased body temperature (such as 95° F [35° C]–98° F [36.7° C] rectally), due to slowed metabolism (unless infection is present).
- Sunken eyes, pinched facial expression (facies).
- Marked oliguria, late.
- Hypotension in all positions, late.
- Cold extremities, late (due to peripheral vasoconstriction to build up central blood volume).
- Stupor or coma, late.
- Laboratory Findings.
    1. Blood urea nitrogen (BUN) rises slowly with a long-standing FVD of sufficient magnitude to reduce GFR (and thus interfere with clearance of nitrogenous wastes).
    2. Hematocrit (Hct) elevated (due to loss of intravascular fluid and subsequent concentration of formed elements of blood).
    3. Urinary specific gravity (SG) high (concentrated urine is the result of renal conservation of needed fluids).

Treatment

- An isotonic electrolyte solution (such as Lactated Ringer's or isotonic saline) is used to treat the hypotensive patient with FVD, since such fluids expand plasma volume.
- As soon as the patient becomes normotensive, a hypotonic electrolyte solution (such as half-strength saline) is indicated because it provides electrolytes and free water for renal excretion of metabolic wastes.
- If the patient with severe fluid volume deficit is oliguric, it is neces-

sary to determine if the depressed renal function is the result of reduced renal blood flow secondary to FVD (prerenal azotemia) or, more seriously, to acute tubular necrosis (ATN) due to prolonged fluid volume deficit. The therapeutic test used in this situation is the *fluid challenge test.* One version of this test, as described by Goldberger, is presented as follows:[1]

1. An initial fluid test volume (200–300 ml in an adult) can be given over a 5 to 10 minute period (provided the CVP is below 15 cm water); the patient should be observed for changes in the CVP, blood pressure, and lung sounds.

   - A CVP line is recommended to adequately assess the patient's response to the fluid load, as is a Foley catheter for urinary output.
   - Fluid used for the fluid challenge is usually a hypotonic saline solution. (Other fluids may be used as indicated by the clinical situation.)

2. If the CVP remains unchanged or does not elevate more than 2 cm or 3 cm above the initial reading, the BP remains stable (or elevated, if hypotension was initially present), and lung sounds remain normal (or no worse), an additional fluid load is given (200 ml in a period of 10 min).

3. If the CVP continues below 15 cm, and if vital signs remain unchanged, the infusion is continued at a rate of 500 ml/hr until the urinary output improves and other parameters (such as CVP and BP) return to normal. It is necessary to monitor the CVP, BP, and lung sounds every 15 minutes.

4. If the problem is prerenal azotemia, the urinary output will increase to more than 20 ml/hr within a few hours. Failure to increase urinary output may indicate the presence of acute renal failure or urinary obstruction.

5. If the patient remains oliguric after the fluid load, and if the CVP and BP have returned to normal, it is then necessary to determine if the oliguria is still prerenal and will respond to further expansion of the blood volume, or whether the volume depletion has led to acute renal failure. If renal failure has occurred, the administration of further fluids can be very dangerous. The physician may elect to use the mannitol infusion test or intravenous injection of furosemide to further evaluate the situation.

- Prompt treatment of FVD is *imperative* to prevent the occurrence of renal damage (ATN).
- If the problem is ATN, the patient requires strict renal management. (See Chap. 22.)
- Maintenance fluid requirements are discussed in Chapter 4, as are replacement requirements for moderate and severe FVD.

Fluid Volume Excess

Fluid volume excess (FVE) is caused by the abnormal retention of both water and electrolytes in approximately the same proportions as they normally exist in the extracellular fluid. It is always secondary to an increase in total body sodium content, which, in turn, causes an increase in total body water. (Remember that sodium retention causes the body to retain water.) Because there is retention of both sodium and water, the serum sodium concentration is essentially normal (unless a concurrent Na^+ imbalance is present).

Causes

Causes of FVE include

- Overzealous administration of sodium-containing parenteral fluids.
- Excessive ingestion of sodium chloride.
- Excessive use of corticosteroid therapy.
- Cushing's syndrome (excessive secretion of steroids).
- Renal failure (failure to excrete body fluids normally).
- Congestive heart failure.
- Cirrhosis of the liver.

Clinical Signs

Clinical signs of FVE include

- Weight gain over a short period of time.
    1. 2% gain in body weight, mild FVE
    2. 5% gain in body weight, moderate FVE
    3. 8% gain in body weight, severe FVE
- Peripheral edema.
- Elevated CVP (> 11 cm of water, assuming the normal CVP in the vena cava is 4–11 cm).
- Distended neck veins. (See assessment of neck veins in Chap. 2.)
- Engorged peripheral veins.
- Slow emptying of hand veins. (See Chap. 2.)
- Bounding, full pulse.
- Polyuria, if renal function is normal (kidneys attempt to rid body of excess fluid if they are functioning properly).
- Ascites, pleural effusion (occurs as excess fluid transudates into body cavities).
- Moist rales in lungs.
- Pulmonary edema.
- Laboratory findings:

1. BUN low, due to plasma dilution with excess fluid. (For example, a BUN of 6–8 mg/dl is frequently associated with states of FVE.)
2. Hct may be decreased (also due to plasma dilution).

Treatment

- Treatment is directed at correction of the pathologic process, if possible.
- Fluid intake is limited, as indicated by the clinical situation.
- Diuretics are administered to promote elimination of excess fluid, as indicated by the clinical situation.

Reference

1. Goldberger E: A Primer of Water, Electrolyte and Acid-Base Syndromes, 6th ed, pp 236–237. Philadelphia, Lea & Febiger, 1980

chapter 6

Third-Space Shifting of Body Fluids

- *Third-space fluid loss* refers to a distributional shift of body fluid into a space from which it is not easily exchanged with the extracellular fluid. Such spaces collect at the expense of, and produce a deficit in, extracellular fluid volume. It is essentially trapped fluid, unavailable for use by the body.
- Fluid can be trapped in potential spaces of the body, such as the pleural cavity, peritoneal cavity, pericardial cavity, and joint cavities. Fluid can also be trapped in the bowel by obstruction or in the interstitial space (as edema) after burns or other trauma.
- Third-space body fluid losses cannot be measured directly, as one would measure fluid losses caused by vomiting or suction.
- The trapped fluid is eventually reabsorbed with bowel obstruction or edema from burns; it can be mechanically removed when the problem is ascites (by paracentesis) or pleural effusion (by thoracentesis).

Clinical Situations Associated With Third-Space Fluid Shifting

A list of clinical situations associated with third-space shift of body fluids is as follows:
- Acute intestinal obstruction. (As much as 5–10 liters or more can accumulate within the lumen of the obstructed bowel.)
- Ascites. (Large quantities of water and electrolytes accumulate in the peritoneal cavity of patients with severe cirrhosis of the liver; removal by paracentesis only causes rapid reaccumulation of the fluid.)

86

- Acute peritonitis. (Inflammatory exudate becomes trapped in the peritoneal cavity; with extensive peritonitis, translocation of 4–6 liters or more, in a 24-hr period, is not uncommon.)[1]
- Acute gastric dilatation.
- Pleural effusion.
- Burns. (In the first 48–72 hr after severe burns, several liters of fluid can be sequestered into the interstitial space, where it becomes temporarily unavailable for use by the body.)
- Crushing injuries.
- Blockage of the lymphatic system (as occurs in malignant involvement of lymph nodes and after extensive resection of lymph nodes during surgery).
- De-clamping phenomenon (follows release of the aortic clamp following surgery on abdominal aorta; fluid accumulates temporarily in the ischemic lower extremities).
- Hypoalbuminemia. (Decreased osmotic pull of plasma causes transudation of fluid into tissue spaces.)
- Retroperitoneal hemorrhage.
- Fractured hip. (A patient may lose 1–2 liters of blood into tissues surrounding the fracture.[2])

Phases of Third-Space Shifting of Body Fluids

Fluid Accumulation

Fluid accumulation refers to the period in which fluid shifts from the intravascular space into the interstitial space or other possible fluid spaces.

Clinical Signs

- Clinical signs occurring with a significant third-space shift of fluids are essentially those of fluid volume deficit (FVD). Even though the fluid is in the body, it is functionally unavailable for use. Signs during this phase include
   1. Tachycardia and hypotension. (Sequestration of fluid in the third-space reduces the effective blood volume.)
   2. Urine output < 30 ml/hr in adult. (Decreased plasma volume causes reduced renal blood flow and thus decreased urine formation.)
   3. High urinary specific gravity (SG) and osmolality. (Kidneys attempt to conserve needed water and thus excrete a concentrated urine.)

4. Elevated hematocrit (Hct). (Loss of intravascular fluid into the third space causes the red blood cells to become suspended in a smaller volume of plasma.)
5. Postural hypotension.
6. Low central venous pressure (CVP).
7. Poor tongue and tissue turgor.
8. Body weight changes are not significant.

- Other clinical signs vary with the cause of third-space fluids. (Bowel Obstruction is discussed in Chap. 18, Cirrhosis in Chap. 19, and Burns in Chap. 24.)

Nursing Assessment Parameters

- Nursing assessment is directed at detecting FVD.
    1. Pulse. (Look for development of weak, rapid pulse.)
    2. Blood pressure (BP). (Look for hypotension, first on position change from lying to sitting, later, in all positions.)
    3. Urine volume. (Look for hourly urine output in adults < 30 ml/hr.)
    4. Urinary SG. (Look for concentrated urine with a high SG.)
    5. Skin turgor. (Look for a decrease in skin turgor.)
    6. Body weight is not helpful in assessing degree of hypovolemia (since the trapped fluid is inside the body).
- Significant findings should be discussed with the physician.

Treatment

- Treatment is directed at
    1. Correcting the cause of the third-space shift of body fluids.
    2. Correcting reduced plasma volume before renal failure develops. The physician may calculate the amount of needed fluid by a formula utilizing the degree of Hct elevation; of course, other factors are also considered.
- Administration of a multiple electrolyte solution (such as lactated Ringer's) is frequently used to correct the FVD. Sometimes albumin or Plasmanate is used to replace lost plasma protein.
- Frequent clinical assessment is needed to determine the patient's response to therapy. With proper management, one would expect to observe
    1. Decrease in pulse rate toward normal
    2. Increase in pulse volume
    3. Rise in systolic BP toward normal
    4. Increased urinary volume and lowered urinary SG
    5. Improved tongue and tissue turgor

Shift of Fluid Back to the Intravascular Space

- Some situations causing third-space shifting of body fluids will resolve themselves as the cause of the fluid leakage is corrected. (For example, injured capillaries caused by burn trauma may eventually heal.)
- During this phase, the extra fluid in the tissue or body spaces shifts back into the intravascular compartment and is excreted through the kidneys.
- It may take days to weeks for this shift to occur (depending on the underlying problem).
- Fluid administration must be curtailed during this phase to prevent circulatory overload.

Nursing Assessment Parameters

- Nursing assessment should include the following parameters and is essentially directed at detecting hypervolemia before serious effects occur:
  1. Urine volume. (Hourly volume may be as high as 200 ml/hr.)
  2. Neck veins. (Look for distention, a sign of volume overload in the intravascular space.) (See Fig. 2-4.)
  3. Lung sounds. (Listen for rales, a sign of pulmonary congestion secondary to fluid overload.)
  4. Ease of breathing. (Shortness of breath can herald FVE.)
  5. BP. (Be alert for hypertension, particularly if parenteral fluid administration is not curtailed.)
  6. CVP. (Be alert for a sharp rise.)

References

1. Schwartz S (ed): Principles of Surgery, 3rd ed, p 1399. New York, McGraw-Hill, 1979
2. Rose B: Clinical Physiology of Acid–Base and Electrolyte Disorders, p 232. New York, McGraw-Hill, 1977

chapter 7

Sodium Imbalances

Facts About Sodium

- Sodium is the most plentiful electrolyte in the extracellular fluid; its concentration ranges from 135 to 145 mEq/liter. Because of this, it is the primary determinant of extracellular fluid concentration.
- Because of its dominance in quantity, plus the fact that it does not easily cross the cell-wall membrane, sodium has a primary role in controlling water distribution throughout the body.
- In addition, sodium is the primary regulator of body fluid volume. A loss or gain of sodium is accompanied by a loss or gain of water.
- A change in the extracellular fluid sodium level has a profound effect on cell size.
    1. Hypernatremia ($Na^+ > 150$ mEq/liter) causes cellular shrinkage.
    2. Hyponatremia ($Na^+ < 130$ mEq/liter) causes cellular swelling.
- Sodium has a major function in the establishment of the electrochemical state necessary for muscle contraction and the transmission of nerve impulses.
- Urinary sodium content varies with sodium intake. The normal range is 80 to 180 mEq/24 hr. The normal ratio of urinary sodium to potassium is 2:1. (Significance of changes in urinary Na^+ excretion is discussed in Table 3-1.)

Sodium Deficit (Hyponatremia)

Hyponatremia refers to a sodium level that is below normal. It may be due to an excessive loss of sodium or an excessive gain of water. It is always due to a relatively greater water concentration (proportionately more water than Na$^+$). This imbalance should not be confused with fluid volume deficit (FVD), which refers to an isotonic loss of sodium and water (resulting in an essentially normal serum Na$^+$ level.)

Causes

The reasons for a drop in the serum sodium level include

- Loss of sodium, as in
  1. Loss of gastrointestinal secretions.
  2. Excessive sweating (particularly in children with cystic fibrosis. Recall that such individuals have high concentrations of salt in their sweat.)
  3. Use of diuretics (particularly in combination with a low-salt diet).
  4. Addison's disease. (Deficiency of aldosterone secretion causes Na$^+$ loss.)
- Gain of water, as in
  1. Excessive administration of dextrose and water solutions (usually D$_5$W), particularly during periods of stress when the patient tends to retain fluids excessively.
  2. Syndrome of inappropriate secretion of antidiuretic hormone (SIADH), described later.
  3. Psychogenic polydipsia. (Rarely, emotionally disturbed persons drink enough water to cause dilutional hyponatremia.)
  4. Induction of labor with oxytocin (discussed later).
  5. Drowning in fresh water. (Hemodilution occurs due to rapid absorption of hypotonic fluid into the circulation.)

Clinical Signs

- Clinical signs of hyponatremia are dependent on the cause, magnitude, and rapidity of onset of hyponatremia.
- Signs in chronic hyponatremia due primarily to sodium loss include
  1. Anorexia
  2. Nausea and vomiting
  3. Abdominal cramps
  4. Weakness
  5. Confusion
  6. Muscular twitching
- Signs of hyponatremia due to acute water overload include

  Previous symptoms plus more severe neurologic signs (such as hemiparesis, grand mal seizures, and coma). Neurologic symptoms are due to brain swelling and are more severe when the serum sodium level falls rapidly and to very low levels.

- Laboratory data
  1. Serum sodium level below 135 mEq/liter (may be quite low, such as 100 mEq/liter in SIADH)
  2. Urinary sodium > 20 mEq/liter in SIADH
  3. Urinary sodium < 10 mEq/liter in hyponatremia due primarily to abnormal sodium losses by nonrenal routes
  4. Urinary specific gravity (SG) usually very low (1.002–1.004) in hyponatremia due primarily to loss of sodium (by nonrenal routes)
  5. Urinary SG usually 1.012 or greater in SIADH

Treatment

- Treatment of hyponatremia due to sodium loss is sodium replacement, by mouth, nasogastric tube, or by parenteral route. If the plasma volume is below normal, Lactated Ringer's solution or isotonic saline (0.9% NaCl) may be used. However, if plasma volume is normal or excessive, it may be necessary to cautiously administer a small volume of hypertonic saline (3% NaCl or 5% NaCl).
- Treatment of hyponatremia due primarily to water overload is presented below in the section on inappropriate secretion of antidiuretic hormone (ADH).

Inappropriate Secretion of ADH

Definition and Pathophysiology

- Syndrome of inappropriate ADH secretion (SIADH) is a kind of hyponatremia that is associated with water excess. It has been found more frequently in the past few years.
- The basic physiologic disturbance is excessive ADH activity, resulting in water retention and dilutional hyponatremia (Fig. 7-1).
- SIADH can be the result of either
  1. Sustained secretion of ADH by the hypothalamus, or
  2. Production of an ADH-like substance from a tumor (aberrant ADH production).
- The term *inappropriate* is used because the secretion of ADH (water-conserving hormone) continues in the face of a hypotonic plasma. (Recall that normally ADH secretion diminishes when the plasma is hypotonic.)
- In the usual situation, hyponatremia and its related extracellular fluid volume reduction stimulates the secretion of aldosterone; however, aldosterone is not operational in SIADH because there is no decline in extracellular fluid volume. Thus, unimpaired by aldosterone's sodium conservative effect, sodium continues to be lost in the

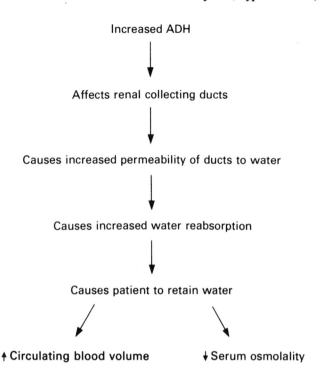

Figure 7–1. Effect of increased ADH.

urine. Because of the low serum sodium level, the plasma osmolality decreases significantly.

- The clinical manifestations in SIADH relate to excessive water retention. Indeed, it is sometimes referred to as *water intoxication.*

Recognizing Patients at Risk

- The nurse needs to be aware of common causes of SIADH. (See Clinical Conditions Favoring or Associated with SIADH.)

- Probably the most frequent cause of SIADH is oat-cell carcinoma of the lung. Other malignancies associated with this syndrome include carcinomas of the duodenum and pancreas, and Hodgkin's disease. Malignant cells from patients with SIADH have, in some instances, been shown to synthesize and release a substance similar to native ADH.

- A wide variety of intracranial disorders can produce SIADH (see Clinical Conditions Favoring or Associated with SIADH); the mechanism is probably one of sustained production of ADH from the neurohypophyseal system. Because the symptoms of SIADH are frequently neurologic in nature, the manifestations can be mistaken for the central nervous system disease itself.

- Endocrine disorders associated with SIADH can include hypothyroidism, Addison's disease, and hypopituitarism.
- Drugs favoring or causing SIADH are listed in this section. (See Drugs Favoring or Causing SIADH.) Patients treated with these medications are particularly likely to develop SIADH in the presence of other factors favoring SIADH (such as one of the conditions listed in Clinical Conditions Favoring or Associated with SIADH, and excessive fluid administration).

   1. High doses of the antineoplastic agents cyclophosphamide (Cytoxan) and vincristine (Oncovin) may cause SIADH. These agents are thought to cause excessive release of ADH. Since fluids are usually forced in patients receiving chemotherapy to prevent uric acid stones, one should be particularly alert for hyponatremia.

   2. Chlorpropamide (Diabenese), an oral hypoglycemic agent, enhances ADH release and potentiates its action at the renal collecting duct level. One study reported that 4% of patients in a clinical population receiving chlorpropamide were hyponatremic (suggesting excessive ADH effect.)[1]

   3. Oxytocin (Pitocin) is a hormone released from the neurohypophysis, capable of initiating uterine contractions; in addition, it has an ADH-like activity. Although oxytocin extracts are much freer of ADH-activity today than in the past, the composition of the two hormones is extremely similar, making women receiving oxytocin susceptible to water retention. Profound hyponatremia is particularly apt to occur when sodium-free intravenous fluids are used excessively.

Clinical Conditions Favoring or Associated With SIADH

Central Nervous System Disorders
- Head injuries (including facial trauma)
- Cerebrovascular disorders
- Brain tumors
- Encephalitis
- Guillain–Barré syndrome

Malignancies
- Oat-cell carcinoma of the lung
- Carcinoma of the pancreas
- Carcinoma of the duodenum
- Hodgkin's disease

Endocrine Disorders
- Hypothyroidism
- Addison's disease
- Hypopituitarism

Pulmonary Disorders
- Tuberculosis
- Bacterial pneumonia
- Status asthmaticus
- Lung abscess

Miscellaneous
- Postoperative states
- Limbic stimulation (pain, fear, major trauma, surgery)
- Prolonged mechanical ventilation

Drugs Favoring or Causing SIADH

Antineoplastic Drugs
- Cyclophosphamide (Cytoxan)
- Vincristine (Oncovin)

Tranquilizers
- Amitriptyline (Elavil, Amitril)
- Carbamazepine (Tegretol)
- Fluphenazine (Prolixin)
- Thioridazine (Mellaril)
- Thiothixene (Navane)

Other Drugs
- Chlorpropamide (Diabinese)
- Oxytocin (Pitocin)
- Morphine
- Barbiturates
- Isoproterenol (Isuprel)

Clinical Manifestations

- Clinical manifestations of SIADH are listed under Common Characteristics of SIADH.
- A high index of suspicion is mandatory to detect SIADH since the clinical manifestations are mainly nonspecific in the early period.
- Patients are usually asymptomatic when the serum sodium level is above 120 mEq/liter.
- When the serum sodium level reaches 115 to 120 mEq/liter, expect to encounter symptoms such as

1. Lethargy
2. Personality changes
3. Anorexia
4. Nausea and vomiting
5. Headache

- When the serum sodium level drops below 115 mEq/liter, expect to see more severe neurologic symptoms such as
  1. Loss of reflexes
  2. Babinski's sign
  3. Seizures
  4. Coma

  These events are almost uniformly seen in patients with serum sodium concentrations in the 90 to 100 mEq/liter range.

- Symptoms are, however, probably as dependent on the rapidity of onset of hyponatremia as they are on the actual serum sodium level. As a general rule, the slower the rate of development, the lower the sodium level before symptoms appear.

Common Characteristics of SIADH

CLINICAL SIGNS

Water Retention

- Intake of fluid greatly exceeds low urinary output (as evidenced by accurate intake–output records).
- Weight gain (reflecting water retention)
- No peripheral edema
- Signs of cerebral edema (see neurologic symptoms below)

Gastrointestinal Symptoms

- Anorexia
- Nausea
- Vomiting
- Abdominal cramps

Neurologic Symptoms

- Lethargy
- Headaches
- Personality changes
- Absent or diminished deep tendon reflexes
- Seizures
- Pupillary changes
- Coma

LABORATORY FINDINGS

Hyponatremia

- Symptoms usually do not appear unless serum Na^+ level is below 120 mEq/liter.
- Plasma osmolality below 280 mOsm/kg (normal is 280–295 mOsm/kg)

Low BUN and Creatinine

- Result of overhydration

Urinary Signs

- Urinary sodium > 20 mEq/liter
- Urinary SG > 1.012
- Urine osmolality is usually higher than plasma osmolality because the urine contains important amounts of Na^+, and the plasma is diluted with water.

Nursing Measures to Detect SIADH

See the summary of Important Nursing Measures to Detect SIADH. Because SIADH is potentially fatal, it is far better to prevent its occurrence than to deal with the condition once it has developed. Recovery is usually rapid if the dilutional state and its cause is recognized early and appropriate measures are instituted. The nurse plays a vital role in this area.

Important Nursing Measures to Detect SIADH

Identify patients at high risk for SIADH (see Clinical Conditions Favoring or Associated With SIADH and Drugs Favoring or Causing SIADH).

- Maintain accurate intake–output records

  Look for fluid intake greatly exceeding output; one should total the I & O for several consecutive days and look at the overall picture.

- Maintain daily body weight measurement records

  Look for sudden weight gain; a liter of fluid weighs approximately 2.2 lb (the typical patient with SIADH retains approximately 3–5 liters).

- Monitor serum Na^+ concentrations

  If the physician fails to order appropriate laboratory tests, pursue whatever route necessary to obtain the needed data.

- Observe for gastrointestinal symptoms

  Be alert for anorexia, nausea, vomiting, and abdominal cramping.

- Observe neurologic status

  Be particularly alert for lethargy, personality changes, decreased or absent deep tendon reflexes, muscle weakness, headache, convulsions, and coma.

Treatment

- Treatment may be divided into two categories.
    1. The underlying cause of excessive ADH secretion
    2. Alleviation of excessive water retention
- If the cause of excessive water retention can be readily corrected (*e.g.,* discontinuation of a drug), clinical and laboratory parameters will return to normal.
- A primary treatment of SIADH is fluid restriction to approximately 500 to 700 ml/24 hr. Fluid is restricted to the extent that urinary and insensible losses induce a negative water balance (*i.e.,* fluid loss exceeds fluid intake). (This treatment may be all that is needed in mild cases.)
- In more severe cases, in addition to fluid restriction, the intravenous administration of a small volume of hypertonic saline (3% NaCl or 5% NaCl) and furosemide (Lasix) is often indicated.

   The simultaneous use of hypertonic saline and furosemide serves to replace sodium and aid in water excretion. (Urinary losses of Na^+ and K^+ must be measured and replaced.) Use of hypertonic saline alone only transiently elevates the serum Na^+ level, since most of the administered sodium is rapidly lost in the urine. In addition, there is significant risk of pulmonary edema when hypertonic saline is not used in conjunction with a diuretic.

- If the underlying cause of SIADH is chronic, other therapeutic measures must be employed. (Fluid restriction is not tolerated well by most patients over a prolonged period.) The most prevalent measure for such patients is the administration of demeclocycline (Declomycin).
    1. Demeclocycline probably acts by interfering with the effects of ADH on the renal tubule, causing increased water excretion and, thus, an elevated serum sodium level.
    2. Its effects take several days to begin; therefore, it is not solely relied on in the acute phases.
    3. Dosage should not exceed 2400 mg/24 hr; renal function should be monitored since nephrotoxicity has been reported.[2]
- See the summary of Nursing Responsibilities During Therapy for SIADH.

Water Intoxication Related to Oxytocin Administration

- Oxytocin has been shown to have an intrinsic antidiuretic effect, acting to increase water reabsorption from the glomerular filtrate.

Nursing Responsibilities
During Therapy for SIADH

Restrict Fluids to the Prescribed Level

- Consider all routes of intake (such as oral fluids, "keep-open" IV fluids, piggyback medications, etc).
- Gain the patient's cooperation, if he is rational, by explaining the need for fluid restriction.
- Place "fluid restriction" signs at the bedside.
- Remove the water pitcher.
- Explain the need for fluid restriction to visitors.
- Space the allotted fluid allowance over the 24-hour period.
- Minimize the risk of accidental over-administration of intravenous fluids by using volume-control devices with microdrip sets.

Maintain an Accurate Intake–Output Record and Study Its Pattern

- A greater urinary output than fluid intake is desired because it indicates a negative water balance. (Remember, the excessive water load must be excreted before significant improvement can occur.)

Maintain an Accurate Body-Weight Record

- With proper therapy one can expect to see an acute decline in body weight (due to excretion of excess water). It should be recalled that a liter of fluid is equivalent to 2.2 lb; for example, a weight loss of 6.6 lb over a period of 1 to 2 days would indicate a loss of approximately 3 liters.

Assess Neurologic Signs

- Assess neurologic signs at regular intervals. With appropriate therapy, one would hope to observe an increased level of consciousness, increased deep tendon reflexes, and increased muscle strength. (Unfortunately, neurologic damage induced by SIADH is not always reversible, particularly if treatment is delayed.)

Initiate Safety Precautions

- For example, side rails should be elevated for the patient with a decreased level of consciousness; be prepared to deal with seizure activity.

Monitor Serum Na+ Levels

- With appropriate therapy, the Na^+ concentration should elevate toward normal.

- Induction of labor with oxytocin can lead to serious water intoxication, particularly when administered by continuous infusion and the patient is receiving oral fluids.
- Because of the danger of water intoxication, it is preferable to mix the oxytocin in a physiologic salt solution, such as Lactated Ringer's or isotonic saline. The usual concentration is 10 mU of oxytocin per ml of solution.

- To avoid accidental infusion of excessive oxytocin, it is important to deliver the solution through a reliable infusion pump, or at least an electronic controller.[4] When oxytocin is infused to induce labor at a rate of 45 mU or more per minute, the effect is equal to maximal doses of vasopressin.[5]
- It is important to restrict water intake (both orally and intravenously) when oxytocin is used to induce labor.
- Symptoms of water intoxication (dilutional hyponatremia) include
  1. Decreased serum sodium level
  2. Behavioral changes
  3. Headache
  4. Blurred vision
  5. Nausea and vomiting
  6. Convulsions
  7. Coma
- Symptoms have been mistaken for eclampsia in some cases until serum sodium levels were determined and hyponatremia was found to be the real cause.[6]
- One case was reported in which a primigravida developed a convulsion during delivery, after receiving 4.5 liters of D5W with oxytocin within a 3½-hour period.
- *Nursing actions during the administration of oxytocin include*
  1. Monitoring the rate of infusion of oxytocin carefully
  2. Evaluating for signs of water intoxication (listed previously)
  3. Monitoring serum sodium levels frequently, particularly if a dextrose in water solution is used, and oral water intake is not restricted

Sodium Excess (Hypernatremia)

Hypernatremia refers to a greater than normal serum sodium level (Na^+ > 145 mEq/liter). It can be caused by a gain of sodium in excess of water, or a loss of water in excess of sodium.

Causes

Causes of sodium excess include

- Decreased water intake, most common in unconscious or debilitated patients unable to perceive or respond to thirst (such as an elderly patient with a recent stroke).
- Decreased water intake in infants, very young children, and unconscious or retarded patients unable to communicate thirst.

- Hypertonic tube feedings without adequate water supplements. (See Chap. 14.)
- Tracheobronchitis (excessive loss of water vapor through coughing).
- Excessive parenteral administration of sodium-containing solutions, such as
  1. Hypertonic saline (3% NaCl or 5% NaCl)
  2. 7.5% sodium bicarbonate in cardiac arrest
  3. 0.9% sodium chloride (isotonic saline)
- Watery diarrhea.
- Hypertonic saline abortions (due to accidental direct introduction of hypertonic saline into maternal circulation or to absorption of sodium from amniotic fluid into the maternal circulation).
- Peritoneal dialysis with hypertonic glucose.
- Greatly increased insensible water loss (as in hyperventilation or in the extensive denuding effects of second- and third-degree burns).
- Diabetes insipidus in postoperative neurosurgical patients (particularly if the patient does not experience, or can't respond to, thirst; or, if fluids are excessively restricted).
- Ingestion of salt in unusual amounts (as in faulty infant formula preparation, or in addition of excessive salt through nasogastric tubes to correct hyponatremia).
- Partial drowning in sea water. (Serum Na^+ level elevates with absorption of salt water into the circulation.)

Clinical Signs

Clinical signs of hypernatremia include
- Dry, sticky mucous membranes
- Tongue rough, red, and dry
- Restlessness, excitement, and delirium
- Lethargy, which may progress to coma

  Level of consciousness depends not only on actual sodium level, but also on rate of development of hypernatremia; for example, a patient may have a serum sodium level of 170 mEq/liter and remain conscious if the imbalance developed slowly.
- Increased deep tendon reflexes
- Nuchal rigidity
- Possible permanent brain damage if severe, especially in children, due to subarachnoid hemorrhages secondary to brain contraction (hypertonic plasma pulls fluid out of cerebral tissue)
- Laboratory findings
  1. Serum sodium level > 145 mEq/liter
  2. Serum osmolality > 295 mOsm/kg

3. Urinary SG > 1.015, as the kidneys attempt to conserve water (provided water loss is from a route other than the kidneys)

Treatment

- Treatment consists in lowering the serum sodium level by infusion of a hypotonic electrolyte solution (usually 0.3% NaCl).
- A hypotonic sodium solution allows a gradual reduction in the serum sodium level (decreasing the risk of cerebral edema). It should be remembered that a rapid reduction in the serum sodium level renders the plasma temporarily hypoosmotic to fluid in the brain (due to the blood–brain barrier); as a result, water can osmose into the brain tissue, causing dangerous cerebral edema.

   The plasma sodium level should not be lowered more than 15 mEq/liter over any 4 to 6 hour period.[7] Correction of hypernatremia usually requires 24 to 48 hours to permit readjustment through diffusion among fluid compartments.
- Serum sodium levels should be monitored frequently (such as every 6 hr).

References

1. Weisman P, et al: Chlorpropamide hyponatremic inappropriate antidiuretic hormone activity. N Engl J Med. 284:65, 1971
2. Hamburger S, Rush D: Syndrome of inappropriate antidiuretic hormone secretion. Crit Care Q, Sept, 1980, p 127
3. Pritchard J, MacDonald P: Williams Obstetrics, 16th ed, p 790. New York, Appleton-Century-Crofts, 1980
4. Ibid
5. Abdaul-Karim R, Assali N: Renal function in human pregnancy: Effects of oxytocin on renal hemodynamics and water and electrolyte excretion. J Lab Clin Med. 57:522, 1961
6. Maxwell M, Kleeman (eds): Clinical Disorders of Fluid and Electrolyte Metabolism, 3rd ed, p 1386. New York, McGraw-Hill, 1980
7. Rose B: Clinical Physiology of Acid-Base and Electrolyte Disorders, p 425. New York, McGraw-Hill, 1977

chapter 8

Potassium Imbalances

Facts About Potassium

- Potassium is the major intracellular electrolyte; in fact, 98% of the body's potassium is inside the cells. The remaining 2% is in the extracellular fluid; it is this 2% that is all important in neuromuscular function.
- Potassium is dynamic—constantly moving in and out of cells according to the body's needs.
- Movement of potassium between the cells and extracellular fluid is controlled by the sodium–potassium pump.
- The normal serum potassium concentration is 3.5 to 5.5 mEq/liter; even minor variations in this range are significant.
- Eighty percent of the daily excretion of potassium from the body is by way of the renal route; the other 20% is lost through the bowel and sweat glands.
- Urinary potassium excretion varies with the dietary intake; normal range is 40 to 80 mEq/24 hr. The normal ratio of urinary sodium to potassium is 2:1.
- The kidneys are not adequately able to conserve potassium, even during times of need; therefore, potassium must be replaced daily. Approximately 40 to 60 mEq/day suffices in the adult if there are no abnormal losses occurring. Dietary intake in the average adult is 50 to 100 mEq/day.

Potassium and Aldosterone

Potassium balance is greatly influenced by aldosterone. Elevation of serum potassium above 6 mEq/liter may stimulate aldosterone secretion. (For example, a rise in serum potassium of 2 mEq/liter can triple aldosterone secretion.) Hypokalemia can inhibit aldosterone secretion if the serum sodium concentration is normal.

Potassium and Heart Function

Potassium influences both impulse conduction and muscle contractility; alterations in its concentration may change myocardial irritability and rhythm. ECG changes produced by serum potassium level variations are illustrated in Figure 8-1.

- Hyperkalemia
    1. Severity of cardiac symptoms from hyperkalemia depends on the rate of increase in serum potassium, with the most severe manifestations being associated with a rapid rise in serum levels.
    2. Also influencing the effects of hyperkalemia on the myocardium is the serum sodium level. (Recall the relationship between potassium and sodium on the resting membrane potential of the cell.) Hyponatremia potentiates the action of hyperkalemia on the heart.
    3. Acidosis and hypocalcemia can also make the effects of hyperkalemia on the myocardium more severe.
    4. The cardiac effects are usually not seen until the serum potassium level reaches 7 mEq/liter, but they may occur at levels as low as 6 mEq/liter. Myocardial effects of hyperkalemia are slowed conduction speed of impulses, and increased speed of repolarization.
    5. In profound hyperkalemia, the heart becomes dilated and flaccid due to a decreased strength of contraction resulting from a decrease in the number of active muscle units.
    6. ECG changes in hyperkalemia are described in the section on Hyperkalemia.
- Hypokalemia
    1. The major cardiac effects of hypokalemia are prolonged cardiac repolarization and decreased strength of myocardial contraction.
    2. Patients with hypokalemia have an increased sensitivity to digitalis toxicity and can be resistant to antiarrhythmics such as lidocaine, procainamide, and quinidine.
    3. Hypernatremia and alkalosis can make the effects of hypokalemia more severe.
    4. ECG changes in hypokalemia are described under the section on Hypokalemia.

Potassium and *p*H

- Hyperkalemia is usually associated with acidosis
    1. Hydrogen ions (H^+) move into the cells in acidotic states and potassium ions (K^+) move out (to maintain electroneutrality).
    2. Serum K^+ generally elevates about 0.5 mEq/liter for each 0.1 unit fall in serum *p*H.

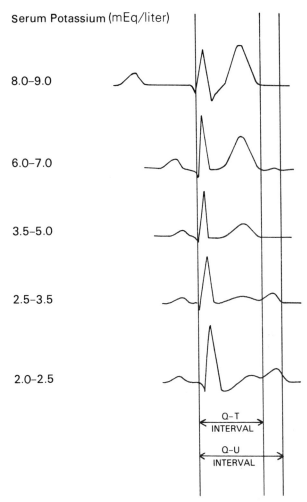

Serum Potassium (mEq/liter)

8.0–9.0

6.0–7.0

3.5–5.0

2.5–3.5

2.0–2.5

Q–T INTERVAL

Q–U INTERVAL

Hyperkalemia initially causes the T wave to increase in magnitude. With increasing serum potassium levels, the ST segment becomes depressed, the U wave disappears, and the QRS duration and P–R interval increase. With serum potassium levels greater than 10 mEq/liter, ventricular fibrillation often ensues.

Hypokalemia broadens and lowers the magnitude of the T wave, increases the magnitude of the U wave, and causes the T and U waves to fuse. It should be noted that, with respect to the heart rate, the Q–T interval and Q–U interval do not increase. With severe hypokalemia, ST segment depression is observed.

Figure 8–1. Effect of serum potassium levels on ECG. (Redrawn from Wilson RF [ed]: Principles and Techniques of Critical Care, Section J. p. 29. Kalamazoo, Upjohn, 1976)

Example:

pH	7.4	7.2
K^+	4.2	5.2

3. In acidotic states, the kidneys excrete more H^+ and thus retain more K^+. (Recall that K^+ and H^+ compete for exchange with Na^+ in the distal convoluted tubules.)

- Hypokalemia is usually associated with metabolic alkalosis
  1. H^+ moves out of the cells in alkalotic states to help correct the high pH and K^+ moves in (to maintain electroneutrality).
  2. Serum K^+ generally falls about 0.5 mEq/liter for each 0.1 unit rise in serum pH.

Example:

pH	7.4	7.6
K^+	4.2	3.2

3. In alkalotic states, the kidneys conserve H^+ and thus excrete more K^+. (Recall that K^+ and H^+ compete for excretion with Na^+ in the distal convoluted tubules.)

Potassium Deficit (Hypokalemia)

Hypokalemia is a common imbalance. As noted below, there are numerous causes of hypokalemia, many of them iatrogenic (treatment-induced).

Causes

- Potassium-losing diuretics (thiazides, furosemide, ethacrynic acid).
- Large doses of corticosteroids (steroids tend to cause the kidneys to excrete K^+ and retain Na^+).
- Prolonged administration of potassium-free parenteral fluids.
- Hyperalimentation with inadequate potassium replacement. (Large amounts of K^+ are needed for tissue synthesis.)
- Severe trauma.
- Parenteral administration of insulin and glucose (causes glycogen formation with temporary shift of K^+ into the cells).
- Prolonged administration of sodium carbenicillin or sodium penicillin. (Infusion of these agents, particularly in volume-depleted patients, can lead to hypokalemia and metabolic alkalosis.)
- Hyperaldosteronism.
- Increased loss of gastrointestinal fluids
  1. Diarrhea (including laxative and enema abuse)
  2. Fistulas

3. Vomiting or gastric suction

4. Villous adenoma of colon

- Dietary decrease in potassium intake, as in anorexia nervosa, alcoholism, or high carbohydrate diet

    The kidneys cannot conserve potassium adequately, even in the presence of starvation. (They may excrete 30 mEq or more of K^+/ day, even when the patient is on a potassium-free diet.)

- Associated with alkalosis (explained in the section on Facts About Potassium).

- Osmotic diuresis (as in uncontrolled diabetes mellitus or mannitol administration).

- Increased loss through sweating, especially in persons accustomed to heat stress. These acclimated persons lose more potassium in their sweat than those not accustomed to heat stress. (See Chap 25.)

- Stress reaction (due to excessive adrenocortical activity).

- Acute myeloid leukemia (partly due to loss of K^+ through kidneys).

- Renal tubular acidosis.

- Early treatment of ketoacidosis (most often seen within the first few hours after treatment is begun). (See Chap. 20.)

Clinical Signs

Clinical signs rarely develop before the serum potassium level has fallen below 3 mEq/liter unless the rate of fall has been rapid; signs may include

- Fatigue
- Anorexia, nausea and vomiting
- Drowsiness, depression
- Muscle weakness (most prominent in the legs, especially the quadriceps; involvement of respiratory muscles soon follows)
- Leg cramps
- Diminished deep tendon reflexes
- Decreased bowel motility, can result in ileus or gastric atony
- Paresthesias or even severely tender muscles
- Postural hypotension
- Premature atrial and ventricular contractions (most common in patients taking digitalis)
- Increased sensitivity to digitalis (digitalis toxicity occurs at lower digitalis levels when hypokalemia is present)
- ECG changes
    1. ST segment depression
    2. Broad, sometimes inverted, progressively flatter T waves
    3. Enlarging U wave sometimes becomes superimposed on the T wave to give the appearance of prolonged Q-T interval

- Impaired renal concentrating ability when hypokalemia is prolonged, causing dilute urine, polyuria, nocturia, and polydipsia (prolonged hypokalemia can damage the renal tubules)
- Impaired glucose tolerance (an adequate K^+ level is required for normal insulin release)
- Flaccid paralysis (late)
- Death in severe hypokalemia due to cardiac or respiratory arrest
- Laboratory findings
    1. Serum potassium level < 3.5 mEq/liter
    2. Often associated with metabolic alkalosis (high pH and low HCO_3^-)
    3. Decreased urine osmolality if hypokalemia has impaired renal concentrating ability; in severe hypokalemia, urine osmolality may decrease to approximately that of plasma
    4. Paradoxical aciduria (increased acid excretion in urine while an alkalotic state exists)

Treatment

- An attempt should be made to correct the underlying cause of hypokalemia, such as discontinuing medications causing excessive potassium loss.
- Treatment of mild hypokalemia is not urgent and often may be handled by dietary increases or oral potassium supplements. Some physicians prefer to reserve potassium supplementation for asymptomatic patients with serum potassium levels below 3 mEq/liter; those with symptoms or with additional risk factors (such as digitalis use) may require supplementation when the serum potassium level drops below 3.5 mEq/liter. Numerous preparations are commercially available. (Table 8-1)
- Salt substitutes contain potassium and can be used to supplement potassium intake for those patients requiring sodium restriction. (Table 8-2)
- If the serum potassium concentration is below 2 mEq/liter, or if serious symptoms such as arrhythmias or paralysis are present, parenteral replacement is urgent.
    1. As much as 40 mEq of potassium per hour may be given by way of solutions that contain no more than 60 mEq/liter.[1] (Constant ECG monitoring is required; in addition, one should monitor muscle strength and serum K^+ levels.)
    2. Sometimes in intensive care units, 10 mEq to 20 mEq of KCl are given via piggyback in a smaller volume of fluid (such as 100 ml); this should be done only with great caution and with continuous ECG monitoring.
    3. Concentrated KCl solutions should be infused through central veins, since they are extremely irritating to peripheral veins.

Table 8–1
Some Commercially Available Potassium Supplements

Liquids	Ingredients	mEq	Per Volume
Kaon Elixir	K gluconate	20	15 ml
Kay Ciel Elixir	KCL	20	15 ml
Kaochlor 10%	KCl	20	15 ml
Kaochlor 20%	KCl	40	15 ml

Tablets	Ingredients	mEq of K^+ Per Tablet
Slow-K (slow-release wax matrix)	KCl	8
K-Lyte (effervescent)	$KHCO_3$, K citrate	25
Kaon	K gluconate	5
K-Lyte/Cl 50 (effervescent)	KCl	50
K-Tab (slow release)	KCl	10

4. KCl should be added to a nondextrose solution (such as isotonic saline) to treat severe hypokalemia, since the administration of KCl in a dextrose solution may cause a small reduction in the serum potassium level (probably as a result of stimulation of insulin secretion by the infusion of dextrose).

5. Rapid administration of potassium is potentially dangerous even in severely hypokalemic patients and should be used only in life-threatening situations.

6. When the serum potassium concentration reaches 2.5 mEq/ liter, and the ECG disturbances of hypokalemia are absent, the rate should be slowed to no more than 10 mEq/hr, using a solution that contains no more than 30 mEq/liter[2].

• Potassium replacement must take place slowly because a rapid rise in the serum potassium predisposes the patient to the effects of hyperkalemia (even when the serum level remains < 5.3 mEq/liter).

Table 8–2
Salt Substitutes

Brand Name	Sodium		Potassium	
	mg/Gm	mEq/Gm	mg/Gm	mEq/Gm
Morton's salt substitute	1	0.044	493	12.62
Co-salt	1	0.044	476	12.18
Adolph's salt substitute	2	0.09	333	8.51
Neocurtasal	100	4.40	469	12.00
Morton's Lite Salt	240	10.54	195	6.15
NaCl (reference)	410	18.00	0	0.00

(Halpern S: Quick Reference to Clinical Nutrition, p 132. Philadelphia, JB Lippincott, 1979)

- Large potassium deficits should be corrected over a period of days; it may take weeks to correct a severe deficit. Fortunately, patients respond well to even partial correction of the potassium deficit.
- According to one source, it takes approximately 40 mEq to 60 mEq of potassium to elevate the serum potassium by 1 mEq/liter.[3] Another source states that if the serum potassium is less than 3 mEq/liter, an infusion of 200 mEq to 400 mEq of potassium is generally necessary to raise the serum potassium by 1 mEq/liter; if the serum potassium is between 3.0 and 4.5 mEq/liter, 100 mEq to 200 mEq will raise the serum potassium by 1 mEq/liter.[4]
- Although KCl is usually used to replace potassium deficits, the physician may prescribe potassium acetate or potassium phosphate. Potassium acetate can be used to treat patients with potassium loss associated with metabolic acidosis (as in renal tubular acidosis, and potassium-losing nephritis); the acetate is metabolized to bicarbonate and thus helps correct the acidosis. Potassium phosphate is used when the patient has deficits of both potassium and phosphate.

Preventive Nursing Measures

- Instruct patients taking potassium-losing diuretics or steroids to take adequate dietary potassium, such as extra fruits and vegetables. Foods high in potassium content are as follows:

  Bananas

  Oranges

  Grapefruit

  Dried figs and dates

  Apricots

  Raisins

  Meats (avoid high sodium varieties such as bacon, sausage, ham and luncheon meats if sodium restriction is necessary)

  Nuts

  Dried lentils

  Sweet potatoes

  Fresh tomatoes

  Fruit juices (such as prune, grapefruit, pineapple, grape, and orange)

  Milk (intake may be limited if more than mild sodium restriction is necessary)

  A well-planned diet can add 40 mEq to 60 mEq of potassium per day to the usual intake.
- In order to detect hypokalemia before it becomes severe, it is wise to monitor serum potassium levels at regular intervals for patients receiving prolonged therapy with medications associated with hypo-

kalemia. This is particularly important in patients receiving digitalis. (Physicians usually like to keep the serum potassium level above 3.5 mEq/liter in digitalized patients receiving diuretics.)

- Instruct the patient as to the reason for the prescribed diuretic and the possible side effects; stress the need for adherence to the correct dosage. Explain the need for potassium supplements if prescribed.

Nursing Responsibilities in Administration of Potassium Supplements

- Dilute liquid supplements as indicated by the manufacturer's directions to avoid gastrointestinal irritation and a saline laxative effect. Instruct the patient to sip the diluted solution over a 5- to 10-minute period.

- Dissolve effervescent tablets in 6 to 8 ounces of water (or as directed by the manufacturer).

- Instruct patients taking slow-release KCl tablets to take them with a full glass of water to help dissolve them.

- Observe patients taking slow-release KCl tablets for gastrointestinal bleeding (since these tablets may cause intestinal and gastric ulceration). These preparations are usually reserved for patients who cannot tolerate liquid or effervescent preparations or for those in whom there is a problem of compliance with these preparations.

- Be aware that the most common adverse reactions to oral potassium salts are nausea, vomiting, abdominal discomfort, and diarrhea. These symptoms are likely due to gastrointestinal irritation and are usually managed by diluting the preparation further, reducing the dose, or by taking it with meals.

- Be aware that hyperkalemia can result from overdosage of these supplements; in fact, some individuals have attempted suicide with their use. Monitor the patient for signs of hyperkalemia.

- Be aware that K^+ supplements are contraindicated in patients receiving K^+-sparing diuretics (e.g. *spironolactone,* triamterene, or amiloride). Dosages of K^+ supplements should be decreased when the patient is using generous portions of K^+-containing salt substitutes.

Potassium Excess (Hyperkalemia)

Hyperkalemia is less common than hypokalemia but is often more dangerous. It seldom occurs in persons with normal renal function. Like hypokalemia, it is often due to iatrogenic (treatment-induced) causes.

Causes

- Severe renal failure, resulting in inability of kidneys to excrete potassium. (Hyperkalemia does not usually occur until the creatinine clearance decreases to approximately 10 ml/min or less.)
- Release of potassium from damaged cells, as in extensive
    1. Burns

2. Crushing injuries
3. Infections
- Chemotherapy, particularly in lymphoma, leukemia, and myeloma (causes rapid lysis of malignant cells).
- Transfusion of aged blood, particularly to patients with diminished renal function (21-day-old blood may contain as much as 30 mEq/liter).
- Frequently associated with acidosis (causes shift of K^+ out of the cells).
- Administration of potassium-sparing diuretics to patients with even mild degrees of renal failure, or to patients with normal renal function ingesting large amounts of potassium (as in potassium supplements or salt substitutes).
- Adrenocortical insufficiency (Addison's disease).
- Too rapid or excessive parenteral administration of potassium fluids.
- Failure to adequately mix a safe dose of KCl in parenteral fluid containers, especially plastic bags (resulting in a bolus of KCl entering the vein).
- Administration of salt substitutes to patients with renal failure. (Recall that salt substitutes contain K^+ as a major ingredient, often as much as 50 to 60 mEq/tsp.)
- Administration of even small amounts of potassium to patients with renal insufficiency, oliguria, or acidosis.
- Artifactual (false) hyperkalemia
  1. Hemolyzed blood sample
  2. Blood sample from extremity with venous stasis (such as caused by prolonged tourniquet placement)
  3. Blood sample drawn from arm vigorously exercised prior to venipuncture (such as repeated fist clenching to make veins more prominent)
  4. Blood sample drawn from an area near the site of an intravenous infusion of potassium

Clinical Signs

- Vague muscular weakness
- Paresthesias of face, tongue, feet and hands are common (may occur at a serum K^+ level of 6 mEq/liter)
- Gastrointestinal hyperactivity
  1. Nausea
  2. Intermittent colic
  3. Diarrhea

- Heart rate may be slow and regular or irregular, or irregular and fast (bradycardia often reported at a serum level of 7 mEq/liter).
- ECG findings
    1. High-tented T waves
    2. Arrhythmias (especially supraventricular tachycardia, premature ventricular contractions, atrial or ventricular fibrillation, or ventricular arrest in diastole)
    3. P waves disappear.
    4. QRS complex widens (a serious sign).
    5. Heart block (may occur at a serum K^+ level exceeding 9 mEq/liter)
- Flaccid muscle paralysis (first noticed in the legs, later in the trunk and arms)
- Death due to cardiac arrest in diastole or respiratory arrest
- Laboratory findings
    1. Serum potassium level > 5.5 mEq/liter
    2. Frequently associated with acidosis (pH < 7.35)

Preventive Nursing Considerations

- Follow rules for safe administration of K^+ solutions
    1. Never administer concentrated K^+ solutions from ampules without first diluting them as directed.
    2. Never give K^+ solutions by IV push.
    3. Limit K^+ concentration in any 1 liter to no more than 20 mEq/liter in usual situations, since accidental rapid infusion is less dangerous when the K^+ concentration is moderate.
    4. Thoroughly mix the KCl when adding to parenteral containers. Squeeze the medicine ports of plastic bottles while they are in the upright position and then mix thoroughly. Never add KCl to a container in the hanging position.
    5. Administer K^+ only after adequate urine flow has been established. A decrease in urine volume to less than 20 ml/hr for 2 consecutive hours is an indication to stop K^+ infusion until the situation is evaluated.
    6. Administer K^+ at a rate no faster than 10 mEq/hr (suitably diluted) for routine maintenance needs.
    7. Administer at a rate no faster than 40 mEq/hr (suitably diluted) in extreme cases of hypokalemia. Monitor the patient and the ECG constantly.
- Avoid administration of K^+-sparing diuretics, K^+ supplements, or salt substitutes to patients with renal insufficiency.
- Caution patients to use salt substitutes sparingly if they are taking other supplementary forms of K^+ or are taking K^+-sparing diuretics.

Treatment

- Restrict dietary potassium intake; discontinue potassium-containing medications.
- Prevent occurrence of serious hyperkalemia in susceptible patients by administration of cation-exchange resins, such as Kayexalate. (One gram of resin removes 1 mEq of potassium from the bowel; see discussion of cation-exchange resins in Chap. 22.)

Severe Hyperkalemia

- Administration of calcium gluconate IV may be necessary when serious cardiac symptoms are present—such as widened QRS complex. Calcium immediately antagonizes the effects of hyperkalemia on the myocardium. From 5 ml to 20 ml of a 10% calcium gluconate solution may be prescribed. The ECG should be continuously monitored during administration; the appearance of bradycardia is an indication to stop the infusion. The myocardial protective effects of calcium are transient, lasting about 1 hour. (Extra caution is required if the patient has been digitalized since parenteral administration of calcium sensitizes the heart to digitalis and may precipitate toxicity.)
- Administration of sodium bicarbonate IV alkalinizes the plasma and causes a temporary shift of potassium into the cells. In addition, $NaHCO_3$ furnishes sodium for antagonizing the cardiac effects of potassium. For example, an ampule of sodium bicarbonate (containing 44 mEq of bicarbonate) may be given as a bolus, or 90 mEq may be added to 500 ml of $D_{10}W$ and given by infusion. The effects are temporary, lasting only a few hours. The patient should be observed for sodium overload.
- Administration of regular insulin and hypertonic dextrose IV causes a temporary shift of potassium into the cells. (For example, a bolus of 50 ml of 50% dextrose with 10–25 U of regular insulin may be given IV.) This form of treatment begins to act in 15 to 60 minutes and the effect lasts 2 to 4 hours. Once begun, insulin–dextrose therapy should not be stopped until the total body potassium has been reduced since dangerous shifts of potassium from the cellular to the extracellular space may occur, with the return of hyperkalemia. Also, rebound hypoglycemia may occur if the hypertonic dextrose solution is abruptly discontinued.
- The above measures are stopgap measures that only temporarily protect the patient from hyperkalemia. If the hyperkalemic condition is not transient, actual removal of potassium from the body is required; this may be accomplished by the use of
    1. Cation-exchange resins
    2. Peritoneal dialysis
    3. Hemodialysis

References

1. Maxwell M, Kleeman C (eds): Clinical Disorders of Fluid and Electrolyte Metabolism, 3rd ed, p 472. New York, McGraw-Hill, 1980
2. Ibid, p 471
3. Wilson R (ed): Principles and Techniques of Critical Care, section K, p 24. Kalamazoo, Upjohn, 1976
4. Maxwell M, Kleeman C (eds), p 471

chapter 9

Calcium Imbalances

Facts About Calcium

- *Calcium* is essential for formation of bones and teeth, and is an essential ion for many enzymes. In addition, it plays a central role in muscle contraction, neural function, and formation of prothrombin for blood coagulation.
- There is very little calcium in serum (4.5 mEq/liter or 8.6–10.5 mg/100 ml). However, even minor changes in the serum calcium level can cause significant symptoms.
- About 50% of serum calcium exists in the ionized form (Ca^{2+}). It is the ionized form that is physiologically active and important for neuromuscular activity. The remainder of serum calcium exists bound to serum proteins (primarily albumin). Most hospitals measure the total serum calcium level only; the ionized fraction must then be estimated in light of simultaneous measurement of serum protein level and arterial pH.
- Clinically it is important to correlate the serum calcium concentration with the serum albumin level. Each fall (or rise) of the serum albumin level by 1 Gm/100 ml (beyond the normal range of 4–5 Gm/100 ml) is associated with a fall (or rise) of serum calcium concentration of approximately 0.8 mg/100 ml. For example, if a patient normally has a total serum calcium level of 10 mg/100 ml and a serum albumin of 4 Gm/100 ml, and then has a decrease in serum albumin to 3 Gm/100 ml, the serum calcium will drop to 9.2 mg/100 ml. Because of this, physicians will often ignore a low serum calcium level in the presence of a similarly low serum albumin level. It should be noted that the ionized calcium level is usually normal in patients with reduced total serum calcium levels and concomitant hypoalbuminemia.
- As the arterial pH increases (alkalosis), more calcium becomes bound to protein; the ionized portion decreases while the total ser-

um calcium remains unchanged. Symptoms of hypocalcemia often occur in the presence of alkalosis. Acidosis (low *p*H) has the opposite effect; that is, less calcium is bound to protein and thus more exists in the ionized form. Rarely will signs of hypocalcemia develop in the presence of acidosis, even when the total serum calcium level is lower than normal.

- Calcium has a reciprocal relationship with phosphorus; that is, a rise in the serum calcium level causes a drop in the serum phosphorus concentration, and a drop in serum calcium level causes a rise in phosphorus concentration.
- *Parathyroid hormone* (PTH) promotes a transfer of calcium from the bone to plasma (raising the plasma calcium level). The bones and teeth are ready sources for replenishment of low plasma calcium levels. PTH also augments the intestinal absorption of calcium and enhances net renal calcium reabsorption.
- *Calcitonin* (produced by the thyroid) promotes transfer of calcium from plasma to the bones, lowering the plasma calcium level. Calcitonin secretion is directly stimulated by a high serum calcium concentration.
- The cardiac effects of calcium include cardiac excitability and myocardial contractility. Alterations in serum calcium level primarily affect repolarization. Arrhythmias and conduction disturbances are uncommon with calcium imbalances when serum potassium and magnesium are normal. The ECG changes produced by serum calcium variations are illustrated in Figure 9-1. Hypocalcemia prolongs the Q–T interval while hypercalcemia shortens it.
- Urinary calcium content is influenced by dietary intake; the normal urinary calcium content is 100 to 250 mg/24 hr (if the patient is on an unrestricted diet).
  1. Urinary calcium is increased in hyperparathyroidism, osteolytic bone disease, and osteoporosis.
  2. Urinary calcium is decreased in hypoparathyroidism.

Calcium Deficit (Hypocalcemia)

Causes

- Surgical hypoparathyroidism (may follow thyroid surgery or radical neck surgery for cancer)
  1. After thyroid surgery, patients with significant parathyroid gland injury usually develop tetany within 24 hours; the parathyroid injury is mainly the result of ischemia caused by dissection and hemostatic maneuvers.

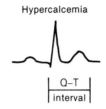

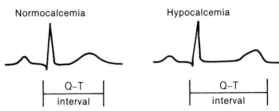

The Q–T interval varies inversely with the blood levels of calcium; *i.e.*, hypocalcemia is associated with a prolonged Q–T interval, and hypercalcemia is associated with a shortened Q–T interval.

Figure 9–1. ECG changes by serum calcium level variations. (Wilson RF [ed]: Principles and Techniques of Critical Care, Section K, p. 26. Kalamazoo, Upjohn, 1976.

    2. Approximately half of the patients who develop postthyroidectomy tetany eventually recover enough for calcium supplements to be withdrawn.

    3. Occasionally, for unknown reasons, hypocalcemia develops between 1 and 30 years after operation.

- Malabsorption
- Vitamin D deficiency
- Prolonged administration of calcium-free parenteral solutions (generally speaking, Ca^{2+} is needed after 1 week of IV therapy using solutions without Ca^{2+})
- Acute pancreatitis (occurs one to several days after onset; it appears that parathyroid hormone secretion is altered in patients with pancreatitis)[1]
- Pancreatic and small intestinal fistulas
- Excessive administration of citrated blood, particularly to patients with liver damage (See Chap. 16)
- Primary hypoparathyroidism
- Alkalotic states (causes decreased Ca^{2+} ionization)
- Hyperphosphatemia, as occurs in renal insufficiency (causes a reciprocal drop in the serum Ca^{2+} level)
- Hypomagnesemia (apparently parathyroid hormone secretion is altered in the presence of hypomagnesemia); in severe hypomagnese-

mia (serum Mg^{2+} < 0.8 mEq/liter), correction of the serum Mg^{2+} may be necessary to normalize the serum Ca^{2+} concentration)[2,3]

- Medullary carcinoma of thyroid (excessive calcitonin secretion)
- Decreased serum albumin level, as in cirrhosis, nephrotic syndrome, and starvation (associated with a decrease in total Ca^{2+} level; however, symptoms rarely occur because the ionized fraction is often normal)
- Decreased ultraviolet exposure
- Maternal diabetes or hyperparathyroidism (can cause transient hypocalcemia in the offspring)
- Cimetidine (Tagamet)
  Interferes with normal PTH function, particularly in patients with renal insufficiency[4]

Clinical Signs

- Numbness, tingling of fingers, toes, and circumoral region
- Cramps in muscles of extremities
- Hyperactive deep tendon reflexes (such as the patellar, Achilles, biceps, and triceps)
- Spasms of muscles in abdominal area (can simulate an acute abdominal emergency)
- Trousseau's sign (Fig. 9-2)
  Refers to carpopedal spasms of hand when blood supply is decreased or nerve is stimulated by pressure; elicited within several minutes by application of blood pressure cuff inflated above systolic pressure
- Chvostek's sign
  Elicited by tapping the facial nerve about 2 cm anterior to the ear lobe, just below the zygomatic process; response consists of spasms of muscles supplied by the facial nerve
- Mental changes include alterations in mood and memory; confusion may be present.
- Convulsions (usually generalized, but may be focal)
- Spasm of laryngeal muscles with airway obstruction
  Seen particularly in early postoperative patients who have undergone thyroidectomy or radical neck dissection; for this reason, it has been recommended that serum calcium levels be measured at 6- to 8-hour intervals in all patients after neck surgery[5]
- Retardation, stunted growth, poor dentition, photophobia, conjunctivitis, and cataracts (may result from chronic hypocalcemia in children)

- Cardiac manifestations
    1. Prolonged Q–T interval and ST segment
    2. Prolonged ventricular systole
    3. Decreased contractility (especially associated with acute losses) leading to heart failure or death in diastole
    4. Decreased sensitivity to digitalis (for example, arrhythmias previously controlled by digitalis may reappear in the presence of hypocalcemia)
- Recurrent episodes of apnea in infants (may be due to hypocalcemia)
- Idiopathic seizure activity in children (may be result of hypocalcemia)
- Papilledema may be noted without an increase in intracranial pressure.[6]

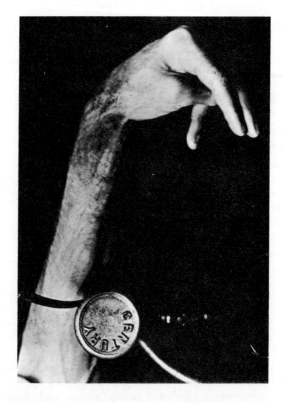

Figure 9–2. Manifestations of hypocalcemia. Trousseau's sign is elicited in a patient with hypoparathyroid tetany. (Ezrin C, Godden JO, Volpe R, Wilson R: Systematic Endocrinology, p. 1510, 2nd ed. Hagerstown, Harper & Row, 1979)

- Laboratory findings
    1. Serum calcium level below 4.5 mEq/liter or 9 mg/100 ml
    2. Ionized calcium level below normal (< 50%)

Note: The presence of hyperkalemia and hypomagnesemia potentiates the cardiac and neuromuscular irritability produced by hypocalcemia; hypokalemia, on the other hand, tends to protect the patient from hypocalcemic tetany.[7]

Treatment

Treatment consists in administration of oral calcium supplements; vitamin D may be given to enhance intestinal calcium absorption.

Acute Management

Acute symptomatic hypocalcemia is a medical emergency requiring prompt treatment with intravenous calcium gluconate (such as 10 ml to 20 ml of a 10% solution over 10 min to 15 min).

Note: Extreme caution is necessary when calcium is given IV to a digitalized patient because calcium increases the sensitivity of the heart to digitalis and may precipitate toxicity.

Postmenopausal Osteoporosis

- Serum concentrations of calcium, phosphorus, and alkaline phosphatase are normal in postmenopausal osteoporosis even though total negative calcium, phosphorus, and nitrogen balance exists.
- There is progressive loss of bone after age 40; by age 80, total bone mass may be reduced to half of what it was at age 40.[8]
- Usually in osteoporosis the rate of bone formation is normal, but the rate of bone resorption is increased.
- Bone resorption is especially increased in women. In fact, one in three postmenopausal women has osteoporosis; one in five will incur a hip or vertebral-crush fracture. Postmenopausal osteoporosis is most common in small-framed white women over 50 years of age.[9,10]
- Back pain is the most frequent symptom, and spinal deformity is the most frequent sign. "Dowager's hump" (a hunchback posture due to severe loss of anterior vertebral height) is a common finding.
- The patient's height may be 1 to 6 inches shorter than the arm span if multiple wedge compression fractures have occurred.
- Involutional bone alterations may reduce cortical bone mass of the femurs by a factor of 30% to 50%.

- Reduced calcium intake seems to play a role in the development of postmenopausal osteoporosis. Also implicated are diminished physical activity, impaired calcium absorption, increased renal calcium loss, increased PTH effect, and reduced secretion of estrogen.[11]
- It seems that oral calcium, vitamin D, and estrogen can reverse negative calcium balance related to menopause and probably delay or prevent the onset of clinical osteoporosis
    1. Calcium supplements help replace the gross deficiency of calcium associated with osteoporosis.
    2. Vitamin D increases calcium absorption from the intestines
    3. Oral estrogen administration has been shown to be the most effective treatment modality; it reduces negative calcium balance and stimulates positive calcium balance. It should be remembered that large daily doses of estrogen have been shown to be associated with sodium retention, hypertension, myocardial infarction, and increased risk of endometrial carcinoma. Many physicians feel that the anticatabolic effect of estrogen (which reduces bone resorption) outweighs the risk of these untoward effects.
- Other therapies for osteoporosis may involve increased exposure to real or artificial sunlight (infrared light) to increase the level of 25-hydroxyvitamin D, and decreased intake of dietary phosphate. (Recent studies suggest that an excess of phosphorus may be an important cause of bone loss.)

Calcium Excess (Hypercalcemia)

Causes

- Hyperparathyroidism
- Malignant neoplastic disease
    1. Solid tumors with metastases (breast, prostate, and malignant melanomas)
    2. Solid tumors without bony metastases (lung, head and neck, and renal tumors)
    3. Hematologic tumors (lymphoma, acute leukemia, and myeloma)
    4. Certain chemotherapeutic regimens (treatment with androgens, estrogens, progestins and antiestrogens)
        - Patients receiving these agents should have frequent serum calcium determinations; the appearance of hypercalcemia is an indication to discontinue their use.
- Paget's disease
- Multiple fractures

- Large doses of vitamin D (causes increased Ca^{2+} absorption from gut)
- Thiazide diuretics (cause a mild increase in serum Ca^{2+} level, usually asymptomatic)
- Overuse of calcium-containing antacids
- Prolonged immobilization
  1. Due to absence of weight-bearing, bone mineral is lost during immobilization, sometimes causing elevation of serum calcium.
  2. Symptomatic hypercalcemia from immobilization is not common; when it does occur, it is often limited to individuals with high calcium turnover rates (as in adolescents during growth spurts or in patients with Paget's disease).
  3. Most cases of immobilization hypercalcemia occur after severe or multiple fractures or after extensive traumatic paralysis (as in quadriplegia).
  4. Immobilization hypercalcemia may remain clinically silent or may be associated with the clinical signs listed in the next section.
  5. Clinical signs usually begin rather abruptly a few weeks or months after injury.[12]

Clinical Signs

Clinical signs of hypercalcemia include
- Tiredness, listlessness, lethargy
- Constipation
- Anorexia, nausea and vomiting
- Decreased memory span, poor calculation, and decreased attention span
- Muscular weakness (Ca^{2+} acts as a sedative at nerve endings)
- Polyuria (due to large renal solute load)
- Polydipsia (follows polyuria)
- Eventual uremia if not treated
- Bone pain, local or diffuse
- Generalized osteoporosis (noted on x-ray film)
- Urinary calcium stones
- Pathologic fractures (due to bone weakening)
- Changes progressing from neurotic behavior to frank psychoses have been documented and shown to be reversible when hypercalcemia was controlled.
- Chronic hypercalcemia can cause itching (due to calcium deposition in skin) and ocular changes (band keratopathy).
- Cardiac manifestations

1. Shortened ST segment resulting in shortened Q–T interval
2. Increased contractility leading to cardiac arrest in systole if hypercalcemia is severe
3. Potentiation of digitalis toxicity (hypercalcemia may aggravate all forms of cardiac toxicity produced by digitalis)
4. Widened and rounded T wave
5. Slightly widened QRS complex and P–R interval

- Laboratory findings

    Serum calcium level > 5.8 mEq/liter or 10.5 mg/100 ml

Note: It should be remembered that prolonged constriction with a tourniquet during phlebotomy can result in a falsely elevated serum calcium level.[13]

- Hypercalcemic Crisis
    1. Occurs often when the serum calcium level exceeds 15 mg/100 ml
    2. Characterized by intractible nausea and vomiting, dehydration, stupor, azotemia, and finally cardiac arrest
    3. Requires prompt therapy to prevent cardiac arrest

Treatment

Acute Hypercalcemia

- Administration of 0.45% NaCl or 0.9% NaCl IV dilutes the serum calcium level and increases urinary calcium excretion. (Recall that sodium inhibits tubular reabsorption of calcium.) Some authorities recommend 250 to 500 ml/hr until the CVP reaches 10 cm of water (provided cardiac function is normal).
- Furosemide (Lasix), such as 20 mg to 40 mg every 2 hours, may be used to prevent volume overloading during saline administration. Also, furosemide increases calcium excretion. Ethacrynic acid (Edecrin) may be used instead of furosemide. Urinary losses of Na^+, K^+, and water should be measured and replaced.
- Calcitonin (Calcimar) will temporarily lower the serum calcium level by 1 to 3 mg/100 ml. It is used only if saline therapy is ineffective; it is not suited for long-term use. (Skin testing should be considered with calcitonin since it is protein in nature.)

Hypercalcemia Associated With Malignancies

- Radiation or chemotherapy to control the tumor.
- Corticosteroids to decrease bone turnover and tubular reabsorption (such as hydrocortisone 250–500 mg q8hr initially, then prednisone 10–30 mg/day) are sometimes administered. These agents are most useful for patients with sarcoidosis, myelomas, lymphomas, and leukemias; patients with solid tumors are less responsive. Due to the

slow onset of action, the serum calcium level may not decrease for 1 to 2 weeks.

- The same measures listed under acute hypercalcemia may be used (*i.e.*, administration of saline and furosemide).
- Mithramycin, a cytotoxic antibiotic, inhibits bone resorption and thus lowers the serum calcium level. This drug must be used cautiously since it has significant toxicities, including thrombocytopenia.

General Measures for Hypercalcemia

- Treatment of the underlying disease may eliminate the source of the hypercalcemia.
- Fluids should be forced to aid in elimination of the high calcium content in the bloodstream.
- A low-calcium diet is indicated when feasible, with a salt intake of 8 to 10 Gm/day.
- Inorganic phosphate salts can be given orally (such as Phospho-Soda [Fleet] or Neutra-Phos), rectally (as retention Fleet's enemas bid), or intravenously. Inorganic phosphate salts lower the serum calcium concentration by inhibiting bone resorption, reducing intestinal calcium absorption, and by forming a calcium–phosphate complex that is deposited in soft tissues and bone. Because of the risk of soft tissue calcification, phosphate therapy should be limited primarily to patients with low serum phosphate levels. (Intravenous phosphate therapy should be used with extreme caution in the treatment of hypercalcemia since it can cause severe calcification in various tissues, including the vein in which it is given.)
- Hemodialysis or peritoneal dialysis can be used to reduce serum calcium levels in life-threatening situations.

References

1. Robertson G, Moore E, Switz D, et al: Inadequate parathyroid responses in acute pancreatitis. N Engl J Med 294:512–516, 1976
2. Chernow B, Smith J, Rainey T, et al: Hypomagnesemia: Implications for the critical care specialist. Crit Care Med 10:193–196, 1982
3. McFadden E, Azloga G, Chernow B: Hypocalcemia: A medical emergency. Am J Nurs 80:226–230, 1983
4. Jacobs A, Lanier C, Canterbury J, et al: Reduction by cimetidine of serum PTH levels in uremic patients. N Engl J Med 302:671–674, 1980
5. McFadden E, Azloga G, Chernow B, op cit
6. Ibid, p 228
7. Ibid, p 229
8. Skillman T: Can osteoporosis be prevented? Geriatrics, Feb 1980, p 95

9. Ibid
10. Ibid, p 96
11. Ibid, p 98
12. Maxwell M, Kleeman C (eds): Clinical Disorders of Fluid and Electrolyte Metabolism, 3rd ed, p 1016. New York, McGraw-Hill, 1980
13. McFadden E, Azloga G, Chernow B, op cit

chapter 10

Magnesium Imbalances

Facts About Magnesium

- Control of magnesium is not well understood; however, many of the factors that regulate calcium balance influence magnesium as well. In addition, magnesium balance is affected by many of the same agents and diseases that influence potassium balance.
- *Magnesium* acts directly on the myoneural junction. An excess diminishes excitability of the muscle cells. Low serum magnesium levels increase neuromuscular irritability and contractility; tetany and convulsions can occur.
- Magnesium acts as an activator for many enzymes and plays a role in both carbohydrate and protein metabolism.
- The normal serum magnesium level is 1.5 to 2.5 mEq/liter (1.8–3.0 mg/100 ml).
- A normal diet supplies approximately 25 mEq of magnesium daily. Requirements for magnesium appear to vary with calcium and phosphorus metabolism. For those with normal bony stores of calcium and a normal serum phosphorus level, approximately 25 to 32 mEq/day will suffice.
- The primary route of magnesium excretion is the kidneys. Thus, patients with impaired renal function have a lower requirement for magnesium than patients with normal function.
- *Hypomagnesemia* can cause hypocalcemia because it interferes with the calcium-elevating effects of PTH; it may also cause what appears to be unexplained hypokalemia.
- Approximately one-third of serum magnesium is bound to protein; the remaining two-thirds exists as free cations (Mg^{2+}).
- Fluid from the lower gastrointestinal tract is richer in Mg^{2+} (10–14

mEq/liter) than is fluid from the upper tract (1–2 mEq/liter).[1] Thus fluid loss from diarrhea and intestinal fistulas is more likely to produce Mg^{2+} deficiency than is gastric suction.[2]

Magnesium Deficit (Hypomagnesemia)

Causes

Causes of magnesium deficit include

- Chronic alcoholism, with or without cirrhosis (the most common cause of symptomatic hypomagnesemia in the United States). Magnesium deficiency in alcoholics may be due to
  1. Inadequate dietary intake
  2. Impaired renal conservation of magnesium
  3. Intestinal malabsorption (due to alcohol itself or liver disease)
  4. Intermittent diarrhea
- Intestinal malabsorption syndromes (especially after small-bowel bypass surgery). In the presence of steatorrhea, magnesium ions are excreted in the stool in the form of magnesium soaps.
- Prolonged diarrhea
- Administration of magnesium-free IV fluids without oral intake or with nasogastric suction for more than a week
- Prolonged malnutrition or starvation (A deficient diet can be tolerated for as long as a month before a deficiency of Mg^{2+} occurs.)
- Hyperalimentation without adequate replacement of magnesium
- Diuretics (can cause excessive urinary losses of Mg^{2+})
- Primary aldosteronism (causes increased urinary and fecal losses of Mg^{2+})
- Treatment of diabetic acidosis (due to increased urinary Mg^{2+} losses and shift of Mg^{2+} into cells)
- Hyperthyroidism (causes deficiency in ionized Mg^{2+} as more Mg^{2+} is bound to protein)
- Following parathyroid surgery in patients with chronic hyperparathyroidism
- Administration of certain drugs, such as gentamicin and cisplatin[3]
- Some causes of renal disease (*e.g.* glomerulonephritis, pyelonephritis, and renal tubular acidosis)
- Acute pancreatitis
- Prolonged gastric suction
- Familial hypomagnesemia in newborns

Clinical Signs

Clinical signs of magnesium deficit include
- Tachycardia
- Neuromuscular and nervous system irritability
    1. Increased reflexes
    2. Coarse tremors (may occur in any extremity, but most commonly in the arms)
    3. Muscle cramps
    4. Positive Chvostek's and Trousseau's signs (See Chap. 9)
    5. Convulsions (usually generalized)
- Paresthesias of feet and legs
- Painfully cold hands and feet
- Cardiac manifestations
    1. Arrhythmias (such as ventricular premature contractions and ventricular fibrillation)
    2. Increased likelihood of digitalis toxicity, since magnesium deficiency potentiates the action of digitalis
    3. Mild hypomagnesemia can cause tall T waves and slight widening of the QRS complex and occasional, mild ST segment depression.
    4. Severe hypomagnesemia (0.2–0.6 mEq/liter) with associated hypokalemia and hypocalcemia can cause prolonged P–R interval, wide QRS complex, slight ST depression, and broad and flat (or inverted) T waves.
- Mental aberrations
    1. Disorientation in memory
    2. Intense confusion
    3. Visual or auditory hallucinations
- Laboratory findings
    1. Serum magnesium < 1.5 mEq/liter or 1.8 mg/100 ml (usually, symptoms do not appear until the serum Mg^{2+} is < 1 mEq/liter)
    2. Hypocalcemia frequently occurs with severe magnesium deficiency.

Treatment

Hypomagnesemia may be treated with magnesium salts by the oral, intramuscular, or intravenous routes.
- Oral magnesium salts (such as magnesium sulfate or magnesium oxide) can be given to counteract continuous magnesium losses.

- For severe hypomagnesemia, 5 Gm (40 mEq) of magnesium sulfate may be added to 1 liter of D_5W or D_5S and given as an infusion.[4] In convulsive states, 1 Gm to 2 Gm (10–20 ml) of a 10% solution of magnesium sulfate may be given by direct IV (IV push) (at a rate of 1.5 ml/min).[5]
- *Nursing considerations in the intravenous administration of magnesium sulfate*
    1. Give the solution at the proper rate; be aware that too rapid administration can lead to cardiac arrest.
    2. Be aware that magnesium sulfate is contraindicated in the presence of heart block.[6]
    3. Be aware that a urine output of at least 100 ml every 4 hours should be maintained to allow adequate renal elimination of magnesium. Oliguric patients should be given only very small doses of magnesium until adequate urine flow is established.
    4. When repeated doses of magnesium are needed, the patellar reflexes should be tested before each dose; if absent, the dose should be withheld.
    5. Be prepared to deal with respiratory arrest should hypermagnesemia inadvertently occur during administration of the magnesium solution.
    6. Be aware that doses of other central nervous system depressants should be reduced when given concurrently with magnesium sulfate.
    7. Monitor serum magnesium levels at appropriate intervals during magnesium replacement therapy.

Magnesium Excess (Hypermagnesemia)

Causes

Causes of magnesium excess include

- Renal failure (particularly when magnesium-containing antacids or laxatives are ingested) (see Commonly Used Medications Containing Magnesium)
- Hemodialysis with excessively hard water or with a dialysate inadvertently high in magnesium content
- Excessive magnesium administration during treatment of eclampsia (causes both maternal and fetal hypermagnesemia)
- Untreated ketoacidosis
- Hyperparathyroidism
- Addison's disease
- Hypothyroidism

Commonly Used Medications Containing Magnesium

ANTACIDS	ANTACIDS (*continued*)	LAXATIVES
Aludrox	Kolantyl	Milk of Magnesia
Camalox	Maalox	Magnesium Citrate
Delcid	Mylanta	Epsom Salts
Escot	Oxaine M	
Gaviscon	Silain-Gel	
Gelusil	Simeco	

Clinical Signs

Clinical signs of hypermagnesemia include

- Early signs (serum Mg^{2+} level of 3–5 mEq/liter)
  1. Flushing and a sense of skin warmth (due to peripheral vasodilatation)
  2. Hypotension (due to blockage of sympathetic ganglia as well as to a direct effect on smooth muscle)
  3. Nausea and vomiting
- Drowsiness, hypoactive reflexes, and muscular weakness (can occur at a serum Mg^{2+} level of 5–7 mEq/liter)
- Depressed respirations (can occur at a serum Mg^{2+} level of 10 mEq/liter)
- Coma (can occur at a serum Mg^{2+} level of 12–15 mEq/liter)
- Cardiac abnormalities
  1. Sinus bradycardia, prolonged P–R, QRS, and Q–T intervals (at serum Mg^{2+} levels of 7.1–10.0 mEq/liter)
  2. Heart block and cardiac arrest in diastole (can occur at a serum Mg^{2+} level of 15–20 mEq/liter)
- Weak or absent cry in newborn

Prevention

- Do not give magnesium-containing medications to patients with acute or chronic renal insufficiency. Caution patients with renal disease to check with their physicians before taking over-the-counter medications.
- Administer magnesium-containing agents at the prescribed proper rate, remaining alert for signs of hypermagnesemia. (See the section on magnesium administration under Hypomagnesemia.)

Treatment

- The source of excessive magnesium intake should be eliminated.
- Calcium gluconate is administered, if necessary, to antagonize the action of magnesium and thus provide temporary relief from hypermagnesemia.

- Use of artificial respiration may be necessary.
- Dialysis is used in severe cases when needed.

References

1. Rude R, Singer F: Magnesium deficiency and excess. Ann Rev Med 32:245, 1981
2. Chernow B, Smith J, Rainey T, Finton C: Hypomagnesemia: Implications for the critical care specialist. Crit Care Med 10:193, 1982
3. McFadden E, Azloga G, Chernow B: Hypocalcemia: A medical emergency. Am J Nurs 80:226–230, 1983
4. Gahart B: Intravenous Medications, 3rd ed, p 129. St Louis, CV Mosby, 1981
5. Ibid
6. Ibid

chapter 11

Phosphorus, Lithium, and Zinc

Phosphorus Imbalances

Facts About Phosphorus

- Phosphorus is the most abundant anion (negatively charged ion) in the intracellular fluid.
- Approximately 80% of the body's phosphorus is combined with calcium in bone and teeth.
- Parathyroid hormone (PTH) plays a major role in phosphate homeostasis because of its ability to vary phosphate reabsorption in the proximal tubule. In addition, PTH causes a shift of phosphate from bone to plasma.
- The normal serum phosphorus level is 2.8 to 4.5 mg/100 ml (1.2–3.0 mEq/liter). Phosphorus exists in the body in the form of inorganic and organic phosphate salts; however, only inorganic salts are present in extracellular fluid.
- Phosphorus plays an important role in the delivery of oxygen to tissues by regulating the level of 2,3 diphosphoglycerate (2,3 DPG), a substance in red blood cells that decreases the affinity of hemoglobin for oxygen. A low level of 2,3 DPG may result in decreased peripheral oxygen delivery.
- An increased serum phosphate level frequently causes hypocalcemia. (Recall that there is a reciprocal relationship between phosphorus and calcium.)

Phosphorus Deficiency (Hypophosphatemia)

Causes

Causes of hypophosphatemia include

- Hyperalimentation (tissue synthesis requires 4 mEq of phosphate per gram of nitrogen; thus, hypophosphatemia is a likely imbalance in total parenteral nutrition [TPN] if inadequate amounts of phosphorus are supplied)
- Treatment of diabetic ketoacidosis (see Chap. 20)
- Malabsorption syndromes
- Overuse of phosphate-binding gels (such as Amphojel, Basaljel, and Dialume)
- Alcoholism
- Prolonged respiratory alkalosis
- Hyperparathyroidism

Clinical Signs

Clinical signs of hypophosphatemia include

- Paresthesia (a frequent, early symptom)
- Profound muscle weakness
- Tremor
- Ataxia
- Incoordination
- Nystagmus, unequal pupils
- Dysarthria
- Dysphagia
- Shallow respiration (may require assisted respiration)
- Disorientation, confusion
- Exquisite pain in the long bones
- Convulsive seizures
- Coma
- Laboratory finding
  Serum phosphorus level < 1.2 mEq/liter (or 2.8 mg/100 ml)

Treatment

- Treatment of phosphorus deficiency varies with the severity of the imbalance.
- For mild to moderate deficiency, administration of oral phosphate supplements (such as Neutra-Phos or Phospho-Soda) is usually adequate.

- For severe deficiency, administration of intravenous phosphate-containing solutions is required.
- Hazards of parenteral phosphorus administration include
  1. Hypocalcemia (serum Ca^{2+} levels should be closely monitored)
  2. Metastatic calcification of soft tissues (should hyperphosphatemia occur)

Phosphorus Excess (Hyperphosphatemia)

Causes

Causes of hyperphosphatemia include
- Renal insufficiency (a major cause, due to abnormal retention of phosphates)
- Hypoparathyroidism
- Increased catabolism
- Vitamin D intoxication
- Myelogenous leukemia
- Lymphoma

Clinical Signs

Clinical signs of hyperphosphatemia include
- Short-term
    Tetany
- Long-term
    Soft tissue calcification
- Laboratory findings
    Serum phosphorus level > 3.0 mEq/liter (or 4.5 mg/100 ml)
    Usually associated with hypocalcemia and reduced serum bicarbonate level

Treatment

- The underlying cause of hyperphosphatemia must be discovered and treated.
- Restriction of dietary phosphorus is indicated to minimize hyperphosphatemia.
- Phosphate-binding gels (*e.g.,* Amphojel, Basaljel, and Dialume) may be used to reduce the serum phosphorus level.

Lithium

Use of Lithium for Affective Disorders

Lithium carbonate is therapeutically effective in the treatment of primary affective disorders of the bipolar type (mania and depression). The manner in which lithium exerts its action is not clearly defined.

Regulation of Dosage

- Lithium dosage is guided by blood serum levels. A lag period of 4 to 10 days is necessary before therapeutic serum levels are reached (0.8–1.2 mEq/liter).
- Initially, lithium determinations should be made twice a week until the desired therapeutic range is reached; then weekly for several weeks, and finally monthly. After the serum lithium level has been stable for 6 to 12 months, lithium determinations can be made less frequently. Lithium blood levels should be drawn 8 to 12 hours after the last dose.
- Since the kidney is the only important route of lithium excretion, dosage should be adjusted according to renal function in the elderly and in patients with known kidney disease. It probably should not be used at all in patients with significant renal damage.
- When possible, lithium should not be used in the first trimester of pregnancy because it can cause fetal malformations and maternal electrolyte disturbances.

Effects on Sodium and Water Balance

- Lithium decreases sodium reabsorption by the renal tubules and can thus lead to sodium depletion.
- Lithium tends to be retained by the kidneys when the serum sodium level is low; thus, lithium toxicity may occur in hyponatremic patients. For example, decreased tolerance to lithium has been reported in patients with protracted sweating and diarrhea. In such instances, supplemental salt and fluid must be administered.
- Long-term use of lithium may have adverse effects on renal function in some individuals. Cases of lithium-induced nephrogenic diabetes insipidus have been reported; symptoms include polyuria and polydipsia.
- Lithium interferes with the synthesis, storage, and release of antidiuretic hormone ADH; because of this, patients taking lithium may experience polyuria. (In fact, lithium has been used in the treatment of patients with excessive ADH activity.)

Factors Causing or Contributing to Lithium Toxicity

Factors causing or contributing to lithium toxicity include
- Overdosage of lithium (accidental or intentional)
- Hyponatremia (as occurs in sweating or diarrhea)
- Dehydration
- Diuretics
    1. Thiazide diuretics reduce renal clearance of lithium and thus increase serum lithium levels; adjustment in lithium dosage is necessary to compensate for this. (Patients taking lithium should use diuretics only under close medical supervision.)
    2. Potassium-sparing diuretics (spironolactone, triamterene, and amiloride) may also cause increased serum lithium levels.
- Significant renal or cardiovascular disease

Clinical Signs

- During the first few days of therapy, mild transient nausea may occur (as may fine tremors of the hand, polyuria, and mild thirst). These symptoms may subside with continued treatment or with a temporary reduction or cessation of the dosage. If they persist, lithium administration should be stopped.
- Serious symptoms of lithium toxicity include
    1. Severe vomiting and diarrhea
    2. Tremors
    3. Marked muscle weakness
    4. Hyperreflexia
    5. Dysarthria
    6. Rigidity
    7. Seizures
    8. Coma
- Laboratory findings include
    Serum lithium level > 1.5 mEq/liter; particularly dangerous when > 2.0 mEq/liter

Treatment

- There is no known antidote for lithium poisoning.
- Early symptoms can usually be managed by reduction or cessation of the drug for 24 to 48 hours (and then resumption of treatment at a lower dosage).
- Severe symptoms require elimination of lithium intake; other measures may include gastric lavage, osmotic diuretics, and perhaps dialysis. (Osmotic diuretics, such as urea and mannitol, increase lithium excretion and thus lower the serum lithium level.)

Nursing Considerations

- Provide adequate fluid intake (such as 2500–3000 ml daily) for patients on lithium therapy, particularly during the initial period. (Recall that during the first few days of therapy, polyuria may occur.)
- Be alert for, and report, persistent polyuria and polydipsia. (Chronic lithium therapy may be associated with diminution of renal concentrating ability, occasionally presenting as nephrogenic diabetes insipidus.) Report urinary specific gravity (SG) less than 1.015 (may indicate diabetes insipidus).
- Encourage adequate salt intake through a normal diet. (Recall that lithium decreases sodium reabsorption by the renal tubules and can lead to hyponatremia.)
- Be alert for, and report, signs of lithium toxicity (particularly when the patient is losing sodium by an abnormal route).
- Caution patients not to take diuretics unless they are under close medical supervision.
- Be particularly alert for signs of lithium toxicity in patients with hyponatremia, because lithium tends to be retained by the kidneys when the serum sodium is low.
- Monitor serum lithium levels at regular intervals.
- Check intake and output, and weigh the patient daily.
- Teach outpatients and their families the toxic symptoms of lithium; instruct them to withhold one dose of lithium and call the physician if toxic symptoms occur.
- Counsel mothers on lithium therapy that it is not wise to breast-feed their infants, because the lithium in mammary milk is one third to one half that in serum.

Zinc Imbalances

Facts About Zinc

- Zinc is necessary for both epithelization and collagen synthesis; therefore, zinc depletion interferes with wound healing. In the presence of zinc deficiency, zinc supplements promote healing. However, wound healing cannot be accelerated by giving zinc supplements to patients with normal zinc serum levels.
- The normal zinc serum level is 77 to 137 $\mu g/100$ ml.
- Zinc is essential for the activity of at least 90 enzymes, which participate in all the major metabolic pathways.
- The small bowel is the major site of zinc absorption; therefore, small bowel disease is a cause of zinc deficiency.

Zinc Deficiency

Causes

Causes of zinc deficiency include

- Hyperalimentation (especially when used longer than 3 wk without zinc supplementation)
- Prolonged poor dietary intake
- Malabsorption syndromes (as in intestinal bypass surgery or regional enteritis)
- Chronic blood loss
- Dialysis
- Diuretics
- Burns

Clinical Signs

Clinical signs of zinc deficiency include

- Impaired sense of taste and smell
- Anorexia (probably related to impaired sense of taste)
- Impaired wound healing
- Depression, poor concentration
- Rash
- Alopecia
- Retarded growth and sexual maturation
- Laboratory finding
    - Serum zinc level < 77 μg/100 ml

Treatment

- Zinc salts can be administered orally, by nasogastric tube, or intravenously; zinc salts have a wide therapeutic index and dosages should be adjusted according to clinical response.
- There is danger of excessive zinc intake when large supplements are added to parenteral fluids. It has been demonstrated that commercially prepared solutions already contain variable amounts of trace elements (including zinc) as contaminants.

Zinc Excess

Causes

Causes of zinc excess include

- Excessive ingestion of oral zinc supplements
- Overdosage of intravenous zinc supplements

Clinical Signs

Clinical signs of zinc excess include
- Anorexia
- Nausea and vomiting
- Dizziness
- Lethargy
- Diarrhea
- Gastric erosions
- Acute renal failure and death (with intravenous overloading of zinc)
- Serum zinc level above normal (> 137 μg/100 ml)

chapter 12

Acid–Base Imbalances

Regulation of Acid–Base Balance

- *p*H is defined as hydrogen (H⁺) ion concentration; the more hydrogen ions, the more acidic is the solution.
- The *p*H range that is compatible with life (6.8–7.8) represents a tenfold difference in hydrogen ion concentration in plasma.
- The body has the remarkable ability to maintain the plasma *p*H within a narrow normal range (7.35–7.45). It does so by means of chemical buffering mechanisms, by the kidneys, and the lungs.
- It has been estimated that metabolism normally produces 13,000 mEq of hydrogen ions each day. Less than 1% of this amount is excreted by the kidneys. It is obvious, then, why renal shutdown can be present for hours or days before life-threatening acid–base imbalance occurs; yet, cessation of breathing for minutes produces critical acid–base changes.
- The best way to evaluate acid–base balance is by measurement of arterial blood gases (ABGs). (Table 12-1)
- Rules for interpreting blood gases are presented in the last section of this chapter, as are nursing responsibilities in obtaining blood samples for ABG studies.

Chemical Buffering Mechanisms

- Chemical buffers are substances that prevent major changes in the *p*H of body fluids by removing or releasing hydrogen ions; they can act within a fraction of a second to prevent excessive changes in hydrogen ion concentration.
- The body's major buffer system is the bicarbonate (HCO_3^-)–carbonic acid (H_2CO_3) buffer system. Normally, there are 20 parts of bicarbonate to 1 part of carbonic acid. If this ratio is upset, the *p*H will change. It is the ratio that is important in maintaining *p*H, not absolute values.

141

Table 12-1
Arterial Blood Gases

Term	Normal Value	Definition—Implications
pH	7.35–7.45	Reflects H+ concentration; acidity increases as H+ concentration increases (pH value decreases as acidity increases) • pH < 7.35 (acidosis) • pH > 7.45 (alkalosis)
$PaCO_2$	38–42 mm Hg	Partial pressure of CO_2 in arterial blood • When <38 mm Hg, hypocapnia is said to be present (respiratory alkalosis) • When >42 mm Hg, hypercapnia is said to be present (respiratory acidosis)

PaO_2	80–100 mm Hg
	(decreases with age)
		Partial pressure of O_2 in arterial blood
		• Any reading above 80 mm Hg (on room air) is considered acceptable
		• In adults < 60 yr (on room air)
		  • <80 mm Hg indicates mild hypoxemia
		  • <60 mm Hg indicates moderate hypoxemia
		  • <40 mm Hg indicates severe hypoxemia
		• Somewhat lower levels are accepted as normal in aged persons, because there is some loss of ventilatory function with advanced age

Standard HCO_3^-	22–26 mEq/liter	HCO_3^- concentration in plasma of blood that has been equilibrated at a $PaCO_2$ of 40 mm Hg, and with O_2 in order to fully saturate the hemoglobin
Base excess (BE)	$-2-+2$ mEq/liter	Reflects metabolic (nonrespiratory) body disturbances, which may be primary or compensatory in nature Always negative in metabolic acidosis (deficit of alkali or excess of fixed acids) Always positive in metabolic alkalosis (excess of alkali or deficit of fixed acids) Arrived at by multiplying the deviation of standard HCO_3^- from normal by a factor of 1.2, which represents the buffer action of red blood cells

Example: In a normal individual

$$\frac{HCO_3^-}{H_2CO_3} = \frac{24 \text{ mEq/liter}}{1.2 \text{ mEq/liter}} = \frac{20}{1} \quad (pH = 7.4)$$

In an individual with chronic obstructive lung disease, one might see

$$\frac{HCO_3^-}{H_2CO_3} = \frac{48 \text{ mEq/liter}}{2.4 \text{ mEq/liter}} = \frac{20}{1} \quad (pH = 7.4)$$

- Carbon dioxide (CO_2) is a potential acid; when CO_2 is dissolved in water, it becomes carbonic acid H_2CO_3 [$CO_2 + H_2O = H_2CO_3$]).
- If either bicarbonate or carbonic acid is increased or decreased so that the 20:1 ratio is no longer valid, acid–base imbalance results. Figure 12-1 demonstrates changes in plasma pH when the bicarbonate:carbonic acid ratio is altered.
- Other buffer pairs in the body include the phosphate buffer system and the protein buffer system.

Kidneys

- The kidneys regulate the bicarbonate level in extracellular fluid; they are able to regenerate bicarbonate ions as well as to reabsorb them from the renal tubular cells.
- In the presence of respiratory acidosis, and most cases of metabolic acidosis, the kidneys excrete hydrogen ions and conserve bicarbonate ions to help restore balance. (The kidneys obviously cannot compensate for the metabolic acidosis created by renal failure.)
- In the presence of respiratory and metabolic alkalosis, the kidneys retain hydrogen ions and excrete bicarbonate ions to help restore balance.
- Renal compensation for imbalances is slow (a matter of hours or days).

Lungs

- The lungs control the carbon dioxide (and thus carbonic acid) content of extracellular fluid. They do so by adjusting ventilation in response to the amount of carbon dioxide in the blood.

A rise in $PaCO_2$ (partial pressure of CO_2 in arterial blood) is a powerful stimulant to respiration. For each mm Hg rise in $PaCO_2$ above normal, the total ventilation increases to about 21 ml/min. The maximal effect of $PaCO_2$ on ventilation is reached at approximately 15% CO_2 concentration. Whenever the $PaCO_2$ is consistently above 50 mm Hg, the respiratory center becomes relatively insensitive to its stimulating effect, leaving the respiratory center

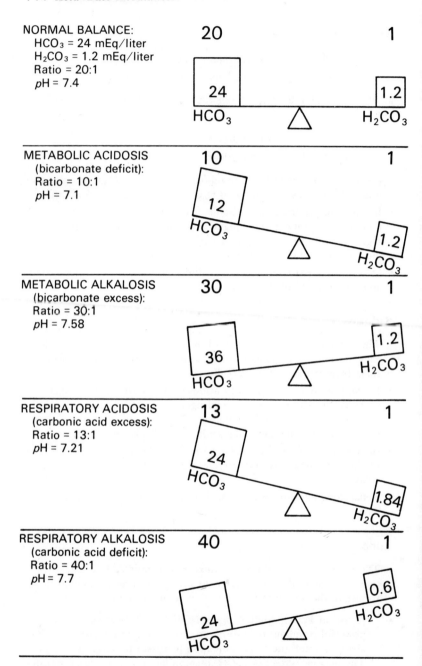

Figure 12–1. Examples of *p*H changes with alterations in the bicarbonate: carbonic acid ratio.

completely insensitive if there is any sudden further increase in $PaCO_2$. (Hypoxemia is then the primary respiratory stimulus).

PaO_2 (partial pressure of O_2 in arterial blood) influences respiration; however, its effect is not as marked as that produced by $PaCO_2$. A fall in PaO_2 to 50 mm Hg (from the normal range of 80–100 mm Hg) usually is necessary before any increase in ventilation takes place.

- Also, the lungs compensate for metabolic disturbances by either conserving or retaining carbon dioxide.
    1. In the presence of metabolic acidosis, respiration is increased, causing greater elimination of carbon dioxide (to lighten the acid load in the body).
    2. In the presence of metabolic alkalosis, respiration is decreased, causing carbon dioxide to be retained (increasing the acid load).
- The lungs can alter plasma *p*H within a matter of minutes.
- Faulty pulmonary function disrupts acid–base balance. For example, hypoventilation due to chronic obstructive pulmonary disease (COPD) causes excessive carbon dioxide retention and respiratory acidosis; hyperventilation due to hysteria causes excessive elimination of carbon dioxide and respiratory alkalosis.
- The lungs are unable to compensate for *p*H disturbances when there is severe pulmonary dysfunction; in these instances, compensation must be accomplished solely by the kidneys.

Acid–Base Imbalances

There are four kinds of acid–base imbalances, as follows:
- Metabolic acidosis (HCO_3^- deficit)
- Metabolic alkalosis (HCO_3^- excess)
- Respiratory acidosis (H_2CO_3 excess)
- Respiratory alkalosis (H_2CO_3 deficit)

Metabolic Acidosis

Causes

- Metabolic acidosis occurs in one of two ways
    1. Loss of alkali, or
    2. Gain of fixed acids (those other than H_2CO_3).
- Metabolic acidosis can be divided clinically into two forms, depending on the values of the serum anion gap:

Anion Gap (AG) = $Na^+ - (HCO_3^- + Cl^-)$ = 12 to 15 mEq/liter

1. There are some unmeasured anions in serum; these include phosphates, sulfates, and anions of organic acids (such as ketones and lactic acid).
2. Normally, these unmeasured anions account for less then 16 mEq/liter of the anion production. (An anion gap > 16 suggests excessive accumulation of unmeasured anions.)
3. Note in the previous equation that the anion gap (unmeasured anions) can be determined by subtracting the sum of the measured anions (HCO_3^- + Cl^-) from the cation concentration (represented by Na^+).

- *High anion gap acidosis* results from excessive, fixed acid accumulation.
  1. Ketoacidosis (as in uncontrolled diabetes mellitus and starvation)
  2. Lactic acidosis (most often due to tissue hypoxia; metabolism switches to anaerobic processes, generating large quantities of lactic acid as a byproduct)
  3. Late salicylate poisoning (causes faulty utilization of glucose and thus excessive ketone formation)
  4. Uremia (accumulation of phosphates and sulfates normally excreted in urine)
  5. Methanol or ethylene glycol toxicity

Example: Patient with lactic acidosis

$$Na^+ = 131 \text{ mEq/liter} \qquad AG = Na^+ - (HCO_3^- + Cl^-)$$
$$HCO_3^- = 9 \text{ mEq/liter} \qquad 131 - (9 + 86) = 36$$
$$Cl^- = 86 \text{ mEq/liter} \qquad \text{mEq/liter (high AG)}$$

- *Normal anion gap acidosis* results from direct loss of bicarbonate, as in
  1. Diarrhea
  2. Urereteroenterostomies (more pronounced in ureterosigmoidostomy because of the large surface area of the bowel to absorb urinary chloride; however, can occur in ileal loop bladders also)
  3. Pancreatic fistulas
  4. Excessive gain of chloride (as in administration of large quantities of isotonic saline or ammonium chloride)
  5. Distal renal tubular acidosis
  6. Acetazolamide (Diamox)

Example: Patient with ureterosigmoidostomy

$$Na^+ = 134 \text{ mEq/liter} \qquad AG = Na^+ - (HCO_3^- + Cl^-)$$
$$HCO_3^- = 10 \text{ mEq/liter} \qquad 134 - (10 + 115) = 9$$
$$Cl^- = 115 \text{ mEq/liter} \qquad \text{mEq/liter (normal AG)}$$

- Metabolic acidosis may be acute or chronic. Expected blood gas changes for uncompensated, partly compensated, and completely compensated metabolic acidosis are presented in Table 12-2.

Clinical Signs

- Headache
- Confusion
- Drowsiness
- Increased respiratory rate and depth (compensatory action by lungs) —may not become clinically evident until bicarbonate is quite low
- Nausea and vomiting
- Peripheral vasodilatation may be present, causing warm, flushed skin.
- Decreased cardiac output when pH falls below 7; bradycardia may develop
- Arterial blood gases
  1. Fall in pH (< 7.35)
  2. Bicarbonate < 22 mEq/liter (primary disorder)
  3. $PaCO_2 < 38$ mm Hg (indicates compensatory action by lungs, which usually begins immediately)
  4. Base excess (BE) always negative

Example: Patient with diabetic ketoacidosis

$$pH = 7.05$$
$$HCO_3^- = 5 \text{ mEq/liter}$$
$$PaCO_2 = 12 \text{ mm Hg}$$
$$BE = -30 \text{ mEq/liter}$$

- Other laboratory data

  Hyperkalemia is frequently present (serum $K^+ > 5.5$ mEq/liter; exceptions—serum K^+ is decreased in renal tubular acidosis, diarrhea, and in use of acetazolamide).

Table 12–2
Expected Directional Changes in Blood Gases in Metabolic Acidosis

Imbalance	pH	HCO_3^-	$PaCO_2$	BE
Uncompensated metabolic acidosis	↓	↓	N*	↓
Partly compensated metabolic acidosis	↓	↓	↓	↓
Completely compensated metabolic acidosis	N*	↓	↓	↓

*N = Normal

Treatment

Treatment is directed toward correcting the metabolic defect. If the cause of the problem is excessive intake of chloride, treatment is obviously directed at elimination of the source. When necessary, bicarbonate is administered. Treatment of Ketoacidosis is described in Chapter 20, Uremia in Chapter 22, and Salicylate Intoxication in Chapter 27.

Metabolic Alkalosis

Causes

- Vomiting or gastric suction (results in loss of H^+ and Cl^-); particularly a problem in pyloric stenosis.
- Hypokalemia, which produces alkalosis in two ways 1. In the presence of hypokalemia, the kidneys conserve potassium and thus increase hydrogen ion excretion. (Recall that these ions compete for renal excretion.) 2. Cellular potassium moves out into the extracellular fluid in an attempt to maintain near normal serum levels (as K^+ leaves the cell, H^+ must enter, to maintain electroneutrality).
- Hyperaldosteronism.
- Cushing's syndrome.
- Diuretics (*e.g.*, thiazides, furosemide, and ethacrynic acid). (These agents promote the excretion of Na^+ and K^+ almost exclusively in association with Cl^-; loss of Cl^- causes increase in HCO_3^-).
- Excessive alkali ingestion (as in bicarbonate-containing antacids; *e.g.*, Alka-Seltzer).
- Excessive parenteral alkali administration (as in administration of sodium bicarbonate during cardiopulmonary resuscitation).
- Abrupt relief of chronically high carbon dioxide level in plasma (*e.g.*, assisted ventilation) results in a "lag period" before the chronically high serum bicarbonate level can be corrected by the kidneys.
- Villous adenoma of the colon.

Clinical Signs

- Symptoms related to decreased calcium ionization caused by alkalosis
    1. Dizziness
    2. Tingling of fingers and toes
    3. Circumoral paresthesia
    4. Carpopedal spasm
    5. Hypertonic muscles
- Depressed respiration (compensatory action by lungs) is usually not clinically evident in an alert patient.
- Arterial blood gases

1. $pH > 7.45$
2. Bicarbonate > 26 mEq/liter (primary disorder)
3. $PaCO_2$ > 42 mm Hg (indicates compensatory action by lungs)

    Compensatory hypoventilation is more pronounced in semiconscious, unconscious, or debilitated patients than in alert patients. The former may develop marked hypoxemia as a result of hypoventilation.
4. Base excess is always positive.

Example: Patient with vomiting

> $pH = 7.62$
> $HCO_3^- = 45$ mEq/liter
> $PaCO_2 = 48$ mm Hg
> $BE = +16$

- Other laboratory data
    1. Hypokalemia is often present (serum $K^+ < 3.5$ mEq/liter).
    2. Serum chloride is relatively lower than sodium.
    3. Urinary chloride concentration is usually > 20 mEq/liter when metabolic alkalosis is due to hyperaldosteronism, Cushing's syndrome, or profound potassium depletion (serum $K^+ < 2.0$ mEq/liter).
    4. Urinary chloride concentration is usually < 10 mEq/liter when metabolic alkalosis is due to vomiting or gastric suction or diuretic use (late).
    5. Decreased ionized calcium level.
- Metabolic alkalosis can be acute or chronic. Expected directional changes in blood gases for uncompensated, partly compensated, and completely compensated metabolic alkalosis are listed in Table 12-3.

Table 12–3
Expected Directional Changes in Blood Gases in Metabolic Alkalosis

Imbalance	pH	HCO_3^-	$PaCO_2$	BE
Uncompensated (acute) metabolic alkalosis	↑	↑	N*	↑
Partly compensated (subacute) metabolic alkalosis	↑	↑	↑	↑
Completely compensated (chronic) metabolic alkalosis	N*	↑	↑	↑

*N = Normal

Treatment

- Treatment is aimed at reversal of the underlying disorder.
- Treatment includes restoration of normal fluid volume by administration of sodium chloride fluids (because continued volume depletion serves to maintain the alkalosis). Sufficient chloride must be supplied for the kidney to absorb sodium with chloride (allowing the excretion of excess HCO_3^-).

Respiratory Acidosis

Respiratory acidosis can be either acute or chronic; the acute imbalance is particularly dangerous. When respiratory acidosis is acute, the bicarbonate level remains in the normal range because renal compensation is very slow. Therefore, the high $PaCO_2$ can quickly produce a sharp decrease is plasma *p*H. When respiratory acidosis is chronic, as in COPD, the kidneys compensate for the elevated $PaCO_2$ by increasing the bicarbonate level. (Compensatory renal HCO_3^- generation takes several hours to several days to develop.)

Causes

- Always due to inadequate excretion of carbon dioxide (inadequate ventilation) resulting in elevated plasma carbon dioxide levels and thus carbonic acid levels. In addition to an elevated $PaCO_2$, hypoventilation usually causes a decrease in PaO_2.
- *Acute Respiratory Acidosis*
    1. Acute pulmonary edema
    2. Aspiration of a foreign body
    3. Atelectasis
    4. Pneumothorax, hemothorax
    5. Overdosage of sedatives or anesthetics
    6. Position on operating table that interferes with respirations
    7. Cardiac arrest
    8. Severe pneumonia
    9. Laryngospasm
    10. Mechanical ventilation improperly regulated
    11. Croup
    12. Hyaline membrane disease
- *Chronic Respiratory Acidosis*
    1. Emphysema
    2. Cystic fibrosis
    3. Advanced muscular dystrophy
    4. Bronchiectasis
    5. Bronchial asthma

- Sleep causes mild hypoventilation and increased $PaCO_2$ (a normal occurrence).
- Factors that interfere with normal pulmonary excursions favor respiratory acidosis.
    1. Obesity
    2. Tight abdominal binders or dressings
    3. Postoperative pain (as in high abdominal or chest incisions)
    4. Abdominal distention from cirrhosis or bowel obstruction

Clinical Signs

Clinical signs are variable between acute and chronic respiratory acidosis.

Acute Respiratory Acidosis

- Sudden development of hypercapnia ($\uparrow PaCO_2$) can cause
    1. Increased pulse and respiratory rate
    2. Increased blood pressure
    3. Feeling of fullness in the head ($\uparrow PaCO_2$ causes cerebrovascular vasodilatation and increased cerebral blood flow, particularly when higher than 60 mm Hg)
    4. Mental cloudiness
    5. Dizziness
    6. Palpitations
    7. Muscular twitching
    8. Convulsions
    9. Warm, flushed skin
    10. Unconsciousness
- Ventricular fibrillation may be the first sign of respiratory acidosis in the anesthetized patient (due to hyperkalemia associated with acidosis). Respiratory acidosis may occur as soon as 15 minutes after the start of anesthesia and is most likely to occur in patients with chronic pulmonary disease.
- Arterial blood gases
    1. $pH < 7.35$ (may reach a level of 7 or less in a few min); pH falls 1/10 for each 20 mm Hg rise in $PaCO_2$ above 40 mm Hg[1]
    2. $PaCO_2 > 42$ mm Hg (may reach 120 mm Hg or higher)
    3. Bicarbonate normal or only slightly elevated, since there has been little time for renal compensation (an exception would be the patient with COPD who chronically has an elevated HCO_3^-; in this situation, it would not be high enough to compensate for the sudden increase in $PaCO_2$)
    4. PaO_2 below normal when patient is breathing room air (result of hypoventilation)

Example:

$$pH = 7.26$$
$$PaCO_2 = 56 \text{ mm Hg}$$
$$HCO_3^- = 24 \text{ mEq/liter}$$

Chronic Respiratory Acidosis

- Weakness
- Dull headache
- Those of underlying disease process
- Arterial blood gases
  1. $pH < 7.35$ or within lower limit of normal (if complete compensation has occurred)
  2. $PaCO_2 > 42$ mm Hg (frequently between 50 to 60 mm Hg or greater)
     - Patients with COPD who gradually accumulate carbon dioxide over a prolonged period (days to months) may not develop symptoms of hypercapnia (listed previously under Acute Respiratory Acidosis) because compensatory changes have had time to occur.

Example: An emphysematous patient kept alive with oxygen therapy for more than 1 year was mentally alert even though his $PaCO_2$ was 140 mm Hg. A rapid rise in the $PaCO_2$ to 140 mm Hg in the normal person would surely produce unconsciousness.

  3. Bicarbonate > 26 mEq/liter (indicates renal, or metabolic compensation)
  4. Base Excess > +2 (indicates renal, or metabolic compensation)

Example:

$$pH = 7.38$$
$$PaCO_2 = 76 \text{ mm Hg}$$
$$HCO_3^- = 42 \text{ mm Hg}$$
$$BE = +14 \text{ mEq/liter}$$

Note: Recall that when the $PaCO_2$ is chronically above 50 mm Hg, the respiratory center becomes relatively insensitive to carbon dioxide as a respiratory stimulant, leaving hypoxemia as the major drive for respiration. Excessive oxygen administration removes the stimulus of hypoxemia, and the patient develops carbon dioxide narcosis (described under Acute Respiratory Acidosis) unless the situation is quickly reversed. (See Considerations for Safe Oxygen Administration in Patients With Chronic Respiratory Acidosis.)

- Expected directional changes in blood gases in uncompensated, partly compensated, and completely compensated respiratory acidosis are listed in Table 12-4.

Table 12–4
**Expected Directional Changes in
Blood Gases in Respiratory Acidosis**

Imbalance	*p*H	PaCO$_2$	HCO$_3^-$	BE
Uncompensated respiratory acidosis (acute)	↓	↑	N*	N*
Partly compensated respiratory acidosis	↓	↑	↑	↑
Completely compensated respiratory acidosis	N*	↑	↑	↑

*N = Normal

**Considerations for Safe Oxygen Administration
in Patients With Chronic Respiratory Acidosis**

- Oxygen should be used intermittently rather than continuously for patients with chronic respiratory acidosis.
- Only devices suitable for low-flow O$_2$ rates should be used (such as a nasal cannula or a Ventimask).
  1. Nasal cannulas supply an O$_2$ concentration of 28% at a flow of 2 liters/min and an O$_2$ concentration of 40% at 6 liters/min.
  2. The Ventimask can deliver O$_2$ concentrations of 24%, 28%, 35%, or 40%.
- No more than a 30% or 40% concentration of O$_2$ in air should be used for intermittent positive pressure breathing (IPPB); a higher concentration may produce serious respiratory depression by removing hypoxemia as a respiratory stimulus.
- It has been recommended that the lowest flow of O$_2$ necessary to provide a PaO$_2$ of 60–80 mm Hg be used. (Some authorities prefer to keep the PaO$_2$ close to 60 mm Hg to avoid eliminating PaO$_2$ as a respiratory stimulus.) When the desired PaO$_2$ cannot be maintained with supplemental O$_2$, it may become necessary to institute mechanical ventilation.

Treatment

- Treatment is directed at improving ventilation; exact measures vary with the cause of inadequate ventilation.
- Pharmacologic agents are used as indicated. For example, bronchodilators help reduce bronchial spasm; antibiotics are used for respiratory infections.
- Pulmonary hygiene measures are employed, when necessary, to rid the respiratory tract of mucus and purulent drainage.
- Adequate hydration (2–3 liters/day) is indicated to keep the mucous membranes moist and thereby facilitate removal of secretions.
- Supplemental oxygen is used as necessary.
- A mechanical respirator, used cautiously, may improve pulmonary

ventilation. Overzealous use of a mechanical respirator may cause such rapid excretion of carbon dioxide that the kidneys will be unable to eliminate excess bicarbonate with sufficient rapidity to prevent alkalosis and convulsions. For this reason, the elevated $PaCO_2$ must be decreased, slowly.

Nursing Actions to Improve Pulmonary Ventilation

- Turn the patient from side to side at frequent intervals to allow for gravitational drainage of mucus from the various lung segments.
- Place the patient in Fowler's position or semi-Fowler's position to allow for greater chest expansion.
- Suction the patient as necessary to rid the respiratory tract of excessive secretions. Give sufficient fluids to keep the mucous membranes moist.
- Before administering any drug that can depress respirations, check the dose carefully as well as the time when the drug was last given. In addition, check the rate and depth of respirations before administering the drug.
- Judiciously administer analgesics to patients whose respiratory efforts are hampered by abdominal or thoracic incisional pain. (Relief of pain allows the patient to breath more deeply and thus excrete CO_2 more effectively.)
- Increase the patient's activity, as outlined by the physician, to promote ventilatory and circulatory improvement.
- Perform other functions, as prescribed, such as O_2 administration, postural drainage, chest cupping, and IPPB treatments.

Respiratory Alkalosis

Causes

- Always due to hyperventilation (which causes excessive "blowing off" of CO_2 and, hence, a decrease in plasma H_2CO_3 content)
- Extreme anxiety (the most common cause of respiratory alkalosis; sometimes referred to as the *hyperventilation syndrome*)
- Hypoxemia (low PaO_2 causes increased respiration)
- High fever
- Thyrotoxicosis
- Excessive ventilation by mechanical ventilators
- Early salicylate intoxication (stimulates respiratory center)
- Gram-negative bacteremia (respiratory alkalosis is often the first sign of this condition; the mechanism for this is not known)
- Central nervous system lesions involving the respiratory center (such as meningitis or encephalitis)

- Pregnancy (high progesterone level sensitizes the respiratory center to CO_2)
- Pulmonary emboli

Clinical Signs of Hyperventilation Syndrome

- Lightheadedness (a low $PaCO_2$ causes cerebral vasoconstriction and thus decreased cerebral blood flow)
- Inability to concentrate
- Numbness and tingling of extremities; circumoral paresthesia (symptoms of decreased Ca^{2+} ionization are more likely to occur if respiratory alkalosis develops rapidly)
- Tinnitus
- Palpitations
- Sweating
- Dry mouth
- Tremulousness
- Precordial pain (tightness)
- Nausea and vomiting
- Epigastric pain
- Blurred vision
- Convulsions and loss of consciousness (may be partly due to cerebral ischemia, caused by cerebral vasoconstriction)

Note: Although hyperventilation is the cause of respiratory alkalosis, it is important to remember that a marked reduction in $PaCO_2$ (hypocapnia) can be present without a clinically evident increase in respiratory effort.

- Arterial blood gases
  1. $pH > 7.45$ (pH rises $1/10$ for each 10 mm Hg fall in $PaCO_2$ below 40 mm Hg in acute disorder)[2]
  2. $PaCO_2 < 38$ mm Hg
  3. Bicarbonate initially unchanged (renal response is slow); if chronic respiratory alkalosis develops, the bicarbonate level will decrease below 22 mEq/liter.
  4. Base excess initially unchanged; if chronic respiratory alkalosis develops, the base excess will decrease below -2 mEq/liter.

Example: Acute respiratory alkalosis

$$pH = 7.52$$
$$PaCO_2 = 30 \text{ mm Hg}$$
$$HCO_3^- = 24 \text{ mEq/liter}$$
$$BE = +2.5 \text{ mEq/liter}$$

Example: Chronic respiratory alkalosis (compensated)

$$pH = 7.40$$
$$PaCO_2 = 30 \text{ mm Hg}$$
$$HCO_3^- = 18 \text{ mEq/liter}$$
$$BE = -5 \text{ mEq/liter}$$

- Other laboratory data

  Decreased ionized calcium level may be present.

- Expected directional changes in arterial blood gases for acute, partly compensated, and completely compensated respiratory alkalosis are presented in Table 12-5.

Treatment

- If cause of respiratory alkalosis is anxiety, the patient should be made aware of his abnormal breathing practices; he should be made to realize that this breathing pattern causes his symptoms. He can be instructed to breath more slowly (to cause accumulation of CO_2) or to breath into a closed system (such as a paper bag). Usually a sedative is required to relieve hyperventilation in very anxious patients. (If alkalosis is severe enough to cause fainting, the increased ventilation will cease and respirations will revert to normal.)

- Treatment for other causes is directed at correcting the underlying problem.

Mixed Acid–Base Imbalances

- The preceding pages have described single acid–base imbalances. These single imbalances do indeed occur. However, one should be aware that in some clinical situations the patient may have two, primary acid–base disturbances simultaneously. Some examples are given at the end of the chapter.

Table 12–5
Expected Directional Changes in Arterial
Blood Gases in Respiratory Alkalosis

Imbalance	pH	$PaCO_2$	HCO_3^-	BE
Uncompensated respiratory alkalosis	↑	↓	N*	N*
Partly compensated respiratory alkalosis	↑	↓	↓	↓
Completely compensated respiratory alkalosis	N*	↓	↓	↓

*N = Normal

- Without a systematic approach to the analysis of acid–base disturbances, it is easy to become greatly confused. (See the following sections.)

Systematic Interpretation of Arterial Blood Gases (ABGs)

The following steps to evaluate ABGs are based on the assumption that the average normal values are

$pH = 7.4$
$PaCO_2 = 40$ mm Hg
$HCO_3^= = 24$ mEq/liter

- Look at the pH. It can be high, low, or normal.

  $pH > 7.4$ (alkalosis)
  $pH < 7.4$ (acidosis)
  $pH = 7.4$ (normal)

  A normal pH may indicate normal acid–base balance, or it may indicate an imbalance that has been compensated. If it is the latter, both the $PaCO_2$ and HCO_3^- will be high, or low. Occasionally, two opposite primary imbalances may be occurring at the same time (sometimes resulting in a normal pH).

- Next, determine the primary cause of the disturbance. This is done by evaluating the $PaCO_2$ and HCO_3^- in relation to the pH.

pH > 7.4 (alkalosis)

1. If the $PaCO_2$ is < 40 mm Hg, the primary disturbance is respiratory alkalosis.

2. If the HCO_3^- is > 24 mEq/liter, the primary disturbance is metabolic alkalosis.

pH < 7.4 (acidosis)

1. If the $PaCO_2$ is > 40 mm Hg, the primary disturbance is respiratory acidosis.

2. If the HCO_3^- is < 24 mEq/liter, the primary disturbance is metabolic acidosis.

- After finding the primary disturbance, determine if compensation has begun. If so, the value other than the primary disorder will be moving in the same direction as the primary value.

  Example:

pH	$PaCO_2$	HCO_3^-
(1) 7.20	60 mm Hg	23 mEq/liter
(2) 7.40	60 mm Hg	37 mEq/liter

  The first set (1) of ABGs indicates acute respiratory acidosis without compensation (the $PaCO_2$ is high, the HCO_3^- is normal). The second set (2) indicates chronic respiratory acidosis. Note that compensation has taken place; that is, the HCO_3^- has elevated to an appropriate level to balance the high $PaCO_2$ and produce a normal pH.

- Use the equations listed in the next section to determine if changes in values other than the primary ones are compensatory in nature or if they indeed indicate a second imbalance. (Recall that mixed imbalances are relatively common.) Some examples of mixed acid–base disturbances are presented after the equations for practice.

(continued)

Key to Equations

If Primary Disturbance is Respiratory

For an acute rise of $PaCO_2$ of 10 mm Hg	HCO_3^- should rise 1 mEq/liter
For a chronic rise of $PaCO_2$ of 10 mm Hg	HCO_3^- should rise 4 mEq/liter
For a fall of $PaCO_2$ of 10 mm Hg	HCO_3^- should fall 2 mEq/liter

If amount HCO_3^- is above expected—metabolic alkalosis
If amount HCO_3^- is below expected—metabolic acidosis

If Respiratory Disturbance is Complicated by Metabolic Acidosis

Calculate anion gap.

Anion gap $= Na^+ - (CO_2{}^* + Cl^-)$, to determine if cause of complicating metabolic disorder is due to increase in organic acids (normal gap is 12 to 15).

If Primary Disturbance is Metabolic Acidosis

Calculate Winter's formula to determine presence of respiratory component.

Winter's formula: $PaCO_2 = (1.54 \times HCO_3) + 8.36 \pm 1$

If $PaCO_2$ higher than predicted	Respiratory acidosis
If $PaCO_2$ lower than predicted	Respiratory alkalosis

If Disturbance is Metabolic Alkalosis

Look at $PaCO_2$ rise.

If $PaCO_2$ exceeds 55 mm Hg	May have respiratory acidosis
If $PaCO_2$ remains low or normal	May have respiratory alkalosis

---

* In this situation, CO_2 refers to "CO_2 content" in the serum. This test reflects total HCO_3^- and a small measure of H_2CO_3.

(Harper R: A Guide to Respiratory Care: Physiology and Clinical Applications, p 157. Philadelphia, JB Lippincott, 1981)

Sample Mixed Acid–Base Problems

Use rules in the previous two sections to analyze the following situations:

Respiratory Alkalosis Plus Metabolic Acidosis

Situation: A patient has taken an overdose of aspirin. Recall that early in salicylate poisoning, the medulla is stimulated to produce hyperventilation (respiratory alkalosis results). Later, due to disrupted glucose metabolism, metabolic acidosis may result.

$pH = 7.4$
$PaCO_2 = 18$ mm Hg
$HCO_3^- = 16$ mEq/liter
$BE = -10$ mEq/liter

- Note in this situation that the simultaneous appearance of respiratory alkalosis and metabolic acidosis has produced a normal pH. (This is not always the case; one imbalance may be stronger than the other and cause an abnormal pH in its favor.)

- If the only problem in this situation were respiratory alkalosis, the HCO_3^- would be expected to be approximately 20 mEq/liter. Note that it is lower than expected (indicating metabolic acidosis).

- If the only problem were metabolic acidosis, the $PaCO_2$ would be expected to be approximately 33 mEq/liter. Note that it is lower than expected (indicating respiratory alkalosis).

Respiratory Acidosis Plus Metabolic Acidosis

Situation: A patient has lactic acidosis secondary to cardiac failure, and hypercapnia secondary to pneumonia.

$pH = 7.10$
$PaCO_2 = 50$ mm Hg
$HCO_3^- = 15$ mEq/liter
$BE = -6$ mEq/liter

- Since both primary disturbances produce acidosis, the pH is quite low.

- If the only problem were acute respiratory acidosis, the HCO_3^- level would elevate, slightly above normal. (Instead, it has decreased.)

- If the only problem were metabolic acidosis, the $PaCO_2$ would decrease. (Instead, it has elevated.)

Metabolic Alkalosis Plus Metabolic Acidosis

Situation: A patient with diabetic ketoacidosis (DKA) has severe vomiting. (Recall that DKA is a form of metabolic acidosis, and that vomiting predisposes to metabolic alkalosis.)

$pH = 7.54$
$PaCO_2 = 49$ mm Hg
$HCO_3^- = 40$ mEq/liter
$BE = +16$ mEq/liter
$Na^+ = 139$ mEq/liter
$Cl^- = 50$ mEq/liter

- The high pH indicates the presence of alkalosis; the high HCO_3^- and BE indicate metabolic alkalosis. (The elevated $PaCO_2$ is a respiratory compensatory change to help correct the alkalosis.)

- The only indication of metabolic acidosis in this situation is the history (severe uncontrolled diabetic mellitus) and the high anion gap (AG).

$AG = Na^+ - (HCO_3^- + Cl^-) = 12{-}15$ mEq/liter
(normally)
$AG = 139 - (40 + 50) = 49$ mEq/liter
(quite high)
The high AG indicates excessive accumulation of metabolic acids (in this situation, probably ketones).

Obtaining Samples for Arterial Blood Gases

- Arterial blood gas (ABG) samples are usually obtained by repeated single arterial punctures. The procedure used in arterial puncture is described below.

Obtaining a Sample by Arterial Puncture

- Explain the procedure to the patient. It is important to prevent unnecessary pain and anxiety which may cause hyperventilation, resulting in a temporary change in blood gases.

- Follow hospital policy for selection of the site for arterial puncture by nonphysicians. The radial artery (at the wrist) is generally considered safest for use by nonphysicians since it is superficially located, has collateral circulation, and is not adjacent to a large vein. The brachial artery (in the antecubital fossa) is also used for arterial puncture because it has considerable collateral circulation. The femoral artery is not used as frequently because it has certain disadvantages. For example, there is not adequate collateral flow if the femoral artery becomes obstructed. Obstruction, therefore, threatens the entire lower limb. Also, it is more difficult to stop bleeding from a large artery.

- Position the site as indicated. If the radial site is to be used, a rolled towel should be placed under the wrist and the hand should be pushed down to obtain wrist extension. If the brachial site is to be used, the arm should be extended and supported with a towel under the elbow.

- Identify the maximal point of pulsation in the artery.

- Cleanse the site with either Betadine or alcohol wipes. (Some operators wear sterile gloves to perform an arterial puncture; the hands, at least, should be thoroughly washed.)

- Flush a 5-ml syringe (using a 19- or 20-gauge needle) with 0.5 ml of 1:1000 heparin solution. Empty the syringe, leaving heparin in the needle to prevent clotting of the blood sample. (For small pediatric patients, a heparinized tuberculin syringe with a 25-gauge needle may be used.) It should be remembered that too much heparin left in the syringe will affect the *p*H of the blood sample and, thus, cause false results. (Many institutions now use commercial ABG sample kits, containing preheparinized syringes.)

- Keep the syringe free of air bubbles.

- Be aware that some authorities prefer a glass syringe to a plastic syringe because a glass syringe has freer movement and, thus, fills more easily. Use of a plastic syringe may necessitate "pulling back" on the plunger to obtain the blood sample. Also, air bubbles adhere more tenaciously to plastic syringes, making them less desirable in this situation.

- Insert the needle at an oblique angle into the artery while stabilizing it with the free hand. (An oblique entrance into the artery minimizes the formation of a hematoma since the puncture in the artery seals better when the needle is removed.) Usually the initial thrust of the needle will bring a pulsating flow of blood into the syringe without pulling out the plunger; this spontaneous pumping of blood into the syringe is proof of entry into an artery. If blood does not enter the syringe, the needle must be redirected; if the needle has gone completely through the artery, it must be slightly withdrawn until blood flows again to obtain the blood sample.

- Withdraw the needle from the artery after obtaining the sample and immediately hand the syringe to a second person (see below). Apply pressure over the puncture site for a period of 5 minutes; a longer time may be necessary if the patient is on anticoagulants.

- Plunge the needle into a cork or rubber stopper to seal it from air. (If air accidentally enters the syringe, the specimen must be discarded.)
- Gently rotate the syringe to mix the blood with the heparin to prevent clotting.
- Quickly place the syringe in a basin or bag of ice slush to slow down O_2 metabolism. (Failure to do so can cause the gas analyser to yield inaccurate results, since O_2 metabolism of blood continues even after it is drawn from the body.)
- Take the specimen immediately to the laboratory after it is properly labeled with the following information:

    Patient's name

    Time of arterial puncture

    Body temperature at time of arterial puncture

    Inspired O_2 level

- It is important that the patient's body temperature be recorded accurately since arterial O_2 and CO_2 increase as body temperature increase; conversely, they decrease as body temperature decreases. Blood gas analyzers are controlled for normal body temperature; thus, the technician must correct the results to the patient's actual temperature (by use of the appropriate nomogram), if severe hypothermia or hyperthermia is present.
- It is important to record the patient's inspired O_2 level on the requisition. (For example: Is the patient breathing room air? Is he breathing 40% O_2? Or, is he receiving O_2 by nasal cannula at a rate of 2 liters/min?) This information is important in intrepreting blood gases. If blood gases are obtained to measure the effectiveness of a new O_2 level, the patient should receive that level of O_2 continuously for at least 20 min prior to obtaining the arterial sample.
- In the critically ill patient, an arterial catheter may be inserted to provide immediate access to arterial blood for blood gas analysis. The procedure for obtaining a sample from an arterial catheter is presented below.

Obtaining a Sample From an Indwelling Arterial Catheter[3]

- Assess the arterial-line system and waveform pattern to ensure correct functioning. Turn off the monitoring alarms.
- Cleanse the exit port of the catheter.
- Aseptically withdraw 3 ml to 5 ml of blood from the exit port and discard the specimen. (This is important because the specimen is diluted with intravenous solution, and analysis of this sample would produce inaccurate results.)
- Aseptically withdraw the desired amount of blood into a heparinized syringe. (Observe the syringe for the presence of air in the syringe; if this occurs, obtain a new specimen.)

(continued)

- Send the specimen to the laboratory with proper labeling (as noted earlier).
- Close the stopcock to the exit port and flush the arterial-line system. Assess the system for correct functioning and accurate stopcock position. Observe arterial waveforms, and reset alarms.

- Complications of arterial lines include tissue necrosis (most likely when the involved artery has inadequate collateral circulation to supply distal tissue) and infection. The risk of infection can be reduced by aseptic handling of the dressing, changing the dressing and intravenous tubing every 24 hours, and using an antibiotic ointment over the puncture site.

References

1. Harper R: A Guide to Respiratory Care: Physiology and Clinical Applications, p 150. JB Lippincott, 1981
2. Ibid
3. Perry A: Fluid Balance in the Patient With Respiratory Disease. In Metheny N, Snively W: Nurses' Handbook of Fluid Balance, 4th ed, p 320. Philadelphia, JB Lippincott, 1983

section three

Selected Clinical Situations and Treatments Associated With Fluid and Electrolyte Problems

chapter 13

Total
Parenteral Nutrition

- *Total parenteral nutrition* (TPN) refers to the infusion of large amounts of basic nutrients sufficient to achieve tissue synthesis and growth. Highly concentrated TPN solutions must be infused through an indwelling catheter into a large central vein, such as the superior vena cava. Initially in adults, 1 to 2 liters is administered daily; this is gradually increased according to the patient's tolerance, to a maximum of 4 to 5 liters daily.
- TPN may be indicated in
    1. Gastrointestinal problems (such as short-bowel syndrome, bowel obstruction, fistulas, pancreatitis, and inflammatory bowel disease)
    2. Severe burns or trauma
    3. Preparation of malnourished patients for surgical procedures
    4. Malignant disease (use for this purpose is controversial, most individuals feel it is indicated if the patient is responding to chemotherapy)
    5. Anorexia nervosa

Composition of TPN Solutions

- TPN solutions infused through a central vein are highly concentrated (osmolality often > 1200 mOsm/kg, as compared to 300 mOsm/kg in plasma). A typical adult solution consists of approximately 25% dextrose, and 3% to 4% protein, supplying close to 1000 calories per liter. (These concentrations may vary.) Dextrose and protein solutions are described in Chapter 4.

- Sometimes fat emulsions are infused through a Y tube ("piggy-backed") near the infusion site simultaneously to provide more calories and prevent fatty-acid deficiency. (Fat emulsions are described in Chap. 4.)
- Because potassium is needed for the transport of glucose and amino acids across the cell membrane, as much as 160 mEq of potassium may be given with each 4000 calories. (Patients with compromised renal function cannot tolerate this amount and require a decreased dose.) Potassium may be administered as the chloride, phosphate, or acetate salt. Serum potassium levels must be closely monitored to adjust potassium intake.
- Magnesium, calcium, sodium, chloride, and phosphate are included as indicated by the patient's clinical condition. (Some commercial preparations are available with electrolytes already added.) Usual requirements as cited by Fischer are[1]
    1. Magnesium, 26–32 mEq/24 hr
    2. Calcium, 10–20 mEq/24 hr
    3. Sodium, 100 mEq/24 hr
    4. Chloride, 50–100 mEq/24 hr
    5. Phosphate, 90–100 mEq/24 hr
  Of course, one must be aware that individual needs for electrolytes are highly variable in patients receiving TPN.
- Water-soluble vitamins should be supplied daily; fat soluble vitamins should be supplied weekly.
- A commercially available solution of trace elements (zinc, copper, manganese, and chromium) may be administered as necessary.
- When necessary to control hyperglycemia, regular insulin can be added to the TPN solution.
- Some authorities favor the addition of heparin (1000 U) to the TPN solution to keep the catheter patent.

Preparation and Storage of TPN Solutions

- The TPN solution should be prepared by a pharmacist when possible, preferably under a laminar-flow hood. (Fig. 13-1)
- The solution should be refrigerated at 4° C until use. The expiration time and date on the label should be checked before administration.
- The solution container should be inspected before use for cracks or punctures. Also, one should hold the solution up to the light and check for clarity (only clear solutions should be used).
- TPN solutions should be used as soon as possible after mixing because glucosamines may form when amino acids remain in contact

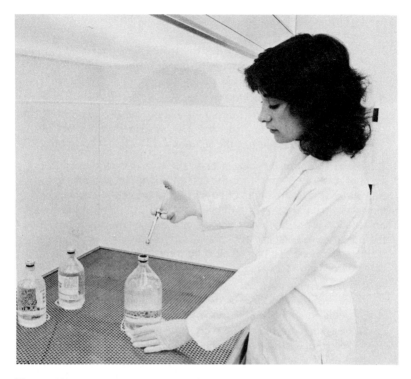

Figure 13–1. Clean-air center (laminar-flow hood) for mixing IV solutions. (Courtesy of Wyeth Laboratories, Philadelphia, PA)

with hypertonic dextrose for a long period of time. (Glucosamines can bind zinc and copper, causing them to be excreted in the urine; deficiencies of these trace elements can then occur.)

- Admixtures that have been delivered to the nursing division for later use should be stored in the medicine room refrigerator.

Metabolic Complications of TPN

Hyperglycemia and Hyperosmolar Syndrome

- Hyperglycemia is a common metabolic occurrence due to the high dextrose concentration in the TPN solution.
- Factors predisposing to hyperglycemia include
  1. Presence of overt or latent diabetes mellitus
  2. Increased age
  3. Stress (recall that during stressful situations there is increased secretion of blood sugar elevating hormones)

4. Hypokalemia (causes glucose intolerance apparently by decreasing insulin secretion)

5. Peritoneal dialysis with dextrose-containing dialysates (imposes a glucose load since some of the glucose from the dialysate is absorbed)

- Preventive measures include

  1. Beginning the TPN infusion at a slow rate (as prescribed), and gradually increasing the rate (as prescribed), allowing the patient time to adjust to the solution

Example: The physician may initiate TPN in the average-weight adult at a rate of 60 ml/hr for the first 24 hours The rate may then be increased by increments of 20 ml/hr at 24- to 48-hr intervals until the maximum desired rate of infusion (such as 165 ml/hr) is achieved (4 liters/day).[2] This gradual increase in flow rate allows the pancreas to establish and maintain the necessary insulin production. (The rate of increase must be more gradual for older patients as compared to younger patients.)

  2. Maintaining a steady rate of infusion around the clock. The most frequent cause of hyperglycemia during TPN is an accidental sudden increase in the infusion rate.

  3. Checking the flow rate every 30 minutes. Do not attempt to "catch up" if the solution falls behind (instead, reset the flow to the desired rate).

  4. Marking the bottle in hourly increments for close monitoring of flow rate

  5. If gravity flow is used, a second clamp should be used to safeguard against accidental free-flow of the solution.

  6. Mechanical infusion devices (such as an IV controller) are more accurate than gravity flow sets and are usually used to administer TPN solutions.

- When severe, hyperglycemia can cause osmotic diuresis and hyperosmolar syndrome. Symptoms include

  1. Frontal headache
  2. Blood sugar > 600 mg/100 ml
  3. Dry skin
  4. Increased longitudinal furrows in tongue
  5. Polyuria (often > 250 ml/hr)
  6. Confusion
  7. Coma and death

- To prevent hyperosmolar syndrome, an attempt is made to keep the blood sugar below 200 mg/100 ml (usually below 140 mg/100 ml) and the urine sugar below 2+ (less than ¼% to ¾%).

- Regular insulin is needed to control excessive hyperglycemia. It may be added to the TPN solution or may be given subcutaneously.

Many authorities prefer to add the insulin to the TPN solution because it provides a more constant serum insulin level than does intermittent subcutaneous injections.

Insulin dosage may be based on the urinary glucose level. This presupposes that the patient's renal threshold and corresponding serum glucose level is known. (Recall that the normal renal threshold is approximately 180 mg/100 ml; however, in patients with renal damage the threshold might be much higher.) Serum glucose levels are more dependable and are indicated when the renal threshold is not known. (Commercial devices are available to allow monitoring of capillary blood sugars; see Fig. 20-1.)

Nursing Assessment for Hyperglycemia and Hyperosmolar Syndrome

- Compare fluid intake and output records at least every 8 hours— look for a urinary output greatly exceeding the intake.
- Measure hourly urine output if urinary losses are large. Recall that urine output may exceed 250 ml/hr in the presence of osmotic diuresis.
- Measure daily body weights. The ideal weight gain for patients receiving TPN is between ½ and 1 lb a day. A gain of over 1 lb/day almost certainly indicates fluid retention; a loss of weight in patients receiving TPN indicates loss of body fluid. (Of course, weights must be measured accurately to be of value; see the section on Body Weights in Chap. 2.)
- Check urine for glucose and acetone content every 6 hr, particularly during the first week of TPN administration. Recall that most physicians favor treating glycosuria of 2+ or more with regular insulin.

  Glycosuria of 3+ or greater requires a stat blood sugar. Be aware of factors that can alter the accuracy of the urine testing agent. (For example: cephalosporins, and high doses of vitamin C and aspirin, can cause false positives in Clinitest tablets.)
- Be alert for the sudden appearance of frontal headaches and confusion; analyze these symptoms in relation to the other clinical signs.
- Monitor vital signs at regular intervals. Look for a rapid, weak pulse and for postural hypotension (a drop in systolic pressure by more than 10 mm Hg after the patient is changed quickly from a lying to a sitting position); both are signs of hypovolemia.
- Monitor skin and tongue turgor. Look for dry skin and longitudinal furrows on the tongue.
- Be aware that patients at greatest risk for hyperglycemia and hyperosmolar syndrome are those with overt or latent diabetes, stress, hypokalemia, underlying renal disease, sepsis, medications that can elevate blood sugar (such as steroids), and produce fluid loss (diuretics).
- Monitor laboratory data. Be alert for elevated blood sugar and serum osmolality.
- Be aware that convulsions and coma are late and serious manifestations of hyperosmolar syndrome.

(continued)

- Treatment of nonketotic hyperosmolar coma includes the rapid infusion of a hypotonic electrolyte solution and administration of regular insulin. If the hyperglycemia is associated with metabolic acidosis, sodium bicarbonate may be needed.

Post-infusion Hypoglycemia

Post-infusion hypoglycemia (due to hyperinsulinism) can occur if the TPN solution is abruptly discontinued. Nursing considerations in the management of this condition are presented below.

Nursing Considerations
Related to Post-infusion Hypoglycemia

- Be aware that TPN solutions should be discontinued *gradually* to allow the pancreas time to adapt to the decreased glucose load. It has been recommended that the infusion rate be slowed by 5 gtt/min each hour when it is time to discontinue the TPN program.[3] Many physicians favor the infusion of a 5% dextrose solution for at least 24 hours after the TPN solution is stopped.
- Be aware that the TPN solution should *never* be discontinued abruptly without allowing for coverage with a dextrose solution.

    If the TPN container empties and the next bottle is not readily available, a dextrose solution (such as $D_{20}W$) should be hung and infused at the same rate as the previous TPN solution until the next bottle can be obtained from the pharmacy.[4] If the patient is receiving regular insulin in the TPN solution, half that amount may need to be added to the dextrose solution. Standing orders from the physician should be available to deal with this situation should it arise.
- Be alert for symptoms of hypoglycemia when TPN solutions are discontinued, particularly if they are stopped abruptly. Be alert for
    1. Occipital headaches
    2. Cold, clammy skin
    3. Dizziness
    4. Rapid pulse
    5. Tingling sensations in extremities and around the mouth
    6. Nervousness
- Obtain a stat blood sugar if these symptoms occur. Be prepared to give intravenous dextrose (such as 20 ml of a 50% solution), if necessary.

Electrolyte Disturbances

- Marked deficiencies of extracellular potassium, phosphorus, and magnesium may occur with TPN if inadequate amounts of these substances are added to the solution; remember, building up of body tissues (anabolism) causes these electrolytes to move into the cells.

- In addition to the other dangers associated with hypokalemia, a glucose intolerance may result, elevating the blood sugar. (Recall that hypokalemia apparently decreases insulin secretion.)
- Hypophosphatemia may decrease the concentration of 2,3 DPG (diphosphoglycerate) in red blood cells, causing them to hang on excessively to oxygen (thereby decreasing tissue oxygenation). Hypophosphatemia may develop within 10 days if a TPN solution with a low or absent phosphate concentration is used.[5]
- Trace element deficiencies are most likely to occur in patients receiving TPN for more than 1 month (particularly zinc and copper).
- Of course, electrolyte excesses will result if too much of these substances are added to the TPN solution.

Nursing Assessment for Electrolyte Disorders

Monitor serum electrolyte levels. Daily blood work usually includes K^+, Na^+, blood urea nitrogen (BUN), creatinine, osmolality, and hemoglobin. (Others are usually measured biweekly.) Report abnormalities to the physician.
Electrolyte imbalances to be particularly alert for include

Hypokalemia

Symptoms of hypokalemia include muscular weakness and cardiac arrhythmias. (Hypokalemia is described in Chap. 8.)

Note: At no time should insulin be given unless potassium supplementation is adequate. (Recall that insulin promotes movement of K^+ into the cells, aggravating any existing hypokalemia.)

Hypophosphatemia

Symptoms of hypophosphatemia include lethargy and slurring of speech, progressing to unconsciousness.[6] Abnormalities in oxygen delivery may also occur. (Phosphorus imbalances are discussed in Chap. 11.)

Hypomagnesemia

Symptoms of hypomagnesemia include tingling in the extremities and around the mouth; agitation is also common.[7] (Magnesium imbalances are described in Chap. 10.)

Zinc Deficiency

Symptoms of zinc deficiency include diarrhea, abdominal pain, impaired sense of taste and smell, mental depression, alopecia, skin eruptions, and retarded wound healing. (Zinc imbalances are described in Chap. 11.)

Copper Deficiency

Symptoms of copper deficiency include anemia, neutropenia, and hypoproteinemia.

Other Metabolic Problems

- Amino acids in the TPN solution can contribute to a high BUN level in patients with renal insufficiency. Renal function tests should be closely monitored.
- Hyperammonemia can be a problem in patients with severe hepatic damage and in infants. Liver function tests are also monitored during TPN administration. (Special TPN solutions are available for patients with renal and hepatic damage.)

Infectious Complications of TPN

- Infections are the most dreaded complication of TPN and pose a constant threat to patients receiving TPN through a central vein.
- Patients requiring TPN are often predisposed to infection as a result of malnutrition, frequent use of broad-spectrum antibiotics, and the presence of concomitant infections in wounds, the lungs, and the urinary tract.
- The indwelling central catheter can serve as an entry port for organisms, and the TPN solution provides a rich culture medium for bacterial growth.

Nursing Considerations
Related to Infectious Complications

PREVENTION

- Maintain aseptic technique in catheter maintenance and in administration of the TPN solution.

  Infectious complications are inversely related to the emphasis placed on aseptic technique. Septicemia rates have been reported to be as high as 33% in institutions in which there is no uniform protocol for infection control; the incidence is much lower (as low as 3%) in institutions using sound infection control practices.

- Do sterile dressing changes every 48 hours, or more frequently if indicated.

  1. Wear a face mask during the procedure; also, provide one for the patient.

  2. Inspect the insertion site for redness or purulent drainage.

  3. Cleanse the catheter exit site and apply an antimicrobial agent.

  4. Visually inspect the catheter for signs of cracking, splitting, kinking, and for clots or leakage.

  5. Apply an occlusive dressing to protect the site.

- Examine the dressing often to detect soiling or wetness.

- Change the administration set every 24 hours. (Have the patient perform the Valsalva maneuver during tubing changes to prevent air emboli; see Chap. 4.)

DETECTION

- As just stated, inspect the insertion site for signs of inflammation.
- Monitor the body temperature. Look for temperature elevations above normal (or, in elderly and chronically ill patients with subnormal temperatures, an elevation to normal). The possibility of catheter-induced septicemia must be considered. An unexplained fever is often an indication to halt TPN administration until the source of the fever is found (or, at least, to start over with a new catheter, tubing, and solution).
- Monitor blood and urinary glucose levels. *The appearance of hyperglycemia in a previously stable patient who has not been spilling glucose generally signals the onset of sepsis.*

  It should be remembered that hyperglycemia may be the *initial* sign of sepsis—occurring up to 12 hours before temperature elevation or other signs of developing sepsis. (Recall that sepsis is a form of stress and can initiate gluconeogenesis and glycogenolysis by stimulating release of ACTH and catecholamines.)

  The appearance of a 4+ glycosuria in a patient previously free of urinary sugar should provoke a search for a source of sepsis before it reaches overwhelming proportions.[8]

References

1. Fischer J: Nutritional management. In Berk J, Sampliner J (eds): Handbook of Critical Care, 2nd ed, pp 377–378. Boston, Little, Brown, & Co, 1982
2. Ibid, p 369
3. Goldberger E: A Primer of Water, Electrolyte and Acid–Base Syndromes, 6th ed, p 427. Philadelphia, Lea & Febiger, 1980
4. Fischer J, p 370
5. Goldberger E, p 425
6. Fischer J, p 383
7. Ibid
8. Ibid, p 382

chapter 14

Tube Feedings

Complications of Tube Feedings

Although tube feedings have proved to be of immense value in many situations, they can cause problems if given incorrectly. Complications of tube feedings include

- Hypernatremia (if hyperosmolar feedings are used)
- Hyponatremia (if abnormal routes of Na^+ loss are present)
- Aspiration pneumonia
- Diarrhea
- Nausea and gastric distention
- Nonketotic hyperosmolar syndrome
- Nasal and pharyngeal irritation (if a large diameter, firm tube is used for nasogastric or nasoenteral feedings)

Hypernatremia

- Hypernatremia is likely to occur in patients receiving hyperosmolar tube feedings with inadequate water supplements. Tube feedings with a high-solute content (osmolality) require extra water to aid in renal excretion of the solute. (The main determinants of osmolality of tube feedings are amino acids, electrolytes, and carbohydrates.) Note in Table 14-1 the wide variance in osmolalities of some commercially available tube feeding mixtures. Recall that the plasma osmolality is approximately 300 mOsm/kg; note that some of these feedings approximate this level while others are much higher. Generally speaking, the higher the osmolality of the feeding, the greater the need for extra water. Actual water needs must be determined individually for each patient considering 1. The osmolality of the

Table 14–1
Osmolality of Some Commercially Available Tube Feedings

Tube Feeding	Osmolality (mOsm/kg Water)
Ensure (Ross)	450
Ensure Plus (Ross)	600
Osmolite (Ross)	300
Vital (High Nitrogen) (Ross)	460
Compleat-B (Bottle) (Doyle)	405
Compleat Modified Formula (Doyle)	300
Formula 2 (Cutter)	435–510
Isocal (Mead Johnson)	300
Meritene (ready to use) (Doyle)	560 (vanilla)
Meritene (powdered, prepared with whole milk) (Doyle)	690 (plain)
Precision LR Diet (Doyle)	525 (orange)
Precision High Nitrogen Diet (Doyle)	557
Precision Isotonic Diet (Doyle)	300
Sustacal (ready to use) (Mead Johnson)	625 (vanilla)
Standard Vivonex (Norwich-Eaton)	550 (unflavored)
	595–610 (flavored)
High Nitrogen Vivonex (Norwich-Eaton)	810 (unflavored)
	850–855 (flavored)

(Data from Ross Laboratories: Enteral Nutrition Ready Reference: The Ross Medical Nutritional System. Division of Abbott Laboratories, Columbus, OH, Oct, 1981)

feeding mixture (listed on the container) 2. Abnormal routes of fluid loss 3. Comparison of intake–output measurements 4. Moisture in mucous membranes 5. Serum sodium, osmolality, and blood urea nitrogen (BUN) levels. (Electrolyte, BUN, creatinine, and glucose levels should be monitored on a regular basis to detect problems before they become severe.)

- If sufficient fluid is not furnished with the hyperosmolar tube feeding mixture, between feedings, or by way of the intravenous route, it is drawn from the patient's intravascular and interstitial fluid (and eventually from cellular stores).

- Extra water is particularly needed for patients with fever, extensive tissue breakdown, and decreased renal concentrating ability (as in the aged). Kidneys in normal persons are able to concentrate urine to an osmolality of approximately 1400 mOsm/kg, and, thus, get by on a minimum of water. However, the ill tube-fed patient (particularly if aged) may not have full renal concentrating ability, and, thus, may need more fluid than a normal individual to excrete renal solutes.

Nursing Measures to Prevent and Detect Hypernatremia and Azotemia

- Monitor intake–output records on all patients receiving tube feedings
  1. Unless abnormal fluid losses are occurring from other routes, the urinary output should roughly equal the total fluid intake.

(continued)

Recorded intake should include fluid given by mouth, by tube, and by the intravenous route. Recorded output should include urine volume (and the volume of liquid feces, vomitus, or other abnormal routes, if present). One should also consider the fluid loss associated with an elevated body temperature, excessive perspiration, and hyperventilation.

2. Be particularly alert initially for a urine output greatly exceeding the fluid intake; later, be alert for a urine output less than 30 ml/hr.

- Give enough supplemental water to maintain a satisfactory urinary output, keep the mucous membranes moist, and maintain normal serum Na^+ levels

1. Be aware that extra water is particularly necessary when the patient has increased production of metabolic wastes (as in fever, or tissue breakdown from decubiti) and decreased renal concentrating ability (as occurs in elderly or severely hypokalemic patients).

2. Add extra water, as indicated, to the tube feeding mixture when the feedings are given continuously over a 24-hour period. When intermittent feedings are used, extra water can be given between feedings.

- Be alert for signs of hypernatremia and azotemia (protein overloading) when hyperosmolar feedings are used

1. Thirst. (An early indicator of need for extra water; unfortunately, the confused or unconscious patient is not able to make this need known.)

2. Sticky tongue and mucous membranes.

3. Urine output greatly exceeds fluid intake (occurs early in solute overloading).

4. Elevated body temperature (may occur).

5. Elevated serum Na^+ and osmolality; elevated BUN.

6. Diminished urine output (occurs later as the patient's body fluids become depleted by the previous polyuria).

7. Confusion.

Hyponatremia

Hyponatremia can be a complication in tube-fed patients if large water supplements are given in conjunction with isotonic or hypotonic tube feedings (particularly when Na^+ losses are incurred by an abnormal route).

Nursing Measures to Prevent and Detect Hyponatremia

- Avoid giving large water supplements to patients receiving isotonic or hypotonic tube feedings, particularly if abnormal routes of Na^+ loss are present (as in diarrhea, vomiting, profuse sweating, renal abnormalities, adrenal insufficiency, and use of diuretics) or water is being abnormally retained (as in SIADH).

- Monitor serum Na$^+$ and serum osmolality levels at regular intervals. Be alert for declines in these parameters.

- Be aware that salt (NaCl) may be given through the feeding tube (as prescribed by the physician) to correct a low serum Na$^+$ level in tube-fed patients. When this is done, one should keep a close watch on the Na$^+$ level to prevent overcorrection.

Aspiration Pneumonia

- Aspiration pneumonia is a dreaded complication of tube feedings. A study of tube feedings and lethal aspiration in 720 neurologic patients found that tube feedings increased the risk of aspiration sixfold.[1]

- Aspiration usually results from reflux vomiting from a distended stomach; it is most likely to occur in comatose and disoriented patients and in those with pulmonary complications. (It appears that the presence of a nasogastric tube inhibits the patient's ability to remove pulmonary secretions.) It is also more likely to occur when the patient is flat in bed and a large volume of feeding solution is given without assessment for gastric distention.

- *Aspiration* is defined as the inspiration of gastric secretions into the lungs; it may be manifested by

  1. Vomiting with sudden intense cough
  2. Cyanosis
  3. Diminished breath sounds, indicative of congestion
  4. Chest x-ray film changes

- Some authors report the addition of 3 gtt of blue food coloring to each tube feeding to make the appearance of gastric secretions in the respiratory tract more visible.[2] The presence of blue secretions in the respiratory tract is indicative of either obvious or silent aspiration of gastric contents, or a fistula between the trachea and esophagus. (It should be noted that aspiration frequently occurs unnoticed in comatose patients.)

Nursing Measures to Prevent Aspiration Pneumonia

- Be aware that it is dangerous to introduce tube feedings into the stomach of unconscious patients and those with decreased protective pharyngeal reflexes.

  It is recommended that feedings be given into the small intestine in such patients. Feedings introduced directly into the small intestine by nasoduodenal or nasojejunal routes or by way of jejunostomy take advantage of the pyloric sphincter in preventing regurgitation.

(continued)

- Confirm that the nasogastric tube is positioned in the stomach.
  1. This is usually accomplished by injecting air (10 cc) into the tube while auscultating the left upper quadrant with a stethoscope.
  2. Aspiration of gastric fluid into a syringe is another method of confirming that the tube is in the stomach.
- Check for gastric retention prior to introduction of feedings.
  1. Aspirate gastric fluid from the nasogastric tube or gastrostomy tube with a syringe (until resistance to pull is felt) prior to each intermittent feeding and every 2 to 4 hours during continuous feedings.
  2. The presence of a sizable volume of the previous feeding (such as > 150 ml in an adult) is an indication to withhold further feedings until the situation is discussed with the physician.
  3. Some physicians recommend stopping the feeding for 2 hours if the aspirant is > 100 ml and then restarting it; if the residual remains high, the feeding rate should be sharply curtailed. (Recall that gastric retention predisposes to regurgitation and aspiration of stomach contents.)

     Note: A 50-ml syringe has been recommended for checking residual volumes or injecting air or medication through mercury-tipped small-caliber nasogastric feeding tubes; use of smaller syringes may cause rupture and separation of the mercury column from the tube.[3]

     One should be aware that soft, small-caliber feeding tubes tend to collapse when aspirated with a syringe, making it difficult to determine the volume of gastric retention. When this occurs, it becomes important to measure abdominal girth serially and palpate the abdomen to detect distention.
- Position the patient in an upright position during tube feedings to minimize the chance of aspiration of fluid into the lungs.
  1. Ambulatory patients should be placed in a sitting position (preferably in a chair) for intermittent feedings.
  2. Bedfast patients should have the head of the bed elevated to a 30° to 45° angle. This position should be maintained for at least 1 hr after the intermittent feeding, or continuously (when possible) for patients receiving feedings around the clock. It has been recommended that the feeding be temporarily stopped during postural drainage maneuvers.[4]
- Keep cuffed tracheostomy or endotracheal tubes inflated during intermittent feedings and for at least 1 hour afterward.

  The cuff should be deflated the minimal amount of time to prevent tracheal complications when the patient is receiving continuous feedings.

Diarrhea

- Diarrhea is a common complication of tube feedings. It can be due to bacterial contamination of the feeding mixture or apparatus, or to too rapid administration of a concentrated feeding. Introduction of a hyperosmolar feeding into the stomach (and especially the small intestine) causes diffusion of fluid into the area to equalize the osmolality of the solution; bloating, hypermotility, and diarrhea may result.

• It is believed that an isotonic feeding mixture (osmolality approximating 300 mOsm/kg) provides optimum benefit with less incidence of diarrhea. Isotonic or hypotonic feedings are usually prescribed for enteral feedings.

Nursing Measures to Prevent Serious Diarrhea

- • Avoid bacterial contamination of the feeding mixture or apparatus.
  1. Keep the tube feeding mixture refrigerated until ready for use (unless, of course, the mixture is in a ready-to-use can).
  2. If continuous tube feedings are used, add small amounts of the mixture at 2- or 3-hour intervals rather than allowing a large volume of the mixture to be exposed to room temperature for a prolonged period of time.
  3. Keep the administration apparatus clean and change the equipment daily.
- • Administer the tube feeding at room temperature.
  One study found that tube feedings administered at room temperature (23° C–26° C or 73.4° F–78.8° F) or at body temperature (36° C–39° C or 98.6° F–102.2° F) were better tolerated than cold tube feedings.[5] (The latter were associated with cramping and diarrhea.)
- • Administer the tube feeding mixture at the desired rate and concentration.
  1. Be aware that continuous feedings should be started out slowly (such as 50 ml/hr in an adult)
  2. Isotonic feedings are sometimes started out at full strength; hyperosmolar feedings should be started out at approximately half strength.
  3. If diarrhea or other untoward effects do not occur, the rate can be increased by 25 ml/hr every 8 to 12 hours until the desired rate is achieved.[6] When the final rate has been achieved, the strength of the solution can be increased as tolerated.
  4. Be aware that an individual intermittent feeding should not usually exceed 300 ml unless it is to be given very slowly. As with continuous feedings, it is necessary to start out with less than a full-strength solution (if the feeding is hyperosmolar), until the patient develops a tolerance for the feeding mixture.
- • Administer the feeding slowly and evenly over the 24-hour period when the feeding tube is positioned in the intestine.
  Remember, the duodenum and jejunum are more sensitive to both volume and osmolality than is the stomach. Slow delivery of the feeding decreases the likelihood of cramping and diarrhea.
- • Be aware that isotonic or hypotonic feedings are generally prescribed when tube feedings are introduced directly into the intestine (as by jejunostomy or nasoenteral feedings).

Nausea and Gastric Distention

Nausea and gastric distention are not uncommon complications of tube feedings. Gastric distention is particularly dangerous because it predisposes to aspiration of gastric contents and can, in severe situations, predispose to gastric rupture.

Nursing Measures to Prevent Nausea and Gastric Distention

- Check gastric contents prior to intermittent tube feedings, or every 2 to 4 hours during continuous feedings, to detect gastric distention.
- Be alert for increased abdominal girth, and complaints of epigastric and left upper quadrant pain. (Gastric rupture secondary to over-zealous tube feeding administration can be lethal.)
- Stop the tube feeding at once if severe nausea occurs; seek medical directives.
  1. It may be helpful to dilute the mixture and administer it more slowly.
  2. A prn medication for nausea may be helpful.
  3. A change in the content of the feeding may be required.

Nonketotic Hyperosmolar Syndrome

- The use of hyperosmolar tube feedings can initiate hyperglycemia and osmotic diuresis, particularly in latent or overt diabetics. For this reason, some authorities recommend monitoring urinary glucose levels every 4 to 6 hours during the first 48 hours. If the urine tests are negative during this period, they can be stopped in nondiabetics. Urine tests should continue for diabetic patients.
- Starting the hyperosmolar tube feedings slowly and in diluted form helps prevent the hyperosmolar syndrome. Slow delivery of a more dilute solution allows the patient time to adjust to the solute load, and the pancreas time to secrete more insulin, as needed.
- The reader is referred to Chapter 20 for a discussion of nonketotic hyperosmolar coma.

General Facts About Administration of Tube Feedings

Routes for Administration of Tube Feedings

- Major routes for administration of tube feedings are depicted in Figure 14-1.
- The use of nasoenteral tube feedings has grown in popularity and is a frequent route of administration, particularly in patients at high

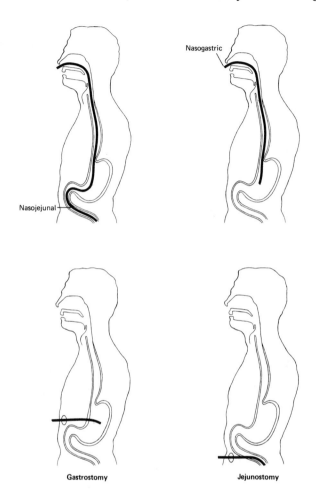

Figure 14-1.   Routes of tube-feeding administration.

risk for aspiration. This route necessitates the use of a weighted feeding tube (such as the Dobbhoff enteric feeding tube).*

- The nasogastric route is best reserved for alert patients requiring relatively short-term nutritional support. This route is generally contraindicated in the presence of unconsciousness or absent pharyngeal protective reflexes.

- The gastrostomy route (tube inserted directly into the stomach through a stab wound) is sometimes used for patients requiring long-term nutritional support; it eliminates nasopharyngeal discom-

---

* Biosearch Medical Products, Somerville, NJ

fort associated with prolonged intubation through the nasopharynx. (However, it should be noted that pliable, small diameter feeding tubes have greatly decreased the complication of nasopharyngeal irritation).

- The jejunostomy route (tube inserted directly into the jejunum through a stab wound) is still a much favored route for unconscious patients. A more recent development is the introduction of a plastic catheter into the jejunum by means of a needle ("needle-catheter jejunostomy").
- The possibility of infection is greater in gastrostomy and jejunostomy than in the nasogastric or nasoenteral routes, because the peritoneum has been entered. Gastrostomy and jejunostomy sites need time to seal and generally are not safe for introduction of feedings until 3 days after insertion.[7]

Tube Feeding Equipment

- Feeding tubes should be as small in diameter as the viscosity of the feeding will allow. A No. 5 or No. 6 French feeding tube can be used for thin feedings; a more viscous solution may require a size 10 French tube.[8] The smaller the diameter of the feeding tube, the less irritating it is to the nasopharynx and esophagus.
- Traditional large-diameter (16–18 French) polyvinyl chloride tubes are extremely irritating with prolonged use and tend to cause pressure necrosis; they also compromise the competency of the gastroesophageal sphincter, possibly increasing the chances of gastric reflux and aspiration.[9]
- Newer feeding tubes are available (such as the Dobbhoff enteral feeding tube, and the Entriflex feeding tube). These small-diameter tubes are made of polyurethane and are very pliable.

  The radiopaque Dobbhoff feeding tube can be used for nasogastric or nasoenteral feedings, depending on the length of the 43-inch tube inserted.[10] The tube is weighted with a mercury bolus to allow entry into the small intestine when desired; spontaneous transpyloric passage of the tip usually occurs within 24 to 48 hours. (The position of the tube should be confirmed by x-ray film after time for passage has elapsed). The radiopaque Entriflex tube is designed for nasogastric feedings; it is weighted with tungsten to facilitate insertion and prevent tube regurgitation. A guidewire stylet is available to aid in insertion of the tube. Proper tube placement can be confirmed by x-ray film. One should be aware that soft, small-caliber feeding tubes tend to collapse when aspirated with a syringe, making it difficult to determine the volume of gastric retention.

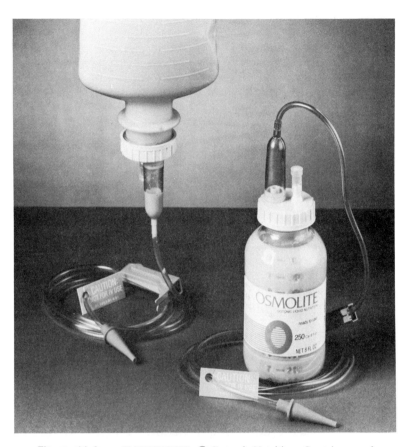

Figure 14–2. FLEXITAINER ® Enteral Nutrition Container and FLEXIFLO ® Gravity Gavage Set. (Courtesy of Ross Laboratories, A Division of Abbott Laboratories, Columbus, OH 43216)

- Disposable gravity-flow gastric feeding units are available commercially and are desirable, since the flow rate of the feeding mixture can be adjusted in drops. One kind of disposable feeding bag is depicted in Figure 14-2.

- An Asepto syringe can serve as a funnel for gravity administration of feeding mixtures of alert patients who can tolerate relatively rapid administration rates. (Devices suited to slower administration rates are preferable to this method.)

- Mechanical pumps are more dependable than gravity-flow equipment for the slow, constant delivery of tube feedings. A mechanical pump assures a constant flow rate regardless of the patient's position or the viscosity of the feeding. An example of a commercial food pump is the Flexiflo Nutrition Pump, shown in Figure 14-3.

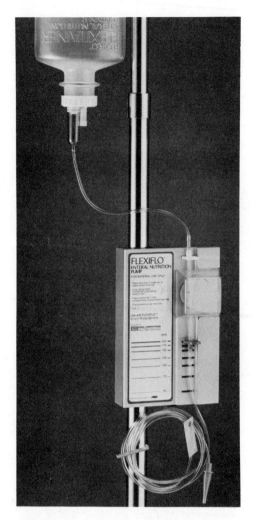

Figure 14–3.   FLEXIFLO ® Enteral Nutrition Pump. (Courtesy of Ross Laboratories, A Division of Abbott Laboratories, Columbus, OH 43216)

References

1. Olivares L, Segovia A, Revuelta R: Tube feeding and lethal aspiration in neurological patients: A review of 720 autopsies. Stroke 5:654–657, 1974
2. Taylor T: A comparison of two methods of nasogastric tube feeding. J Neurosurg Nurs 14, No. 1:53, 1982
3. Orr M, Shinert J, Gross J: Acute Pancreatic and Hepatic Dysfunction, p 107. Bethany, CT, Fleschner Publishing, 1981
4. Ibid, p 106
5. Kagawa-Busby K, Heitkemper M, Hansen B, et al: Effects of diet temperature on tolerance of enteral feedings. Nurs Res 29, No. 5:278, 1980
6. Calaldo C, Smith L: Tube Feedings: Clinical Application, p 22. Ross Laboratories, Division of Abbott Laboratories Columbus, 1980
7. Condon R, Nyhus L (eds): Manual of Surgical Therapeutics, 5th ed, p 207. Boston, Little, Brown & Co, 1981
8. Calaldo C, Smith L, p 16
9. Ibid
10. Ibid

chapter 15

Postoperative Period

The surgical patient is a likely candidate for numerous fluid and electrolyte problems. Nursing care must be directed toward preventing or minimizing these imbalances when possible. Effective nursing care is facilitated by reviewing the physiologic changes brought about by surgical stress.

The Body's Response to Surgical Trauma

Endocrine Response (Stress Reaction)

Period of Fluid Retention and Catabolism

- The stress reaction described by Selye, representing a response to surgical trauma, is present for the first 2 to 5 days. The intensity of changes depends on the severity and duration of the trauma.
- Postoperative apprehension and pain enhance the stress reaction, and extreme preoperative apprehension can initiate the stress reaction before surgery. Anesthesia also constitutes a form of stress.
- The endocrine responses are briefly outlined as follows (Fig. 15-1):
    1. Increased adrenocorticotrophic hormone (ACTH) secretion from the anterior pituitary.
    2. Increased mineralocorticoid and glucocorticoid secretion from the adrenal cortex (in response to stimulation by ACTH).
        - Mineralocorticoids (primarily aldosterone) cause
        a. Sodium retention

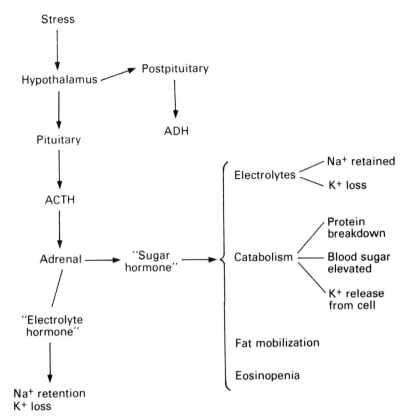

Figure 15–1.   Effect of stress on electrolyte excretion. (Adapted from Statland H: Fluids and Electrolytes in Practice, 3rd ed, p 45. Philadelphia, JB Lippincott, 1963)

    b. Chloride retention

    c. Potassium excretion

  • Glucocorticoids (mainly hydrocortisone) cause

    a. Sodium retention

    b. Chloride retention

    c. Potassium excretion

    d. Catabolism

       Protein breakdown

       Gluconeogenesis and elevated blood sugar

    e. Fat mobilization

    f. Drop in eosinophil count

  3. Increased antidiuretic hormone (ADH) secretion from posterior pituitary, causing decreased urinary output.

4. Vasopressor substances (epinephrine and norepinephrine) secreted from the adrenal medulla to help maintain blood pressure, a response stimulated by fear, pain, hypoxia, and hemorrhage.

- The body's response to stress appears purposeful. For example, sodium retention and increased ADH secretion help maintain blood volume. Glucocorticoids cause protein breakdown, make amino acids available for healing at the site of trauma, and cause conversion of protein and fat to glucose (gluconeogenesis), creating a ready supply of glucose for use during the stress period. (The elevated blood sugar is sometimes mistaken for diabetes mellitus.)

- As a result of stress, the early postoperative patient loses more urinary nitrogen than normal, even though his protein intake is low or absent. Because the body nitrogen losses exceed intake, the patient is said to be in negative nitrogen balance.

- The changes described above are normal responses to trauma and do not usually require corrective measures.

Period of Diuresis and Anabolism

- After the second to fifth postoperative days, adrenal activity is decreased and a mild water and sodium diuresis occurs. The body also begins to retain potassium.

- Anabolism, the building up of body protein, usually begins by the seventh to the tenth postoperative day. At this time the urinary nitrogen losses are decreased even though the patient is consuming protein foods. Thus, the renal excretion of nitrogen no longer exceeds the nitrogen intake, and the patient begins to gain weight, provided oral intake is adequate.

Tissue Injury

- Trauma results in the formation of edema in and about the operative site for the first few days after surgery. The fluid closely resembles plasma; its volume is roughly proportionate to the amount of tissue trauma.

- Traumatic edema is functionally sequestered as a third space and cannot be readily mobilized to meet body needs. While the amount of fluid lost in edema is not itself significant, it may enhance the extracellular fluid volume deficit created by peritonitis, hemorrhage, or other complications. The edema fluid is eventually reabsorbed and excreted during the diuretic phase of the stress reaction. Fluid overload in the postoperative period can increase the local edema at the surgical site; when a bowel anastomosis has been performed, the edema fluid can cause a temporary obstruction.

Starvation Effect

- Most patients eat inadequately, or not at all, during the first few postoperative days; thus, a starvation effect is induced. The majority of patients receive only routine dextrose and electrolyte fluids during this period, furnishing limited calories. (For example, a liter of 5% dextrose solution contains only approximately 170 cal.)
- Accompanying starvation is a daily weight loss of about ½ lb, reflecting a decrease in lean and fatty tissue mass. Renal nitrogen excretion is increased as a result of lean tissue catabolism.
- The weight loss following surgery is generally constant as revealed by accurate weighing procedures. A weight gain, in the face of starvation, indicates fluid retention.
- Usually the patient will regain his normal weight in 2 to 3 months if preoperative nutrition was good, if operative trauma was of only moderate severity, and if there was prompt return of gastrointestinal function.

Prevention of Postoperative Complications by Preoperative Assessment and Preparation

Nutritional Status

- A patient in good nutritional condition preoperatively is able to withstand postoperative negative nitrogen balance and early starvation without serious effects.
- On the other hand, the nutritionally depleted patient goes to surgery under a serious handicap—a poor tolerance for operative stress.
    1. Increased susceptibility to infection results from diminished ability to form antibodies and from the superficial atrophy in the mucous membrane linings of the respiratory and gastrointestinal tracks that often accompanies malnutrition.
    2. Hypoproteinemia follows negative nitrogen balance and increases susceptibility to shock from hemorrhage.
    3. Diminished supplies of protein and vitamin C retard wound healing.
- The nurse has primary responsibility for the following areas related to nutrition:
    1. Working with the physician and clinical dietitian to promote optimal nutrition; particularly when a problem exits.

2. Assisting the patient to eat a therapeutic diet.
3. Describing actual food consumption, including any provided by nonhospital sources, and noting inadequate intake (sadly, most hospital charts contain little information about what the patient eats or doesn't eat).
4. Determining height and weight on admission and measuring weight at regular intervals throughout hospitalization.
5. Measuring all routes of fluid intake and output.

- In addition to these duties, the nurse may be responsible for measuring other parameters of nutritional status, such as skinfold thickness and mid-upper-arm circumference. From these two measurements, the lean muscle mass can be estimated.
- Other methods for evaluation of nutritional status include certain laboratory tests (such as 24-hr urinary creatinine excretion, serum albumin, total white blood cell count, white blood cell differential, and total iron-binding capacity) and skin antigen testing.

Medications

Steroids

- The adrenals normally produce about 25 mg of hydrocortisone a day; under stress, however, they may need to increase their output as much as 10 times this amount.
- A patient's ability to withstand surgical stress depends upon his adrenals' ability to secrete extra hydrocortisone.
- When a patient is on steroid therapy there is less need for adrenal secretion and the glands tend to atrophy from disuse; after withdrawal of steroid therapy, the adrenals gradually resume their function. However, if steroids are suddenly withdrawn and the patient is subjected to massive trauma, such as a major surgical procedure, the atrophied adrenal glands may be unable to respond to the stress signal; adrenocortical failure may follow cessation of adrenocortical substitution therapy. Symptoms include hypotension, nausea and vomiting, a thready pulse, subnormal temperature early on with hyperpyrexia later on, hallucinations, confusion, stupor, or coma. The reaction usually occurs in the first 24 hours.
- The actual degree of adrenal suppression caused by steroid administration is related to the dose and the duration of use. Although there are variations in degrees of adrenal suppression, patients who are currently receiving steroids, or have received steroids for more than 1 to 2 weeks within the preceding 6- to-12-month period, are usually considered as having adrenal suppression. Such patients require appropriate perioperative glucocorticoid supplementation. One should remember that short-term excess of glucocorticoids is

relatively harmless, but short-term deficiency during stress can be fatal!

- It is wise to ask all surgical patients routinely if they have taken cortisone or any other steroid preparation for more than 1 to 2 weeks within the past 6 to 12 months.

- Today, because of medical specialization, one patient may have two or three physicians prescribing medications at the same time. The fact that the patient may have recently received steroids is often overlooked, and, in fact, many patients do not know what medicines they have taken. If this is the case, the nurse can ask for a description of the medication and why it was given. Conditions for which steroid therapy is often used should be kept in mind (for example, rheumatoid arthritis, asthma, dermatitis, and ulcerative colitis). When in doubt, one should check with those who prescribed medications for the patient.

Drug Allergies or Idiosyncrasies

- One should always ask the prospective surgical patient if he knows of any drug allergies, sensitivities, or idiosyncracies he may have. If the patient does not understand the question, typical symptoms of sensitivity, such as urticaria, asthma, and the like, can be mentioned. Also, the patient should be asked if a physician has ever cautioned him to avoid a specific medication because of an unusual reaction to it. In questioning the newly admitted patient, the nurse will do well to remember that he is often upset and may have difficulty remembering.

- Allergies must be discovered before the patient is sedated or anesthetized; it is too late to ask him when he is unconscious or semireactive after surgery. Remember that most medications are ordered in the immediate postoperative period. Failure to ascertain the presence of allergies may be disastrous. The most extreme and dreaded allergic reaction is anaphylactic shock; other less dangerous reactions include skin eruptions and asthma.

- Idiosyncratic reactions to drugs deserve consideration. For example, a narcotic may cause more depression in one patient than in another of equal weight and age. A dose creating the desired effect in one patient may overwhelm another patient. The aged are particularly sensitive to narcotics and should generally be given smaller doses than younger adults; this is important to aged surgical patients.

Chronic Illnesses

- Certain chronic illnesses greatly increase the hazard of postoperative water and electrolyte imbalances. Such illnesses include
  1. Renal disease
  2. Pulmonary disease

3. Diabetes mellitus
4. Adrenal insufficiency
5. Heart disease

- Patients with chronic renal disease may present a number of problems during surgery. Preoperative preparation should include adequate fluid administration to furnish the greater-than-normal urine volume required by the diseased kidney to excrete metabolic wastes, taking care not to overload the circulatory system and cause congestive heart failure. Anemia and abnormal coagulation factors should be corrected preoperatively as should metabolic acidosis. (Respiratory acidosis caused by anesthesia may be lethal when added to an already existing metabolic acidosis.) Hemodialysis may be required the day prior to surgery in patients with severe renal damage. Hyperkalemia due to tissue trauma, increased catabolism, and failure of the diseased kidneys to adequately remove potassium should be anticipated following surgery.

- Patients with pulmonary disease are at special risk for such complications as atelectasis, pneumonia, and pulmonary embolism. Some indicators of significant operative risk in patients with pulmonary disease include

  1. Elevated preoperative $PaCO_2$ ($>$ 45 mm Hg)
  2. Severe hypoxemia before operation (PaO_2 $<$ 55 mm Hg)
  3. Maximal volume ventilation less than 50% of predicted or less than 50 liters/min[1]

- Obviously, patients with diabetes mellitus should be in the best possible balance at the time of operation. Therefore, unless the operation is urgent, it should be delayed until the diabetic state, nutritional status, hydration, and electrolyte status are controlled. Serum glucose levels should be closely monitored during the perioperative period. Close cooperation between the internist, surgeon, and anesthesiologist is required for optimal management of diabetic surgical patients. One should also remember that the diabetic patient is particularly susceptible to gram-negative septic shock.

- Patients with known or suspected adrenal insufficiency should be supplied with adequate doses of hydrocortisone preoperatively to prevent precipitation of adrenal crisis during the stress of surgery. (This subject was discussed earlier.)

- Patients with cardiovascular disease are more susceptible to the cardiovascular stresses associated with general anesthesia and surgery (*e.g.*, hypotension, hypoxemia, sepsis, and thromboembolism). Review of the electrocardiogram (ECG) and a careful history and physical examination are required to identify factors that predict serious postoperative cardiac complications.

Preoperative Fluid Therapy

- The preoperative evaluation and correction of existing water and electrolyte disorders are an integral part of care of the surgical patient.

- A common problem in preoperative patients is fluid volume deficit (FVD); general guidelines for the treatment of this problem include reversal of the signs of volume deficit, combined with stabilization of the blood pressure and pulse, and an hourly urine volume of 30 to 50 ml/hr. In order to prevent operative and postoperative renal insufficiency, it is best that the preoperative urine volume be at least 50 ml/hr.

- Potassium administration for the correction of hypokalemia should be started only after adequate urine output has been established. It should be remembered that potassium deficiency is frequent in patients who are taking potassium-losing diuretics, in those receiving cathartics for bowel preparation, and in those with gastric outlet obstruction. *It is important that hypokalemia be corrected prior to surgery to reduce the risk of an intraoperative arrhythmia.*

Operative Fluid Therapy

- If preoperative fluid replacement has been inadequate, the patient may promptly develop hypotension upon induction of anesthesia; unfortunately, fluid replacement after induction is far less effective in dealing with hypotension and decreased renal blood flow.

- Blood lost during the surgical procedure should be replaced steadily. It is usually unnecessary to replace a blood loss of less than 500 ml; however, after the loss exceeds this amount, replacement should begin. Allowance should also be made for administration of a balanced electrolyte solution to replace extracellular fluid that is trapped in the surgical site as edema in major operations. Other measurable losses should also be replaced.

Postoperative Gains and Losses of Water, Electrolytes, and Other Nutrients

Need for Intake–Output Measurement

- Surgery often brings into play abnormal routes of fluid loss, such as gastric or intestinal suction, vomiting, or drainage from an ileostomy or colostomy.

- Failure to measure the amounts and kinds of fluids lost makes adequate replacement therapy almost impossible, and, without an accurate account of gains and losses, the early discovery of water and electrolyte imbalances is unlikely. The nurse should automatically place postoperative patients on the intake–output list and make a conscientious effort to keep the intake–output record accurate.
- The 8-hour and 24-hour totals are significant in assessing fluid balance in general; it is equally important to know the types and amounts of fluids making up the total. When in doubt about the patient's fluid balance status, it is helpful to total the intake–output for several consecutive days to obtain an overall picture.
- A summary of the electrolyte content of gastrointestinal fluids (explaining expected imbalances with significant losses) is presented in Chapter 17.

Urinary Output

- In health, the daily urinary output is roughly equal to the volume of liquids taken into the body. However, during postoperative stress reaction, the urine volume may tend to be low regardless of the amount of fluids taken in. Additional secretion of aldosterone and antidiuretic hormone (ADH) during the early phase of stress causes a reduction in renal ability to excrete excessive water and sodium loads.
- Following a major surgical procedure the 24-hour output may be only 750 ml to 1200 ml for the first few postoperative days. Preferably, urinary output should be at least 30 to 50 ml/hr; an hourly urinary output below 25 ml should be investigated. It is important to differentiate between the decreased urinary output of stress reaction (healthy physiologic response to surgery) and pathologic developments.
- Factors contributing to decreased urinary volume in the postoperative patient include
  1. Hypovolemia resulting from fluid loss incurred in surgery
  2. Preoperative FVD
  3. Subtle accumulation of fluid at the surgical site (third-space effect)
  4. Disturbance in myocardial function, causing decreased blood flow to the kidneys and, thus, decreased urine formation
  5. Renal failure (a serious cause of postoperative oliguria)
- Measurement of urinary specific gravity (SG) is a simple method of evaluating renal function. A low fixed urinary SG (1.010–1.012) in the presence of oliguria indicates acute renal failure; however, a high urinary SG (1.026–1.036) suggests water conservation by a

healthy kidney. The oliguria of renal failure does not respond to increased intravenous fluid administration or to diuretics. The serum creatinine level rises rapidly in renal failure whereas the urinary creatinine level remains low (indicating a disturbance in excretion of body wastes). Renal failure is discussed in Chapter 22.

Fluid Intake

- The need for immediate postoperative fluid replacement is determined by assessment of vital signs and urinary output in the recovery room. Unrecognized deficits of extracellular fluid during this period may be manifested by circulatory instability (*e.g.*, hypotension, tachycardia, and a urinary output < 30–50 ml/hr). Circulatory instability is most commonly due to underestimated initial losses, or to insidious, concealed, continuous losses (as in sequestration of edema fluid at the surgical site). For the patient with circulatory instability, further volume replacement, within reason, often resolves the problem. Persistence of symptoms despite fluid replacement necessitates prompt investigation of the problem.
- The presence of nausea, postoperative ileus, and gastrointestinal suction contraindicates oral fluids; intravenous fluids must then be relied on to furnish the body's maintenance needs and to replace lost fluids.
- Fluid volume replacement usually involves the replacement of measured losses (such as from vomiting or gastrointestinal suctioning) and estimated insensible losses. (An accurate daily intake and output record and weight chart should be kept to aid in the determination of the volume and type of fluid replacement required.) In the adult, the insensible loss averages 600 ml to 900 ml daily; this may be increased, of course, by the presence of fever and hyperventilation. Approximately 1 liter of fluid is given daily to replace the volume of urine necessary to excrete the metabolic end products. (Urine loss is not replaced on a milliliter-for-milliliter basis.)
- After adequate renal function has been established, potassium is given daily to prevent potassium deficit. Many surgeons prefer to wait until the first postoperative day to begin potassium replacement because the serum potassium may be temporarily elevated on the operative day due to tissue trauma.

  Daily renal potassium excretion usually ranges from 40 to 100 mEq/day. A daily maintenance allowance of 40 to 80 mEq of potassium will cover minimal needs; unusual losses of potassium-rich fluids necessitates the administration of greater amounts. Inadequate replacement of potassium can prolong the usual postoperative ileus and can contribute to the development of metabolic alkalosis. (Nursing responsibilities in the administration of potassium solutions are discussed in Chap. 4.)

- Negative nitrogen balance can be reduced by the administration of dextrose solutions, and they frequently are used for this purpose in the postoperative period.
- Hypotonic multiple electrolyte solutions are used for maintenance water and electrolyte needs; specific electrolyte replacement solutions are used as indicated by the patient's condition.
- After the fluid retention phase of stress has subsided, a larger amount of fluid can be given. If oral intake is still prohibited, an attempt must be made to supply body needs solely with parenteral fluids.
- Parenteral vitamin preparations of the B-complex group and vitamin C should be given daily when parenteral therapy is necessary for more than 2 days.
- Magnesium replacement is necessary when parenteral fluid administration is prolonged, particularly when magnesium losses have resulted from gastric suction.

Postoperative Problems in Water and Electrolyte Balance

Water Intoxication

- Water excess (Na^+ deficit) is also referred to as *water intoxication* or *hyponatremia*, an imbalance most likely to occur in the first day or two postoperatively, while the water-retention effect of stress (transient, excessive ADH secretion) is still present. Excessive administration during this period of water-yielding fluids, such as D_5W, predisposes to this condition. Symptoms of water excess (Na^+ deficit) include the following
  1. Behavioral changes
     - Inattentiveness
     - Drowsiness
     - Confusion
  2. Acute weight gain
  3. Neuromuscular changes
     - Isolated muscle twitching
     - Weakness
     - Headache
     - Blurred Vision
     - Incoordination
     - Elevated intracranial pressure may occur with hypertension, bradycardia, decreased respiration, projective vomiting, and papilledema
     - Convulsions

- The nurse should suspect water excess (Na^+ deficit) when several of these symptoms occur in the early postoperative period. Behavioral changes are usually noticed first, particularly drowsiness.

- A case was reported in which a young boy was inadvertently given only D_5W during the early postoperative period (a physician's order to alternate isotonic saline with D_5W was transcribed incorrectly). The boy developed cerebral edema and permanent brain damage. This tragic event could have been averted had the nursing staff understood the danger of excessive water administration in the early postoperative period.

- Prevention of body fluid disturbances demands the study of daily accurate body weight measurements; a sudden weight gain in the early postoperative period is an indication to decrease fluid intake. Fluid intake during the water-retention phase of stress should not greatly exceed body fluid losses.

- Mild water excess can be corrected by prohibiting further water intake; however, brain damage or death may result if the condition is allowed to go untreated. (See the section on inappropriate antidiuretic hormone syndrome in Chap. 7.)

Fluid Volume Excess

Overadministration of sodium-containing fluids (such as 0.9% NaCl) during the first few postoperative days can cause fluid volume excess (FVE) with pulmonary edema and increased local edema at the surgical site. In intestinal surgery, the increased edema may be sufficient to cause partial or complete obstruction.

Diminished Postoperative Ventilation and Respiratory Acidosis

- The surgical patient may develop decreased ventilation and respiratory acidosis for one or several of the following reasons:
  1. Depression of respiration by anesthesia
  2. Blockage of oxygen–carbon dioxide exchange in the lungs owing to atelectasis, pneumonia, or bronchial obstruction
  3. Depression of respiration with too frequent or too large doses of narcotics
  4. Shallow respiration because of abdominal distention and crowding of the diaphragm
  5. Shallow respiration as a result of pain in the operative site or large cumbersome dressings

- Pulmonary complications are a frequent cause of death and morbidity in the period following anesthesia and surgery. Inadequate pulmonary ventilation is common in postoperative patients and can lead to respiratory acidosis, hypoxemia, and atelectasis.

- Patients having high abdominal or thoracic incisions are particularly prone to develop ventilation problems. For example, vital capacity may be decreased by 40% of normal on the day after subtotal gastrectomy. However, this loss of reserve still allows for adequate oxygen and carbon dioxide exchange if preoperative ventilation was normal.
- Other factors that increase the risk of postoperative ventilatory problems include advanced age, heavy cigarette smoking, obesity, extended length of operation, and, of course, pre-existing pulmonary conditions.
- Indiscriminate use of oxygen in the postoperative period increases the chance of overlooking respiratory acidosis. Cyanosis is one of the criterion for detecting inadequate ventilation; oxygen therapy may prevent cyanosis and keep the skin color pink even though respiratory acidosis is progressing.
- In addition to causing respiratory acidosis, decreased ventilation interferes with the correction of metabolic acidosis. (Recall that the lungs attempt to compensate for metabolic acidosis by eliminating more carbon dioxide.)

Nursing Interventions to Improve Postoperative Ventilation

- Encourage the patient to breathe deeply at regular intervals. Deep breathing is helpful in loosening secretions for coughing, helps improve venous return to the heart, and is thought to help stimulate the production of surfactant (which helps prevent alveolar collapse).[2] With the patient positioned comfortably in the semi-Fowler's position, he should be instructed to inhale slowly, hold his breath for a few seconds at end-inspiration, and exhale passively.
- Encourage the patient to cough at regular intervals as indicated by his condition. Coughing has long been considered an integral part of chest therapy. However, one should keep in mind that coughing is most needed for patients whose respirations are noisy, which suggests secretion retention secondary to atelectasis. (The patient who is able to maintain an adequate alveolar ventilation without secretion retention does not need to cough routinely to prevent the development of atelectasis.)[3] See the next section dealing with measures in assisting the patient with coughing.
- Position the patient in high Fowler's position, when permitted, to allow for better gas distribution throughout the lungs. (The sitting position allows for greater lung expansion, because gravity pulls the abdominal organs away from the diaphragm.)
- Ambulate the patient as early as possible. (More than just a few steps to a bedside chair are necessary to prevent deep vein thrombosis and pulmonary embolism.)

- Teach the patient to do leg exercises to prevent deep vein thrombosis and pulmonary embolism. (The nurse should perform passive leg exercises for the patient if he is unable to do active exercises.)
- Assure that dressings or restraints are not tight enough to interfere with normal respiratory excursions.
- Assist the patient in the use of devices prescribed by the physician to improve ventilation (such as blow bottles, and so forth).

Assisting the Patient to Cough

Techniques to Position Patient for Maximal Chest Expansion
For ambulatory patient

- Sitting on side of bed with legs dangling and arms supported

For patient on bed rest

- Place bed in semi-Fowler's position
- Support the patient's head slightly flexed
- Instruct patient to rotate shoulders slightly inward
- Support arms with pillows
- Instruct patient to flex his knees

Techniques to Support Wound Incision to Minimize Pain

- Gently and firmly place one or both hands over incision dressing, or
- Have patient support the incision site with pillows and fold his hands over the pillows with coughing

Technique for Forceful Coughing

- Take in a slow, deep breath through the nose with the mouth slightly open
- Hold breath for the count of one
- Lean forward and contract the abdominal, thigh, and buttock muscles to build up abdominal pressure on the count of two
- Forcefully cough twice on the count of three (once to raise secretions and the second time to expel them)

Staged Coughing Technique
For patients with vascular disorders and painful chest incisions

- Take slow, deep breath through the nose with mouth slightly open
- Lean forward, cross arms over pillows to contract abdominal muscles
- Cough in several short bursts

Huff Technique
For patients with increased sputum production who have difficulty expectorating mucus

- Take a slow, deep breath through the nose with mouth slightly open
- Lean forward and cross arms over pillows
- Whisper word "huff" several times while rapidly exhaling
- Stop exhalation when most of the inspired air is expelled.

(continued)

Note: Do not encourage patient to exhale beyond his functional residual capacity; to do so would raise airway pressures and precipitate small airway collapse.

(Harper R: A Guide to Respiratory Care, p. 274. Philadelphia, JB Lippincott, 1981)

Imbalances Associated With Loss of Gastrointestinal Secretions

- Metabolic alkalosis is most commonly seen in surgical patients as a result of the loss of large amounts of gastric secretions, either through vomiting or gastric suction. It is closely associated with potassium deficit, produced by the excessive loss of potassium-rich intestinal secretions or by prolonged parenteral therapy without potassium replacement. Potassium deficit is enhanced by the stress reaction in the early postoperative period.
- Metabolic acidosis follows excessive loss of alkaline intestinal secretions, bile, and pancreatic juice.
- See Chapter 17 for a more detailed discussion of imbalances associated with the loss of gastrointestinal secretions.

Ileus

- Peristalsis is inhibited for 2 to 4 days after intraabdominal manipulation, and, for this reason, patients with abdominal surgery usually receive nothing by mouth (NPO) until bowel sounds indicate the return of peristalsis. Active bowel sounds are rare or absent for 1 to 3 days after abdominal surgery. An abdomen silent for longer than 3 to 4 days should be brought to the attention of the physician. (See assessment of bowel sounds in Chap. 18.)
- Gastric intubation is often employed early in the postoperative period for patients having abdominal surgery to prevent the bowel from becoming greatly distended with swallowed air and gastrointestinal fluids.
- If paralytic ileus persists longer than normal, the patient must be maintained on intravenous fluids and gastric suction until peristalsis returns (as evidenced by passage of flatus). Factors that may prolong the expected period of ileus postoperatively include bacterial and chemical peritonitis, excessive handling of the intestines during surgery, advanced age, actual mechanical obstruction, or hypokalemia.
- Oral intake should be withheld in any postoperative patient who is vomiting persistently; the nature of the vomitus, as well as the degree of abdominal distention, should be noted and reported to the physician.

- Large amounts of water and electrolytes may be sequestered into the bowel. The amount of fluid lost in this manner is not revealed by body weight changes. Clinical signs of FVD, decreased urinary volume, and increased urinary SG help indicate the amount of fluid trapped in the bowel. (See the discussion of bowel obstruction in Chap. 18.)

Shock

- Shock is a clinical syndrome indicating inadequate tissue perfusion secondary to decreased effective circulating blood volume.
- Shock in surgical patients is usually of the hypovolemic type; however, a number of associated problems may cause other forms of shock. A commonly accepted classification of shock is as follows:
  1. Hypovolemic shock
  2. Neurogenic shock
  3. Septic shock
  4. Cardiogenic shock

Hypovolemic Shock

- Hypovolemic shock is due to actual internal or external blood loss (hemorrhage), acute FVD (as in severe diarrhea or vomiting), or third-space losses of fluid (as in intestinal obstruction or severe burns).
- If blood volume is reduced by more than one-third and the deficiency is maintained, the shock that ensues will be fatal.[4]
- Clinical signs of hypovolemic shock include
  1. Poor capillary refill time (may take seconds) as noted by pressing the fingernails.
  2. Anxiety (due to hyperactivity of the sympathetic nervous system when blood loss is sustained).
  3. Decreased systolic blood pressure of 10 mm Hg or more when position is changed from lying to standing or sitting (early sign).
  4. Decreased pulse pressure—often less than 20 mm Hg (when blood pressure changes are due to blood loss, the systolic pressure falls more rapidly than the diastolic pressure).
  5. Eventually, blood pressure remains low even in flat position as the body's compensatory mechanisms are unable to cope with hypovolemia. (Systolic pressure generally does not fall significantly until a blood volume deficit of at least 15% to 25% has been sustained.)

6. Collapse of neck veins. (See discussion of neck veins in Chap. 2)

7. Respiratory alkalosis in early shock (caused by hyperventilation).

8. Pulse rate usually increased by more than 20 per minute when position changed from lying to standing or sitting; later, pulse remains rapid and thready in all positions. (With slow hemorrhage, as much as 1000 ml of blood may be lost without significant increase in pulse rate as long as the supine position is maintained.[5])

9. Blood loss is indicated by a decreased hematocrit level. (Recall that it takes 4 to 6 hours for interstitial fluids to move into the bloodstream after a hemorrhage; therefore, a hematocrit value will not reflect a very recent hemorrhage.)

10. Decreased urinary output. (Increased ADH and aldosterone secretion causes water and sodium retention by kidneys.)

11. Pale, cool, moist skin (caused by peripheral vasoconstriction).

12. Metabolic acidosis in late shock (resulting from lactic acid build-up caused by hypoxia).

13. Central venous pressure (CVP) usually below normal. (Isolated CVP reading may mean relatively little, but, the change caused by fluid replacement is quite significant.)

14. If treatment of shock is delayed, or unsuccessful, the patient may develop the lethal combination of metabolic acidosis and respiratory acidosis (due to respiratory failure).

Positioning

Elevating the legs to a 45° angle temporarily releases about 500 ml of blood and, thus, partially relieves the effect of moderate hypovolemia; in severe hypovolemia there may be insufficient blood in the extremities to be of value. Trendelenburg position is contraindicated in most instances because it allows the abdominal viscera to interfere with respirations by pressing against the diaphragm; also, the abdominal contents press against the vena cava, interfering with venous return.

Fluid Replacement

- Many physicians feel that hypovolemia in hemorrhagic shock is best treated with a judicious mixture of electrolyte solutions and blood. As a rule, resuscitation is begun with a balanced isotonic electrolyte solution and blood is added when serial measurements show that the hematocrit has fallen below 30%. This approach provides as effective balance between electrolytes, red-blood-cell mass, and colloid to allow adequate volume distribution and oxygen-carrying capacity. Parameters to be monitored during fluid resuscitation include

1. Blood pressure
2. Pulse
3. Urinary output
4. Hematocrit
5. CVP

- A typical regimen for replacement therapy in the emergency room (for the patient in hemorrhagic shock) allows for the rapid infusion of lactated Ringer's solution (such as 1000–2000 ml in 45 min in adults); if the blood pressure, pulse, and urinary output return to normal, and if the hematocrit is greater than 30%, the solution is continued at a maintenance level. If, however, blood loss has been severe or hemorrhage is continuing, the rise in blood pressure and decrease in pulse rate will usually be transient; at this point, blood should be readied for administration. (Blood should be drawn for typing and crossmatching at the same time the electrolyte solution is started.) The amount of blood to be given is dependent on the degree of blood loss.
- Considerable controversy exists about the choice of fluids for treatment of hypovolemia. The physician has a choice of electrolyte solutions, whole blood or packed cells, dextran, and plasma or albumin. A brief review of each of these substances follows.

Electrolyte Solutions

- Many physicians advocate the use of a balanced electrolyte solution (such as lactated Ringer's) as a principal fluid in initial shock therapy. Among its advantages are the following:
  1. It is readily available and inexpensive.
  2. It is virtually free of reactions.
  3. It appears to reduce the amount of blood needed for correction of blood loss.
  4. It is effective in controlling hypovolemia.
  5. It does not aggravate other fluid and electrolyte problems that may be present.
- Since balanced electrolyte solutions diffuse out of the intravascular space rather rapidly, it may take 3 to 4 times as much solution to replace the volume of lost blood. Care must be taken to observe the hematocrit during fluid administration to avoid diluting the red cell mass beyond the level needed for normal peripheral oxygenation.
- Electrolyte solutions are only substitutes for blood in the initial resuscitation of the patient in hemorrhagic shock; major blood loss always necessitates the administration of blood as soon as it is available. Only minor blood loss can be completely replaced by the sole use of electrolyte solutions.

- Isotonic solution of sodium chloride (0.9% NaCl) is not a first choice because it is not a balanced or physiologic solution. It contains approximately 10 mEq of sodium and 40 mEq of chloride in excess of the normal extracellular concentrations. (Isotonic solution of NaCl contains 154 mEq of Na^+ and 154 mEq of Cl^-.)
- Electrolyte-free solutions (such as D_5W) *should never* be used for emergency fluid replacement in hypovolemia; these solutions may rapidly dilute the contracted extracellular fluid, producing the threat of "water intoxication." Contributing to the threat of water overload is the increased ADH secretion associated with acute stress states. (Water intoxication is described earlier.)

Blood

Whole blood (or more often packed red blood cells) remains a mainstay for fluid replacement in major hemorrhage. The major advantage of blood is its oxygen-carrying capacity. As stated earlier, blood administration is usually preceded by the administration of a balanced electrolyte solution, allowing time for typing and cross-matching, and actually decreasing the volume of blood needed for resuscitation. Blood transfusions are limited by

1. Cost
2. Limited availability
3. Delay required for typing and crossmatching, and
4. Most importantly, a number of potentially serious complications

(The reader is referred to Chap. 16 for discussion of nursing responsibilities in blood administration.)

Dextran

Regular dextran (dextran 70) is an effective plasma volume expander. Fifty percent of the infused particles are still circulating by 24 hours.[6] Problems that may be associated with its use include dilutional hypoproteinemia and lowered hematocrit, occasional allergic reactions, defects in the clotting mechanism, and interference with crossmatching. Because of these possible complications, it is recommended that no more than 1000 ml of dextran solution (6%) be given in 24 hours in the average adult.

Plasma and Albumin

The risk of hepatitis transmission from plasma has severely limited its use as a volume expander. In addition, the effect of plasma or albumin as a blood volume substitute is transient. Human serum albumin is free of the risk of hepatitis transmission and is readily available. However, when given in large volumes during resuscitation, transudation into the pulmonary interstitium may occur, perhaps contributing to adult respiratory distress syndrome (ARDS). Those who advocate the use of colloids in resuscitation strive to give enough to restore hemodynamics without overloading the pulmonary interstitium.

Oxygen Administration

Oxygen is not used routinely in the treatment of shock because the oxygen saturation in most patients with uncomplicated hypovolemic shock is generally normal. However, in hypovolemic patients in whom the oxygen saturation is not normal, the initial use of oxygen may be very important. This may occur in patients with obstructive lung disease; but more frequently, the oxygenation problems arise from the patient's injury (such as pneumothorax, pulmonary contusions, or aspirated gastric secretions). When necessary, oxygen is best administered through a loose-fitting face mask; if a controlled airway is indicated for other reasons, an endotracheal tube is an ideal route. When in doubt as to the adequacy of arterial oxygenation, it is best to administer oxygen until the patient can be assessed more thoroughly.

Neurogenic Shock

- This form of shock follows serious interference with the balance of vasodilator and vasoconstrictor influences to both arterioles and venules. In neurogenic shock there is dilatation of arterioles and venules; shock is produced by the decrease in effective circulatory blood volume.

- This is the shock seen with high spinal anesthesia, syncope upon sudden exposure to unpleasant stimuli, and even acute gastric dilatation.

- Clinically, it is manifested quite differently from hypovolemic shock (*i.e.*, while the blood pressure may be extremely low in both hypovolemic shock and neurogenic shock, the pulse rate is often slower than normal in the patient with neurogenic shock, as opposed to the rapid pulse in the hypovolemic patient). Also, the patient with neurogenic shock often has dry, warm, and even flushed skin. In neurogenic shock, the blood volume is apparently normal, but its distribution pattern is changed (due to an increased reservoir capacity in the arterioles and venules). As a result of peripheral pooling of blood, the venous return to the right side of the heart is diminished causing a reduction in cardiac output.

- Failure to treat neurogenic shock can cause all the ravages of hypovolemic shock, including damage to the kidneys and brain due to decreased blood flow. Fortunately, neurogenic shock is usually rather easy to treat. For example, high spinal anesthesia shock can be treated with a vasopressor to produce peripheral vasoconstriction; acute gastric dilatation can be relieved by insertion of a nasogastric tube. In milder forms of neurogenic shock, as in fainting from emotional stimuli, merely removing the patient from the source of the unpleasantness may be sufficient. Similarly, relieving acute pain usually corrects the state of shock due to this cause. (The patient in neurogenic shock may respond favorably to a head-down position.)

- Of course, when neurogenic shock is due to injury, its treatment becomes more complicated. For example, if neurogenic shock is due to spinal cord transection with associated loss of blood and cerebrospinal fluid in the injured area, it becomes necessary to balance fluid replacement with vasopressor therapy.

Septic Shock

- Septic shock is the result of peripheral pooling of blood in capacitance (primarily venous) vessels, causing decreased effective circulating blood volume (without actual blood loss).
- This form of shock is most often due to gram-negative bacterial sepsis; other causative organisms may include gram-positive bacteria and, far less often, viruses, parasites, and fungi.
- The highest incidence of septic shock occurs during the seventh and eighth decades of life.
- It is more likely to occur in patients debilitated from other conditions, such as malignancies, diabetes mellitus, uremia, and cirrhosis. The mortality remains in excess of 50%
- Despite improved antimicrobial therapy, there has been an increase in the incidence of gram-negative sepsis; this increase has been attributed to
    1. More extensive operations on elderly and poor-risk patients
    2. More frequent use of steroids, immunosuppressive, and anticancer drugs
    3. Widespread use of antibiotics, with the resultant development of more virulent and resistant organisms
- Gram-negative sepsis as a cause of shock is more serious than that seen with gram-positive sepsis. (Gram-positive infections can be more effectively controlled by antibiotics.)
- The most frequent source of gram-negative infections is the genitourinary system; many patients have had as associated operation or instrumentation of the urinary tract.
- The second most frequent site of origin is the respiratory tract.
- Septic shock also follows peritonitis, biliary tract infections, burns, septic abortions, and postpartum infections. Indwelling catheters for hyperalimentation and CVP monitoring are used more frequently than in the past and are an increasing source of contamination. (Nursing responsibilities in caring for central venous catheters are described in Chap. 4.)
- Gram-negative organisms that most commonly produce septic shock include *Escherichia coli, Klebsiella, Aerobacter, Proteus, Pseudomonas,* and *Bacteroides.*[7]

- Clinical manifestations of septic shock include
  1. Mild hyperventilation with respiratory alkalosis and change in sensorium. (These symptoms may precede the usual symptoms by several hours to several days.)
  2. A temperature above 101° F (38.3° C) and chills.
  3. Hypotension.
  4. Peripheral vasoconstriction and cold, clammy extremities. (Earlier in its course, however, there may have been warm, dry extremities.)
  5. Generally the white blood cell (WBC) count is elevated; however, it may be normal or low in debilitated patients, in patients on immunosuppressive drugs, or in patients with overwhelming sepsis. (There is usually a left shift in the WBC differential.)
  6. Deterioration of pulmonary function with severe hypoxemia and resultant metabolic acidosis (late).
- The best treatment of septic shock is prevention; that is, treating the infection before the onset of shock. Use of appropriate antibiotics is mandatory, as is the debridement or drainage of the site of the infection when possible.
- Fluid replacement varies with the patient's underlying disease process. If the patient has incurred third-space losses (from peritonitis, burns, or bowel obstruction, for example) the replacement of fluid should be a balanced electrolyte solution. Enough fluid must be given to maintain an effective circulating volume; however, care must be taken to avoid fluid overload. This is best accomplished by monitoring at least the CVP; if function of the left side of the heart is in question, a Swan-Ganz catheter should be inserted for measurements of pulmonary artery (PA) and pulmonary capillary wedge (PW) pressure. Many patients show improvement after appropriate fluid replacement and antibiotic therapy.
- The use of corticosteroids in the treatment of septic shock is controversial. Although there is no direct evidence that steroids are helpful in these cases, some improvement in cardiac, pulmonary, and renal function has been reported. Some physicians elect to use steroids only if the patient fails to respond adequately to antibiotics and fluid therapy.
- Since many patients with sepsis may develop significant pulmonary insufficiency, the use of a controlled airway (such as an endotracheal tube) and assisted ventilation may be required.
- Fever in the septic patient can contribute to the deleterious cellular metabolic effects of shock and thus must be treated; the patient should be kept at normal temperature by use of a hypothermic blanket and administration of aspirin or acetaminophen.

Cardiogenic Shock

- This form of shock is due to failure of the heart as a pump, with resultant inadequate output and inadequate tissue perfusion, in spite of normal blood volume.
- Causes may include primary myocardial dysfunction (as in myocardial infarction or serious arrhythmias); mechanical restriction of cardiac function or venous obstruction, such as occurs in cardiac tamponade, tension pneumothorax, or vena cava obstruction; and myocardial depression (resulting from a variety of causes).
- Management of postoperative patients with this type of hypotension must be accompanied by accurate hemodynamic monitoring. Typically, the classic findings in cardiogenic shock are a CVP that is elevated or rises briskly with fluid administration and a depressed cardiac output that fails to respond to fluid administration.
- Primary treatment of cardiogenic shock must be directed at the cause; for example, cardiac tamponade should be relieved by pericardiocentesis.

References

1. Luce J: Preoperative evaluation and perioperative management of patients with pulmonary disease. Postgrad Med Jan, 1980, p 203
2. Harper R: A Guide to Respiratory Care: Physiology and Clinical Applications, p 270. Philadelphia, JB Lippincott, 1981
3. Ibid, p 272
4. Schumer W, Nyhus L: Treatment of Shock: Principles and Practice, p 23. Philadelphia, Lea & Febiger, 1974
5. Shires T, Canizaro P, Carrico C: Shock. In Schwartz S, Shires T, Spencer F, Storer E (eds): Principles of Surgery, 3rd ed, p 137. New York, McGraw-Hill, 1979
6. Pestana C: Fluids and Electrolytes in the Surgical Patient, 2nd ed., p 49. Baltimore, Williams & Wilkins, 1981
7. Shires T, et al., p 175

chapter 16

Blood Administration

- Blood replacement therapy is often lifesaving; but without careful attention to its hazards, it may also be lethal. Nurses need to be familiar with complications associated with blood transfusion since it is the nurse's responsibility to monitor patients receiving blood.
- Before discussing the possible complications, it is helpful to review general rules for safe blood administration.

Rules for Safe Blood Administration

- Inspect the blood bag for tears or inadequately sealed closures; if present, discard the solution.
- Inspect the blood for discoloration or gas bubbles; if these are present, the blood is probably contaminated and should be discarded. (Look for a purple, red, or brown color to the plasma—indicative of hemolysis.)
- Cross check the blood identification data against the patient's identification to assure giving the right blood to the right patient.

  1. Always check the data with another registered nurse or a physician for validation.

  2. Check the assigned blood bank number on the unit with the number on the requisition (in addition to the blood type, Rh factor, and the expiration date).

  3. Identify the patient by asking for a name and by checking the wrist identification band.

     - Use particular caution when the patient has a common name; many errors have occurred by not checking beyond the last name and first initial.

- Be aware that a unit of blood (500 ml) can usually be administered to an adult in a 1½- to 2-hour period.

(continued)

- Unless the patient is severly hypovolemic, blood should not be given faster than 500 ml in 30 minutes.

- Patients with cardiac, renal, or liver damage may require a much slower than normal rate (such as 1 unit over a 3–4-hr period).

- Use a small container of isotonic saline (0.9% NaCl) as a starter solution for blood administration.

   Do not use dextrose and water solutions, since they cause aggregation of red blood cells (rouleaux formation); do not use calcium-containing solutions (such as lactated Ringer's), since they may cause formation of clots in the blood administration set.

- Collect baseline data (temperature, pulse, respiration, lung sounds, complaints of pain, and urine color) prior to starting the transfusion. This data serves as a comparison for changes occurring during the transfusion.

- Start blood within 30 minutes after it is delivered to the nursing division—do *not* store it in the ward refrigerator.

   Rapid deterioration RBCs occurs after blood has been exposed to room temperature for more than 2 hours. If the patient requires a slow administration rate, the time of exposure to room temperature becomes critical. Ward refrigeration is inadequate for storage of blood because it is not controlled and has no alarm to fluctuations in temperature (accidental freezing of blood renders it unsuitable for use).

- Give blood through a blood filter to remove the particulate matter formed during storage.

   1. All blood administration sets are equipped with standard filters to trap particulate matter.

   2. A condition known as posttransfusion lung syndrome can occur when multiple units have been transfused through a standard (170-μ) filter (allowing particulate matter to enter the pulmonary system).

   3. It has been recommended that microaggregate filters (40 μ and smaller) be used when more than 2 units of blood are to be given in 1 day, and in all patients with respiratory insufficiency.

- The first 15 minutes of the transfusion should be carefully monitored for adverse reactions since this is a critical time. (See the section dealing with complications, which is discussed later). After the first 50 ml has infused with no adverse reactions, the rate may be increased to allow the blood to infuse in a 1½- to 2-hour period (unless a slower rate is indicated).

- Check the patient at least every 30 minutes (including vital signs) throughout the transfusion for adverse reactions; if a reaction occurs

   1. Stop the transfusion.

   2. Keep the vein open with the starter solution (usually isotonic saline).

   3. Contact the physician to report the patient's condition and to receive treatment orders.

   4. Other procedures vary with the type of reaction (discussed later).

- Be aware that a unit of blood should not be infused longer than 4 hours.

If the transfusion of 1 unit of blood will take longer than 4 hours, the unit should be divided by the blood bank into two containers to avoid prolonged exposure to room temperature.

- Warm blood, when indicated, by means of a mechanical blood warmer (Fig. 16-1) or a warm-water coil apparatus. *Never* use hot water to warm blood since excessive heat destroys red blood cells.
- Change the IV tubing after every 2 to 4 units of blood and not less often than every 24 hours.[1]
- Record vital signs immediately posttransfusion and 30 minutes later.

Complications of Blood Transfusion

- Changes in blood induced by storage are referred to as the *storage lesion*. This term refers to both chemical and physical changes in blood brought about by preservatives and prolonged storage. These changes in stored blood can cause electrolyte problems and other complications.
- Of course, one must always remember that a blood transfusion is similar to a tissue transplant because blood is a human tissue with specific antigen markings.
- Blood transfusion is a complex and potentially hazardous procedure, demanding skill and knowledge on the part of the nurse.

Stored Blood and Potassium Excess

- Continual destruction of red blood cells occurs when blood is stored; the plasma potassium level increases to 10, 20, and 23 mEq/liter at 1, 2, and 3 weeks of storage. (K^+ is released from the destroyed RBCs and is also transferred from intact red blood cells into the surrounding plasma.)
- When the stored blood is infused, the blood cells are reoxygenated and take up the leaked potassium; however, before this occurs, there can be a transient hyperkalemia if blood administration is extremely rapid.
- Aged blood poses no problems for the normal patient but should not be given rapidly to patients with oliguria or anuria since there is a danger of causing potassium excess with cardiac arrest. It is best to use fresh blood or packed cells for oliguric individuals (fresh blood has a low potassium level, packed red blood cells have much of the plasma [and thus electrolytes] removed).
- The incidence of cardiac arrest during surgery has been correlated with the rapid administration of large quantities of aged blood.

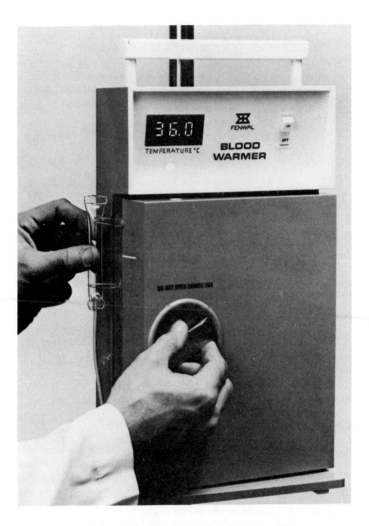

Figure 16–1. FENWAL Dry Heat Blood Warmer. (Illustration provided by Fenwal Laboratories, Division of Travenol Laboratories, Inc, Deerfield, IL)

Citrated Blood and Hypocalcemia

- Acid-citrate-dextrose (ACD) and citrate-phosphate-dextrose (CPD) preserved bank blood contain excessive citrate ions. The excess citrate normally causes no difficulty since the liver rapidly converts it to bicarbonate; another safeguard to protect the body from hypocalcemia is the rapid mobilization of calcium from bone when needed. However, in some situations, the excess citrate may combine with ionized calcium in the recipient's blood, producing a deficiency of ionized calcium. (Recall that it is ionized calcium that controls neuromuscular irritability).

- Calcium defiency caused by administration of citrated blood is rare; however, it can occur in

  1. Infants receiving exchange transfusions
  2. Adults with severe liver disease
  3. Patients with inadequate bone stores (such as young children, elderly, or osteoporotic patients, those with bony tumors, or those who have been immobilized for a prolonged period)
  4. Normal adults given rapid transfusions (such as more than 2000 ml in 30 min or less)

- Hypocalcemia has been implicated in the cardiac output and blood pressure fall induced by multiple blood transfusions.

- The presence of potassium excess, hypovolemia, and hypothermia increases the danger of mortality in citrate toxicity.[2]

- If citrate toxicity is anticipated or suspected, the patient may be given calcium gluconate in a vein remote from the infusion site. Authorities disagree as to when (and indeed if) calcium should be administered.

  1. Some recommend the slow administration of 10 ml of a 10% calcium gluconate solution if more than 2000 ml of blood must be given rapidly.
  2. Other sources recommend calcium gluconate prior to or concomitant with whole blood transfusions exceeding 1000 ml in 4 hours.
  3. In patients with known hepatic insufficiency, some authorities recommend that citrate toxicity be prevented by use of washed, frozen–thawed, or heparinized blood.[3]

*p*H Changes with Citrated Blood

- The *p*H of 2- to 3-week-old bank blood falls to approximately 6.8 (due to leakage of lactate and pyruvate into the plasma as a result of red cell hypoxia during storage). The decreased *p*H of stored blood usually has no markedly adverse effects on healthy adults since it is diluted by the patient's own blood.

- However, rapid transfusion of citrate preserved blood in patients with pre-existing acidosis (such as premature infants with respiratory problems) can make the acidosis more severe. Patients with pre-existing acidosis who are to receive rapid transfusions (such as 1 unit of blood every 20 min or faster) may require the administration of sodium bicarbonate.

- Although an initial metabolic acidosis may occur during rapid transfusion, metabolic alkalosis (with concomitant hypokalemia) may eventually result due to rapid metabolism of citrate to bicarbonate. (Each unit of CPD blood eventually generates 22.8 mEq of bicarbonate.)[4] Rarely does the alkalosis require treatment.

- It is wise to monitor the patient's pH during multiple blood transfusions.

Hyperammonemia

- Ammonia concentration in stored blood begins to rise after 5 to 7 days and reaches fairly high levels after 2 to 3 weeks of storage.
- Normal patients can tolerate the extra ammonia with no difficulty; however, patients with severe liver disease should be given blood not more than 5 to 7 days old, or receive packed red blood cells only (since the excessive ammonia is in the plasma).

2,3 Diphosphoglycerate

- 2,3 Diphosphoglycerate (2,3 DPG) is a substance in red blood cells important in regulating the amount of oxygen released by hemoglobin to the tissues. Storage decreases the level of 2,3 DPG in blood.
- Blood with 2,3 DPG depletion is initially less efficient in releasing oxygen to the recipient's tissues. However, red cells begin to regenerate 2,3 DPG after transfusion (half is restored in 4 hr).[5]
- The PaO_2 may be temporarily decreased as a result of rapid transfusion of 2,3 DPG depleted blood; however, it is only significant in critically ill patients unable to compensate for hypoxemia with increased cardiac output.
- When patients require transfusions of near normal 2,3 DPG, fresh, frozen red cells or blood less than 1 week old should be used.

Circulatory Overload

- Many physicians order monitoring of central venous pressure (CVP) during the rapid administration of blood. (The procedure for CVP measurement is described in Chap. 2.) More extensive hemodynamic monitoring may be required for patients with cardiac disease.
- The onset of circulatory overload is usually gradual, with symptoms becoming more severe as the transfusion continues. Symptoms include cough, shortness of breath, neck vein distention, and pulmonary congestion.
- If symptoms of circulatory overload occur, the transfusion should be stopped and the vein kept open with isotonic saline while the physician is notified.
- Administration of packed cells instead of whole blood decreases the likelihood of fluid overload (recall that packed cells are obtained by

centrifuging whole blood and drawing off approximately 200 ml to 225 ml of plasma).

Febrile Reactions

- Febrile reactions are probably the most frequently encountered adverse responses to blood transfusions.
- Minor febrile reactions are characterized by fever (rarely exceeding 103° F [39.5° C]), flushing, chills, and headache.
- Usually febrile reactions are due to sensitivity to platelets and plasma proteins in patients who have received numerous transfusions in the past; they may also result from contamination of the administration setup.
- Febrile reactions will usually occur within 6 hours after the transfusion is initiated (often within the first hour).[6] However, they may occur up to 24 hours after transfusion.[7] Symptoms do not usually occur until more than 250 ml of blood has been transfused.
- It is sometimes impossible to differentiate a simple febrile reaction from the initial stages of a hemolytic transfusion reaction and from that of bacterial contamination. In such instances, the transfusion should be halted and the physician notified. Blood samples should be obtained from the transfusion set and recipient for culture and recrossmatching.
- Therapy for a simple febrile reaction is essentially symptomatic; aspirin is usually used to relieve the fever. Usually, another unit of blood is started.

Allergic Reactions

- Anaphylactic reactions are apparently quite rare; urticarial reactions may occur as often as once in every 20 transfusions.[8]
- The allergic reaction results from transfer of donor antigen to a sensitive recipient; frequently the donor has just ingested the antigen prior to donating blood.
- Reactions are much more apt to occur in patients with a history of asthma, hay fever, or atopy.
- Symptoms include flushing, chills, pruritis, and urticaria; rarely bronchospasm and vascular collapse.
- Symptoms usually do not occur until 250 ml of whole blood or 125 ml of packed cells have been transfused; they may not occur until the transfusion has been completed (or up to 24 hr after transfusion).
- If an allergic reaction occurs, stop the transfusion, keep the vein open with the starter solution, and notify the physician. The transfusion may be continued if the physician deems the reaction minor.
- Antihistamines usually relieve mild symptoms; but the rare, more

severe reactions require epinephrine or steroids. Sometimes antihistamines are used prophylactically when an allergic reaction is considered likely.

Hypothermia

- Cold donor blood is quickly warmed as it mixes with the recipient's blood during usual infusion rates.
- However, infusion of 3 to 5 units of cold blood (at refrigerator temperature) within 1 to 2 hours may lower the recipient's body temperature by 5° F to 7° F (3° C to 4° C).
- Generalized hypothermia can result in decreased heart rate, diminished cardiac output, and abnormal cardiac conduction. Local cardiac hypothermia caused by rapid transfusion of cold blood through a central catheter can produce cardiac arrest even though generalized hypothermia is not present.
- During rapid blood replacement, it is necessary to warm the blood to 37° C either by a mechanical blood warmer (see Fig. 16-1) or by passing it through a coil of tubing submerged in a basin of 37° C (98.6° F) water. (Warming should never be attempted by placing the container in warm water!)
- It may also be necessary to warm blood (even 1 unit) for patients with pretransfusion hypothermia, for those who are paralyzed, or for some other reasons unable to maintain their own body temperature.

Bacterial Contamination

- It has been reported that approximately 0.1% of all units of whole blood are contaminated with cold-growing organisms.
- Blood grossly contaminated with gram-negative organisms can cause the recipient to develop severe symptoms (such as high fever, intense flushing, vomiting, diarrhea, headache, and fatal shock) after as little as 50 ml has been administered.
- Occurrence of the above symptoms is an indication to halt the transfusion immediately, keep the vein open with a slow saline drip, and quickly summon any physician for help. Blood samples should be cultured and the patient treated with appropriate antibiotics, steroids, and fresh uncontaminated blood.
- The best treatment of this potentially fatal condition is prevention. Before blood is used it should be inspected for signs of bacterial growth, such as discoloration or gas bubbles. (Look for a purple, red, or brown color to the plasma—a sign of hemolysis.) Do not use any container with inadequately sealed closures. The presence of air bubbles is cause for suspicion of contamination.

Coagulation Problems

- Platelets are almost totally destroyed after 24 hours of blood storage. Massive transfusion therapy can, therefore, produce dilutional thrombocytopenia. Platelets are transfused as indicated by laboratory data and the patient's clinical status.

- Fresh frozen plasma (FFP) is rich in clotting factors V, VIII, and XI. (Recall that prolonged storage of blood decreases the levels of Factors V and VIII substantially.) FFP is indicated in the treatment of postsurgical hypovolemia with a hemorrhagic diathesis associated with a clotting factor deficiency.

Serum Homologous Hepatitis

- It has been reported that either serologic or clinical evidence of hepatitis occurs in approximately 5% to 10% of recipients of blood transfusions.

- Hepatitis A is rarely (if ever) transmitted by blood transfusion. Hepatitis B transmission in blood transfusions has been reduced significantly since blood banks have developed the capability to test for its presence. The majority of cases of transfusion hepatitis are thus apparently of the non-A, non-B variety; unfortunately, blood banks are not able to test for the presence of this virus (or viruses).

- Persons known to have had hepatitis are not allowed to give blood; however, many persons who have had hepatitis have not been diagnosed (many cases are subclinical in nature).

- Statistics indicate that paid donors cause many more cases of serum hepatitis than do unpaid donors. (The Food and Drug Administration has indicated that blood from paid donors must be labeled as such in bold letters.)

- When serum hepatitis occurs in a blood transfusion recipient, it is necessary to trace the donors who contributed the blood and to eliminate them from the donor pool.

- The incubation period for serum hepatitis is from 35 days to 120 days; symptoms include malaise, anorexia, vomiting, abdominal discomfort, enlarged liver with tenderness, diarrhea, headache, fever, and jaundice.

- Serum hepatitis is particularly serious in the very young and elderly and in patients already debilitated by serious illness or injury.

- Pooled plasma products (such as Factor VIII concentrate, and Factor II–VII–IX–X complex) may be associated with a greater risk of hepatitis transmission than is whole blood or packed red cells.

- Blood components virtually free of hepatitis transmission include albumin, plasma protein fraction, and immunoglobulin preparations.

Acute Hemolytic Reaction

- The most dreaded reaction to blood transfusion is that caused by the administration of grossly incompatible whole blood or packed cells.
- A transfusion of incompatible blood causes rapid cell agglutination in the recipient and eventual intravascular hemolysis.
- Incompatible transfusion is most often caused by the careless administration of the wrong blood; it can also be the result of inadequate cross matching or improperly handled blood (such as leaving the blood exposed to room temperature for an excessive time). It has been estimated that as many as two-thirds of the cases of hemolytic reactions occur as a result of clerical error.[9]
- Onset of hemolytic reaction may be delayed when factors of less magnitude than those involving the ABO system are present.
  1. For example, a patient who has received multiple transfusions in the past may have become sensitized to Rh or to the minor factors, and his hemolytic reaction might have a delayed onset.
  2. The slow-developing reaction might be characterized by jaundice appearing hours or days after the transfusion.

Nursing Actions Related to Hemolytic Blood Reactions

- Follow rules for safe blood administration presented at the beginning of the chapter.
- Be alert for symptoms of an acute hemolytic reaction. They usually occur during the transfusion of the first 50 ml to 100 ml (sometimes as little as 5–10 ml) of blood and include
  1. Throbbing occipital headache
  2. Lumbar and flank pain (due to precipitation of hemoglobin in renal tubules)
  3. Flushed face, then cyanosis
  4. Tachypnea and tachycardia
  5. Diaphoresis and cold, clammy skin
  6. Feeling of constriction in the chest or precordial pain
  7. Distended neck veins
  8. Urge to defecate or urinate
  9. Fever and severe shaking chills (a hemolytic reaction may cause considerable fever and general toxicity; these effects, in the absence of renal failure, are rarely fatal)
  10. Hemoglobinuria and acute tubular necrosis
  11. Jaundice (usually does not occur unless more than 3000 ml of blood is hemolyzed in less than a day)
- Observe the unconscious patient for tachycardia and bleeding from wounds or incisions. The only signs of serious transfusion reaction in

an anesthetized or otherwise unconscious patient may be increasing tachycardia, shock, and oozing of blood at the site of operation. (Bleeding may follow disseminated intravascular coagulation, which presumably follows the release of erythrocytic thromboplastic substances.)

- Be aware that hemolytic reactions are more prone to occur in patients who have had past transfusions and tend to be proportionate to the number of such transfusions, regardless of when given.
- Observe the patient closely during the administration of the first 50 ml to 100 ml of blood.
  1. If the transfusion is stopped early, acute tubular necrosis (ATN) and death rarely occur.
  2. If more than several hundred ml are infused, renal shutdown and death are common (renal shutdown is due to blockage of the tubules with hemoglobin and to the powerful vasoconstriction caused by toxic substances released from the hemolyzed blood).
- Stop the transfusion immediately when a hemolytic reaction is suspected—keep the vein open with isotonic saline. Notify the physician.
- Refrigerate the blood bag and administration set so that compatibility tests can be made.
- Save all urine and send to the lab for analysis for hemoglobin content (look for port-wine discoloration).
- Be prepared to administer fluids to promote diuresis—alkaline fluids, such as $1/6$ molar sodium lactate, may be prescribed to increase the solubility of hemoglobin and aid in its excretion. An osmotic diuretic (usually mannitol) may be given. (Further treatment is dependent on the amount of renal damage present.)

Acquired Immune Deficiency Syndrome

- Acquired Immune Deficiency Syndrome (AIDS) is a problem of increasing concern in patients receiving blood and blood products because it is suspected that the disease can be transmitted from blood donors with AIDS. Physicians suspect that an infectious agent such as a virus is responsible for AIDS, although the cause is yet to be determined.
- Through the first quarter of 1983, more than 1100 cases of AIDS has been reported. Two to three new cases of AIDS are being reported to the Centers for Disease Control (CDC) every day.[10] The mortality rate for this disease is extremely high, ranging from 40% to 70%, according to one source[11] Other sources report even higher mortality rates. Almost 90% of AIDS deaths have occurred in men less than 45 years of age.[12]
- AIDS is primarily a disease of homosexual men (72%); other groups considered at risk are intravenous drug abusers (17%), Haitian immigrants (4%), and hemophiliacs (1%).[13] Hemophiliacs are thought to be at high risk because of their frequent need for antihemophiliac factor (AHF), a blood by-product prepared from numerous blood

donors. For reasons that are still unclear, Haitians appear to be at greater risk for AIDS than non-Haitian hemophiliacs.

- Alarmingly, cases of AIDS have been reported in young children who have received blood transfusions. Although most of the AIDS victims are homosexual or bisexual, there are reported cases of AIDS in heterosexual men and women.

- Presenting signs and symptoms may include malaise, fever, night sweats, constant tiredness, decreased appetite, weight loss, shortness of breath, dry cough, enlarged lymph nodes, and painless purple-colored papules or nodules on the body or inside the nose, mouth, or anus.[14] Because of their weakened cellular immunity, AIDS victims are susceptible to a wide range of devastating infections. Certain kinds of cancer, most notably Kaposi's sarcoma, are more likely to occur in AIDS victims.

- In March 1983 the Public Health Service urged that several groups considered at risk for AIDS (including homosexual men, intravenous drug abusers, and Haitian immigrants) refrain from donating blood due to the possibility of transmitting the disease through blood transfusions.

References

1. Mertes Y: Care of the patient receiving transfusion therapy. In Rutman R, Miller W: Transfusion Therapy, p 191. Rockville, Aspen Systems Corp, 1982
2. Harris E: Massive transfusions. In Rutman R, Miller W: Transfusion Therapy, p 286. Rockville, Aspen Systems Corp, 1982
3. Ibid
4. Ibid
5. Ibid, p 287
6. Miller W: Adverse effects of blood transfusion. In Rutman R, Miller W: Transfusion Therapy, p 304. Rockville, Aspen Systems Corp, 1982
7. Hyatt C, Miller W, Rutman R: Transfusion reactions. In Rutman R, Miller W: Transfusion Therapy, p 356. Rockville, Aspen Systems Corp, 1982
8. Miller W: Adverse effects of blood transfusion. In Rutman R, Miller W: Transfusion Therapy, p 302. Rockville, Aspen Systems Corp, 1982
9. Ibid, p 307
10. Apuzzo-Berger D: A.I.D.S.: Could you be at risk? RN, Feb, 1983, p 67
11. Ibid
12. Ibid
13. "Battling a Deadly New Epidemic." *Time*, March 28, 1983, p 53
14. Apuzzo-Berger, p 68

chapter 17

Loss of Gastrointestinal Fluids

Character of Gastrointestinal Secretions

- The average daily volume of gastrointestinal secretions is approximately 8000 ml, as compared to a plasma volume of 3500 ml. Most of these secretions are reabsorbed in the ileum and proximal colon; only about 150 ml of relatively electrolyte-free fluid is excreted daily in the feces.
- Gastrointestinal secretions consist of saliva, gastric juice, bile, pancreatic juice, and intestinal secretions. Their average values are listed in Table 17-1. The electrolyte content of gastrointestinal secretions are summarized in Table 17-2.
- With the exception of saliva, the gastrointestinal secretions are isotonic with the extracellular fluid. In addition, material entering the gastrointestinal tract tends to become isotonic during the course of its absorption. Many liters of extracellular fluid pass into the gastrointestinal tract, and back again, as part of the normal digestive process. This movement of water and electrolytes is sometimes referred to as the *gastrointestinal circulation.*
- Loss of gastrointestinal fluids is the most common cause of water and electrolyte disturbances. This fact becomes evident when one considers the large volume of fluids in the gastrointestinal tract and the many ways in which they can be lost. Vomiting, gastrointestinal suction, diarrhea, fistulas, and drainage tubes are some of the abnormal ways in which these fluids can be lost. Fluids trapped in the gastrointestinal tract, as in intestinal obstruction, are physiologically outside the body (third-space effect). Any condition that interferes with the absorption of fluids from the gastrointestinal tract can cause serious water and electrolyte disturbances.

Table 17–1
Gastrointestinal Secretions and Their Usual *p*H

Secretion	*p*H
Saliva	6.0–7.0
Gastric juice	1.0–3.5
Pancreatic juice	8.0–8.3
Bile	7.8
Small intestine	7.8–8.0
Large intestine	7.5–8.0

Table 17–2
Approximate Electrolyte Composition of Gastrointestinal Secretions

Secretion	Usual Maximum Volume/ Day	Sodium (mEq/ liter in adults)	Chloride (mEq/ liter in adults)	Potassium (mEq/ liter in adults)
Normal				
Saliva	1000	100	75	5
Gastric juice (*p*H < 4.0)	2500*	60	100	10
Gastric juice (*p*H > 4.0)	2000*	100	100	10
Bile	1500	140	100	10
Pancreatic juice	1000	140	75	10
Succus entericus (mixed small-bowel fluid)	3500	100	100	20
Abnormal				
New ileostomy	500–2000	130	110	20
Adapted ileostomy	400	50	60	10
New cecostomy	400	80	50	20
Colostomy (transverse loop)	300	50	40	10
Diarrhea	1000–4000	60	45	30

*Nasogastric suction volume is usually much less than this unless pyloric obstruction exists.
(Condon R, Nyhus L: Manual of Surgical Therapeutics, 5th ed, p 174. Boston, Little, Brown & Co, 1981)

Vomiting and Gastric Suction

Fluid and Electrolyte Disturbances

- Major electrolytes in gastric juice are hydrogen, chloride, potassium, and sodium. Gastric juice is the most acid of the gastrointestinal secretions, having a *p*H of 1.0 to 3.5 (sometimes higher). Imbalances most often associated with the loss of gastric juice include

1. Fluid volume deficit (FVD)
2. Metabolic alkalosis
3. Potassium deficit
4. Sodium deficit
5. Magnesium deficit

- *Fluid Volume Deficit.* If vomiting is prolonged, and fluid replacement therapy is inadequate, FVD will result. Since many liters of fluid can be removed daily by gastric suction (depending on the underlying disease process), FVD is likely to occur unless fluid replacement therapy is adequate. One should be alert for signs of FVD (such as dry skin and mucous membranes, longitudinal furrowing of the tongue, oliguria, and acute weight loss).

- *Metabolic alkalosis.* Excessive loss of gastric juice by vomiting or gastric suction causes metabolic alkalosis, because hydrogen and chloride ions are lost from the body. Loss of chloride ions cause a compensatory increase in bicarbonate ions. The bicarbonate side of the carbonic acid–bicarbonate ratio is increased and the plasma pH becomes more alkaline. The nurse should be alert for signs of metabolic alkalosis when the patient has sustained a prolonged loss of gastric juice by gastric suction or vomiting, particularly if the patient has been allowed to drink water or ingest ice chips in large amounts. Symptoms can include muscle hypertonicity and tetany (caused by decreased Ca^{2+} ionization), depressed respirations (usually not clinically evident in an alert patient), personality changes, and disorientation.

Note: Vomiting of gastric contents alone is not as common as vomiting of gastric fluid plus duodenal fluid. Vomiting of alkaline duodenal fluid in addition to gastric juice can cause plasma pH to remain essentially normal or even cause metabolic acidosis if the loss of alkaline fluid exceeds loss of acidic gastric juice. Contributing to metabolic acidosis is ketosis of starvation (if caloric intake is inadequate).

- *Potassium Deficit.* A prolonged loss of gastric fluid frequently leads to potassium deficit, particularly if potassium replacement therapy is inadequate. Hypokalemia frequently accompanies metabolic alkalosis. Typical symptoms of hypokalemia include muscular weakness, gaseous intestinal distension, muscle cramps (particularly in the legs), irregular pulse, and in severe cases, cardiac or respiratory arrest.

- *Sodium Deficit.* Gastric suction or prolonged vomiting can lead to sodium deficit, especially if plain water is ingested. Symptoms to be alert for include anorexia, weakness, abdominal cramping, and confusion.

- *Magnesium Deficit.* Prolonged vomiting or gastric suction can result in magnesium deficit, an imbalance not as common as those listed above. The magnesium concentration in gastric juice is only 1.4 mEq/liter. In addition, the body conserves magnesium well. However, its continued loss by suction or vomiting, plus its dilution with magnesium-free replacement fluids, can result in symptoms of magnesium deficit. These include tachycardia, arrhythmias, increased neuromuscular irritability, and paresthesias.

Gastric Suction

- Many physicians prefer that patients undergoing gastrointestinal suction receive nothing by mouth; others allow ice chips sparingly to relieve a dry mouth. The word sparingly is open to interpretation by the staff, and more ice chips may be given than was intended by the physician, because of the patient's constant plea for more ice.
- Drinking plain water causes a movement of electrolytes into the stomach to make the solution isotonic; before the water and electrolytes can be absorbed they are removed by the suction apparatus. This process can deplete the body of valuable electrolytes, primarily sodium, chloride, and potassium.
- Profound states of metabolic alkalosis and of sodium deficit have been caused by the unwise practice of giving plain water to a patient undergoing gastric suction.
- If ice chips are to be given, they should be limited carefully. (Some physicians allow approximately 1 ounce of plain ice chips per hr to relieve oral dryness.) Ice chips made from an isotonic electrolyte solution are recommend by some physicians to avoid electrolyte washout.
- Most authorities favor the use of isotonic saline to irrigate gastrointestinal tubes rather than plain water, because the latter can promote electrolyte loss from the gastrointestinal mucosa (electrolyte washout).
- Sometimes there are specific orders for the frequency of the irrigation and the amount of solution to be used; other times the order is general and leaves the frequency of the irrigation and the amount of fluid to the nurse's discretion. In either event, the amount of fluid instilled should be removed; and the difference between the amount instilled and the amount removed should be recorded on the intake–output record.
- Some physicians prefer hourly irrigation of the tube with 30 ml to 50 ml of air to prevent clogging; air does not interfere with electrolyte balance or with intake–output record keeping. However, when blood or thick secretions are occluding the tube it may be necessary to use isotonic saline. In addition, the use of air may lead to uncomfortable abdominal distension and bloating.

• The patient with gastric suction or persistent vomiting requires parenteral fluid therapy to prevent FVD and other imbalances. Fluid lost from gastric suction or vomiting may be replaced with a special gastric replacement solution (such as Electrolyte #3, Isolyte G, or Ionosol G). Frequent serum electrolyte levels should be obtained to guide parenteral therapy.

Nursing Measures for the Patient With Vomiting

• Discourage the intake of plain water if vomiting is severe. Either encourage the frequent intake of small volumes of fluids containing electrolytes, or discourage oral intake altogether to promote rest of the stomach. (Discuss the situation with the physician.)

• Report vomiting early so that appropriate treatment can be started before water and electrolyte losses become severe. The physician will likely prescribe a medication to relieve nausea, and prescribe parenteral fluids.

• Administer prn medications, as indicated, to relieve nausea.

• Measure, or estimate as accurately as possible, the amount of vomitus so that lost water and electrolytes can be replaced by the parenteral route. In fact, all fluids lost and gained from the body should be measured.

• Measure body weight daily to detect significant changes in fluid balance. Daily weights are helpful in detecting FVD, particularly if vomitus has not been measured. A patient on a starvation diet should lose about a half pound a day, whereas a loss in excess of this amount probably implies a FVD. A weight gain implies fluid volume excess (FVE) if the patient is on a starvation diet. (Recall that routine IV fluids are low in caloric intake; for example, a liter of 5% dextrose solution contains only 170 cal.)

• Be alert for the imbalances associated with loss of gastric fluid (see the preceding discussion).

Nursing Measures for Patients With Gastric Suction

• Irrigate the suction tube with isotonic NaCl solution (or other electrolyte solution as prescribed by the physician). Realize that plain water is not recommended as an irrigating solution because it can wash out electrolytes.

• Record the irrigating solution volume as intake and whatever is recovered as output.

• Prohibit the intake of large quantities of water or ice chips by mouth, since water washes electrolytes from the stomach. Be aware that some physicians prefer that patients with gastric suction receive nothing by mouth; others allow small quantities of ice chips (such as one ounce per hour). Occasionally physicians will prescribe ice chips made from electrolyte solutions for patients experiencing great discomfort (dry mouth and/or thirst).

• Measure and record the amount of fluid lost by suction, as well as all other fluid losses and gains.

(continued)

- Measure daily weight variations to help detect early fluid volume deficit (or excess).
- Be alert for imbalances associated with loss of gastric fluid (see previous discussion).

Diarrhea, Intestinal Suction, and Ileostomy

- Abnormal loss of intestinal fluid occurs in the presence of diarrhea, intestinal suction, and ileostomy. Imbalances likely to occur in these situations include
  1. FVD
  2. Metabolic acidosis
  3. Potassium deficit
  4. Sodium deficit
- *Fluid Volume Deficit.* Intestinal hypermotility shortens the opportunity for absorption of intestinal fluids and, thus, results in increased fluid loss in bowel movements.
  1. The liquid stools expelled as a result of hypermotility contain water and electrolytes derived from secretions, ingested food and fluids, and extracellular fluid brought into the bowel to render ingested substances isotonic.
  2. Many liters of fluid can be lost daily as diarrheal fluid. Obviously, prolonged diarrhea is a serious threat to water and electrolyte balance. The amount of fluid lost in intestinal suction averages around 3000 ml daily.
  3. Thus, the nurse should be alert for symptoms of FVD when the patient has sustained large fluid losses from the intestinal tract. (These can include dry skin and mucous membranes, furrowing of the tongue, oliguria, and acute weight loss).
- *Metabolic Acidosis.* Intestinal secretions are alkaline, including pancreatic juice and bile, which are mixed with intestinal juices in the intestines. The chief electrolytes in intestinal juice include sodium, potassium, bicarbonate, and chloride. Intestinal secretions are alkaline because of the preponderance of bicarbonate ions. Loss of bicarbonate results in a decreased *p*H. Symptoms of metabolic acidosis can include hyperventilation (compensatory action by lungs) and confusion.
- *Potassium Deficit.* Relatively large amounts of potassium are contained in the intestinal fluid; therefore, potassium deficit occurs frequently with diarrhea, prolonged intestinal suction, and recent ileostomy. Symptoms to be alert for include muscular weakness, gaseous intes-

tinal distention, muscle cramps (particularly in the legs), irregular pulse, and in severe cases, cardiac or respiratory arrest.

- *Sodium Deficit.* Intestinal secretions have a relatively high concentration of of sodium; excessive loss of these secretions can result in sodium deficit. Symptoms to be alert for include anorexia, abdominal cramping, and confusion.

Nursing Measures in the Care of Patients With Diarrhea

- Discourage intake of irritating foods that are likely to stimulate peristalsis. Encourage frequent intake of small volumes of fluids containing electrolytes. In the presence of severe diarrhea, the physician may recommend that oral intake be discontinued altogether to allow the intestine to rest.

- Report diarrhea early so that appropriate treatment can be instituted before water and electrolyte losses become severe. The physician will likely prescribe medication to reduce peristalsis; in severe cases, parenteral fluids may be necessary to allow the intestinal tract to rest. (Be aware that less gastrointestinal fluids are formed when the intestinal tract is at rest.)

- Administer prn medications, as indicated, to prevent diarrhea from becoming severe.

- Measure, or estimate as accurately as possible, the amount of liquid feces lost so that lost water and electrolytes can be replaced by the parenteral route. (All fluids lost and gained by the body should be recorded on the intake–output record.)

- Measure body weight daily to detect significant changes in fluid balance. Daily weights are helpful in detecting FVD, particularly if liquid stools have not been measured.

- Be alert for imbalances associated with the loss of intestinal fluid (see the previous discussion).

Nursing Considerations and Actions for the Patient With Recent Ileostomy

- Be alert for symptoms of water and electrolyte disturbances in the immediate postoperative period; K^+ deficit is the most frequent inbalance. Other imbalances may include Na^+ deficit and FVD. (These imbalances have been described previously.)

  Patients with ileostomies are more likely to develop water and electrolyte disturbances when their stomas first begin to function, as shown by a comparison of the amount of water and electrolytes lost in a 24-hour period in the early postoperative period, with the amounts lost in a similar period after the ileostomy has adapted. Fluid loss from a recent ileostomy may be as high as 4000 ml in 24 hours; each liter of fluid may contain 130 mEq of Na^+ and 20 mEq of K^+ (sometimes as much as 70 mEq). An *adapted* ileostomy usually loses no more than 500 ml in 24 hours; each liter of fluid may contain 50 mEq of Na^+ and 10 mEq of K^+.

- Measure and record the fluid lost by ileostomy, as well as other fluid losses and gains by the body.

- Various parenteral fluids are available to replace intestinal fluid losses. Duodenal replacement fluid (Na^+ 138 mEq/liter, K^+ 12 mEq/liter, HCO_3 50 mEq/liter, and Cl 100 mEq/liter) may be used. Loss of ileostomy fluid is usually replaced with lactated Ringer's solution to which sufficient K^+ has been added to maintain a normal serum K^+ concentration.
- Nursing considerations in the care of patients with intestinal suction are the same as those listed earlier for gastric suction.

Fistulas

- Fistulas are abnormal communications between the intestine and the skin (external fistula) or another hollow viscus (internal fistula).
- Fistulas may result from trauma, or they can occur spontaneously as a complication of pancreatitis, inflammatory bowel disease, neoplasia, or other gastrointestinal disorders.
- The amount and kind of fluid lost through a fistula depends on its location. An educated guess as to the content of the fluid and imbalances likely to accompany its loss can be made by reviewing the usual electrolyte content of the fluid in the region of the fistula. (See Table 17-2.) For example, one would expect metabolic alkalosis to occur if the fluid loss is mostly gastric or, metabolic acidosis if the fluid loss is duodenal or pancreatic. Jejunal fistulas are not common and usually do not result in a serious acid–base problem. When in doubt as to the origin of the fistula, the fluid's pH and electrolyte content can by analyzed by the laboratory.
- In addition to pH changes, fistulas can cause a serious contraction of extracellular fluid volume. For example, a gastric, duodenal, or jejunal fistula may drain 3 to 6 liters daily; a pancreatic fistula may drain 2 liters daily. Pancreatic fistulas are particularly serious since pancreatic enzymes may digest the abdominal wall and cause peritonitis.

Nursing Considerations in the Care of Patients With Fistulas

- Attempt to measure the fluid loss from the fistula. Usually, a stoma bag can be applied over the stoma's exit. If the drainage cannot be directly measured, an estimate should be made of the volume. Statements as to how much of a dressing is saturated, as well as the extent of gown and linen saturation, help in planning fluid replacement therapy.
- Encourage adequate nutritional intake. Be aware that fistulas may close spontaneously with proper nutritional support (such as elemental diets or intravenous hyperalimentation). (Examples of elemental

feedings include Vivonex [Norwich-Eaton] and Flexical [Mead Johnson].) At times, surgical intervention is required.
- Be alert for symptoms of imbalances likely to be associated with the fistula. (See previous discussion.)

Prolonged Use of Laxatives and Enemas

- The prolonged use of laxatives and enemas can result in serious water and electrolyte disturbances, particularly potassium deficit. Other possible inbalances include sodium deficit and FVD.
- *Cathartics* increase the water and electrolyte output through the fecal route by hastening the excretion of fecal contents and, thus, reducing absorption time. Irritant cathartics cause hypermotility of the bowel by irritating the bowel mucosa. Saline cathartics draw water form the extracellular fluid into the bowel. The distended bowel produces mechanical stimulation and the large amount of fluid is propelled out of the bowel. A large fluid volume deficit can result from continued use of saline cathartics, which interfere with electrolyte absorption from the intestines.
- It has been reported that persons over 70 years of age take laxative twice as often as those in the 40-to-50 age group. Unfortunately, the use of laxatives in the aged often becomes habit forming, requiring larger and more frequent doses to achieve results.
- *Enemas* can also affect electrolytes; repeated tap water enemas can result in hyponatremia due to absorption of the water by the colon. On the other hand, use of commercial hypertonic cleansing enemas (containing sodium biphosphate and sodium phosphate) can result in some degree of sodium absorption and, thus, should be avoided in patients requiring sodium restriction.
- Nursing actions

  1. Teach patients to avoid the repeated use of cathartics and enemas.
  2. Encourage patients to increase fluid and bulk intake and to exercise regularly.
  3. Encourage the ingestion of a glass of warm fluid first thing in the morning to stimulate the evacuation reflex.
  4. Inform patients that stool softeners are useful physiologic tools to promote evacuation and are less apt to disrupt fluid balance than are more harsh laxatives.
  5. Use an isotonic solution of sodium chloride when repeated bowel irrigations are needed. A roughly isotonic sodium chloride solution can be easily made by adding 1 tsp of table salt to 1 liter of tap water (furnishes approximately 120 mEq of Na^+ and 120 mEq of Cl^-).

Imbalances Associated With Preparations for Diagnostic Studies of the Colon

- Standard colon cleansing techniques for diagnostic studies usually include several days of dietary restrictions (such as low-residue or liquid diets), purgatives (such as castor oil or magnesium citrate), and numerous cleansing enemas (either saline or tap water).
- Studies have indicated that the rigorous catharsis associated with this kind of preparation leads to significant shifts of fluids among body compartments—if the shifts occur rapidly in an elderly person with cardiovascular disease, the results can be dangerous.
- The elderly patient undergoing rigorous bowel preparation should be observed closely for adverse reactions. Care should be taken to perform the procedure correctly the first time to avoid the need for repeated roentgenograms (requiring more cathartics, more enemas, and more fluid restriction). The elderly patient can ill afford to undergo one test after another without a rest period in between; it is frequently up to the nurse to intervene in this area on the patient's behalf.
- The already reduced glomerular filtration rate (GFR) in the elderly potentiates the hazards of any further decrease in extracellular fluid volume (as occurs in vigorous catharsis). The aged patient having gastrointestinal roentgenograms should probably receive intravenous fluids during the preparation period of reduced oral intake and increased fluid loss by catharsis and enema.
- It should also be noted that the use of radiocontrast agents in diagnostic radiology (particularly in large doses) has been incriminated as predisposing to acute renal failure, especially when the patient has a severe FVD.

chapter 18

Intestinal Obstruction

- *Intestinal obstruction* is said to exist when there is interference with the normal progression of intestinal contents. It may be complete or incomplete, acute or chronic, intermittent or continuous, mechanical or functional.
- A *mechanical obstruction* is defined as an actual physical barrier (such as adhesions, hernia, tumor, intussusception, or diverticulitis) blocking normal passage of intestinal contents.
- *Functional obstruction* (paralytic ileus) refers to the inability of the bowel to propel intestinal contents forward due to inhibition of the neuromuscular apparatus; it occurs temporarily after abdominal surgery and is commonly associated with peritonitis.
- It has been estimated that 20% of surgical admissions for acute abdominal conditions are for intestinal obstruction.
- A *strangulated obstruction* is one in which there is occlusion of the blood supply to a segment of the bowel, in addition to obstruction of the lumen. Surgical intervention is needed immediately to prevent gangrene, perforation, and peritonitis. A *simple obstruction* is one in which the intestinal blood supply is not compromised.

Simple Mechanical Obstruction

- Mechanical obstruction of the bowel causes hyperperistalsis; severe cramping pain occurs concurrently with the hyperperistaltic movements. The quiescent period between attacks of pain is short in high intestinal obstruction (approximately 5 min); these periods are longer the lower the obstruction.
- Episodes of pain are associated with bowel sounds that are loud,

high-pitched and metallic, and occur in bursts or rushes. The abdomen is quiet between attacks of colic. Hyperperistaltic movement of the bowel attempts to propel intestinal contents past the point of obstruction; it may be so strong that it actually traumatizes the bowel wall and contributes to its edematous state. Eventually, the hyperperistalsis subsides as distention becomes more severe and the colicky pain is replaced by steady abdominal discomfort. Steady severe pain with no quiescent periods is usually indicative of strangulation.

- Intestinal gas accumulates proximal to the obstruction and is largely the result of swallowed air. Bacterial fermentation of intestinal contents also contributes to the gas volume. Distention may be absent, or confined to certain loops of the bowel (forming a so-called *ladder pattern*). Diffuse distention may occur late and is often indicative of colon obstruction. Serial abdominal girth measurements will detect progressive abdominal distention.

- Fluid accumulates proximal to the obstruction and is derived mainly from normal gastrointestinal secretions; the fluid is isotonic in nature. (Large quantities of water and electrolytes continue to be secreted into the bowel lumen by the edematous gut wall even in the absence of oral intake.) The fluid contained in the bowel forms a third space, which is temporarily lost to the body. (See Chap. 6 for a discussion of third-space body fluids.)

- Vomiting of the accumulated fluid is common in bowel obstruction and is an early manifestation. Vomiting is more frequent and copious in high obstructions than it is in low small bowel obstruction. The vomitus is frequently malodorous (feculent) due to stagnation of fluid and associated bacterial action. Vomiting does not usually occur in colon obstruction until retrograde distention involves the small bowel. If the ileocecal valve remains competent, small bowel distention may not occur in large intestinal obstruction.

- Decreased selectivity of the edematous intestinal membrane allows plasma proteins to enter the intestinal lumen and gut wall. Because of distention, the mucosa of the bowel above the obstruction is stimulated to secrete more fluid. The edematous bowel wall is not able to absorb the large fluid volume; thus, the distention becomes progressively worse and the patient progressively more volume depleted.

- Sustained elevated intraluminal pressures compromise lymphatic and capillary circulation and can lead to patches of ischemic gangrene and perforation.

Fluid and Electrolyte Imbalances

- *Fluid volume deficit* (FVD) occurs due to accumulation of trapped fluid in the intestines (as much as 5–10 liters or more), vomiting, and gastrointestinal suction. The FVD may be severe enough to cause hypovolemic shock and death. The hematocrit rises roughly in proportion to the fluid loss, it has been estimated that each point of elevation in the hematocrit represents the loss of approximately 100 ml of plasma.[1] The blood urea nitrogen (BUN) elevates due to decreased urinary output. Oliguria and high urinary specific gravity (SG) are results of the FVD. It is important to remember that a sustained FVD can cause renal failure and thus requires prompt treatment. (See treatment of FVD in Chap. 5.)

- Plasma concentrations of electrolytes are initially preserved because the fluid lost is primarily isotonic.
- In the untreated patient, there is a tendency for sodium-free fluid (derived from cellular catabolism and oxidation of fat) to partially correct the FVD. The plasma osmolality decreases below normal as the sodium level is decreased by endogenous production of water. The previously noted rise in the hematocrit can be halted (or even reversed) and urine production is increased. Potassium excretion is increased as cellular potassium is released by increased catabolism. Thus, the untreated patient with bowel obstruction must be closely monitored for both sodium and potassium deficiencies.
- A low intestinal obstruction (as in a distal segment of the small intestine) causes the patient to vomit larger volumes of alkaline intestinal fluids than it does acidic gastric juice. Thus, metabolic acidosis can result. Ketosis of starvation (often present in patients with intestinal obstruction) contributes to the metabolic acidosis; mild ketonuria is frequently noted.
- Metabolic alkalosis is infrequent in bowel obstruction; when present, usually it is primarily due to loss of acidic gastric fluid.
- If the upper small intestine is obstructed, the patient may vomit approximately equal volumes of intestinal and gastric fluids, preventing a serious *p*H disturbance.
- A markedly distended abdomen presses against the diaphragm and embarrasses respiration, resulting in carbon dioxide retention and respiratory acidosis.

Fluid Replacement Therapy

- Fluids used for replacement of lost fluids include balanced isotonic and hypotonic electrolyte solutions, and plasma protein fraction or human serum albumin. Blood is used if blood loss into infarcted bowel segments is significant, or if blood loss has occurred with carcinoma of the bowel or diverticular disease. If loss of gastric juice has been prominent, isotonic saline (0.9% NaCl) may be used. Otherwise, equal volumes of lactated Ringer's solution and D_5W may be used to replace fluid losses and cover maintenance needs; extra potassium is needed.
- Potassium and sodium administration must be guided by frequent monitoring of serum levels. Potassium should not be given until uri-

nary output is adequate. (Nursing responsibilities in the administration of potassium are discussed in Chap. 4.)

- Treatment of FVD is guided by clinical assessment of hourly urinary output, skin and tongue turgor, blood pressure and pulse changes, and laboratory data (primarily hematocrit and BUN). (Clinical assessment and treatment of the patient with FVD is described in Chap. 5.)

- Fluid administration before surgery aims at stabilizing the vital signs sufficiently to withstand the stress of surgery. In general, the longer the obstruction has existed, the greater the FVD and thus the longer it takes to prepare the patient for surgery.

- Ideally, fluid–electrolyte normality should be achieved before surgery. However, when the possibility of strangulation exists, there is not sufficient time to fully correct all imbalances because this is an emergency situation requiring quick surgical intervention. In any event, it is highly desirable that the hourly urine output be at least 50 ml before surgery. Correction of imbalances may have to extend over into the postoperative period.

- The rate of fluid administration is best guided by monitoring the central venous pressure (CVP), particularly when severe FVD is present and there is a history of cardiovascular instability.

- Electrolyte-free solutions should *never be used alone* in the previous situation because water intoxication (dilutional hyponatremia) can be easily precipitated. The patient has lost both water and electrolytes, and needs replacement of both. Excessive administration of water-yielding fluids (such as D_5W) dilutes the contracted plasma and causes serious hyponatremia; it also dilutes other plasma electrolytes such as postassium.

- Chronic bowel obstruction may necessitate the use of total parenteral nutrition (TPN) through a central catheter to maintain or build up the patient's nutritional status. If the obstruction is high in the small bowel, the physician may elect to insert a jejunostomy tube for feeding purposes (introduction of feeding solution below the point of obstruction). (See Chap. 14 for a discussion of tube feedings.)

- About the third postoperative day, the fluid sequestered in the intestine is picked up by the vascular compartment. At this time, fluid administration must be slowed down to avoid overloading the intravascular space, causing congestive heart failure and pulmonary edema.

Nursing Assessment of the Patient With Intestinal Obstruction

Assess the Patient for Pain

Be aware that patients with early simple mechanical obstruction frequently have intermittent colicky pain occurring with bouts of hyperperistalsis. They are pain free between attacks of hyperperistalsis. Later, the pain subsides and becomes a generalized abdominal discomfort. The presence of a steady severe pain can indicate strangulation and is a significant finding that should be reported to the physician.

Assess Bowel Sounds at Regular Intervals

- Use the diaphragm of the stethoscope to listen to bowel sounds in order not to miss the usually faint peristaltic sounds. Lightly place the stethoscope on the abdomen and listen for several min over each quadrant. It is best to begin at the *auscultation center*, which is approximately 1 cm to 2 cm below and to the right of the umbilicus. (Bowel sounds, particularly from the small intestine, are most audible at this point.)

- Be aware that normal bowel sounds are soft, gurgling, or rushing sounds that occur approximately every 5 to 15 seconds.[2]

- If abnormal sounds are detected, try to locate the point of greatest intensity.

- Be aware of significant findings.

  1. Very slow (one or less per min) or absent bowel sounds. Slow bowel sounds occur in adynamic ileus. Totally absent bowel sounds are uncommon; this phenomenon may occur in a strangulated obstruction. Before deciding bowel sounds are absent, listen at the auscultation center for at least 5 minutes.

  2. Waves of high-pitched "tinkling" sounds (as occurs with the hyperperistalsis seen in early mechanical bowel obstruction). Sounds may be heard occurring every 2 to 3 seconds (hypermotility).

Assess the Vital Signs at Frequent Intervals

- A fall in blood pressure with an increased pulse rate indicates further contraction of the plasma volume. With appropriate fluid replacement, blood pressure should rise and the pulse rate should decrease and the pulse volume become stronger. An elevated temperature can be indicative of bowel strangulation or perforation.

Monitor Hourly Urine Volume and SG

- Prior to initiation of parenteral fluid administration, oliguria is present and the urinary SG is high (*i.e.*, 1.025 or above). With appropriate fluid replacement therapy, hourly urine flow should increase, and the SG should decrease. If the patient is being readied for surgery it is vital that hourly urine flow be at least 50 ml/hr (to prevent hypotension and renal insufficiency after induction of anesthesia).

Assess the Patient for Other Signs of FVD

- Signs are dry skin and mucous membranes, furrowed tongue, sunken eyes, and pinched facial expression. These signs should improve with adequate replacement therapy.

Measure all Routes of Fluid Gains and Losses

- Effective fluid replacement therapy depends on accurate intake–output records. Other factors the physician considers in fluid replacement therapy are clinical signs (discussed above) and the hematocrit and BUN levels.

(continued)

Monitor Central Venous Pressure (CVP) During Fluid Replacement Therapy

- Fluids can usually be given aggressively (as indicated by the physician) as long as the CVP remains < 10 cm to 12 cm of water (or the limit designated by the physician). A sudden rise in CVP is an indication to cut back on the rate of fluid administration. (Nursing responsibilities in monitoring CVP are described in Chap. 2.)

Check the Nasogastric Tube for Patency

- Also, check the suction apparatus. (Gastric suction must be maintained to keep the gastrointestinal tract as decompressed as possible.)

References

1. Pestana C: Fluids and Electrolytes in the Surgical Patient, 2nd ed, p 54. Baltimore, Williams Wilkins, 1981
2. Orr M: Assessment of the abdomen. In Orr M Shinert J Gross J: Acute Pancreatic and Hepatic Dysfunction, p 5. Bethany, CT, Fleschner Publishing, 1981

chapter 19

Cirrhosis of the Liver and Acute Pancreatitis

Cirrhosis of the Liver

Pathophysiology

- *Ascites* represents the accumulation of fluid in the peritoneal cavity. In the cirrhotic patient, ascites results from a combination of factors:
    1. Mechanical obstruction to venous outflow from the cirrhotic liver
    2. Hypoalbuminemia
    3. Hyperaldosteronism
- Decreased plasma albumin, combined with the increased capillary pressure produced by portal hypertension, allows albumin-rich fluid to shift into the peritoneal cavity. It is thought that portal hypertension is more important in the pathogenesis of ascites than is hypoalbuminemia, because ascites disappears when portal hypertension is relieved by surgical shunting procedures.
- Ascites caused by liver disease is a transudate with a relatively high protein content (varying from approximately 20%–50% of the protein level in plasma).
- The decrease in plasma volume (associated with ascites formation) stimulates aldosterone secretion, which in turn causes renal retention of sodium and water. This mechanism explains the fluid volume excess associated with cirrhosis of the liver.

237

Clinical Manifestations

Clinical manifestations of cirrhosis are variable but may include the following:

- Hypoglycemia (glycogen metabolism impaired).
- Spider angiomata (small dilated superficial vessels resembling small bluish-red spiders may appear in the skin of the face, forearms, and hands; the "spider" disappears when pressure is applied to the central point with a pencil, and reappears when the pressure is removed).
- Palmar erythema (bright, red color on the palms of the hands, due to increased circulation; redness disappears when pressure is applied).
- Enlarged breasts and atrophy of testicles in males (caused by the liver's inability to deactivate estrogen).
- Embarrassed respirations if the amount of ascites is large. (However, if a high ammonia level is present, hyperventilation may occur.)
- Enlarged abdominal veins, esophageal varices, and internal hemorrhoids caused by portal hypertension. (These may be sites of profuse hemorrhage.)
- Increased bleeding tendencies resulting from vitamin K deficiency and decreased prothrombin formation.
- Deficiency of vitamins A, C, and K caused by inadequate formation, utilization, and storage by the liver.
- Intensified reaction to certain medications, caused by the liver's inability to detoxify them. (Morphine and barbiturates should be avoided in patients with advanced liver disease.)
- Greater susceptibility to infection, secondary to decreased formation of antibodies by the liver.
- Elevated ammonia level at first causes dullness, drowsiness, loss of memory, slow, slurred speech, and personality changes—later these may progress to confusion and disorientation and, finally, to stupor and coma.
- *Hepatic flap* is caused by a high ammonia level; this tremor is characterized by spasmodic flexion and extension at the wrists and fingers, as well as lateral finger twitching. (Hepatic flap can be elicited in susceptible patients by elevating the arms, hyperextending the hands, and spreading the fingers.)

Associated Water and Electrolyte Disturbances

Water and electrolyte disturbances that can occur in patients with cirrhosis of the liver and ascites are summarized in Table 19-1.

(Text continues on p. 242.)

Table 19–1
Water and Electrolyte Disturbances in Cirrhosis of the Liver With Ascites

Disturbance	Cause and Comments
Fluid volume excess (Na^+ and water retention)	Secondary hyperaldosteronism Recall that aldosterone causes Na^+ retention, which in turn causes water retention. (The liver normally inactivates aldosterone and prevents its excessive buildup in the body; however, the cirrhotic liver does not perform this function well.)
Mild hyponatremia is quite common (Severe hyponatremia, < 125 mEq/liter, is a serious prognostic sign)	Impaired renal water excretion (pathogenesis not clear) Perhaps related to excess ADH secretion or inability of liver to deactivate ADH. Severe dietary Na^+ restriction Frequent paracentesis Na^+ moves into the cells to replace K^+ loss (K^+ loss is the result of diuresis, malnutrition, or hyperaldosteronism).
Decreased plasma protein level (Hypoalbuminemia results in reduced plasma osmotic pull)	Albumin lost in ascites (1 liter of ascitic fluid contains as much albumin as 200 ml of whole blood) Normally, blood flowing into and out of the liver has the same volume. In cirrhosis, however, the volume of hepatic venous outflow is reduced by half. This mechanical blockage leads to the formation of an ultrafiltrate of blood, which escapes into the abdominal cavity. Decreased albumin synthesis results from liver dysfunction. The normal albumin–globulin ratio of approximately 3:1 is disrupted—the ratio may be reversed so that globulin levels are higher than albumin. (Recall that globulin exerts a weaker osmotic pull than albumin, and that it is produced mainly by lymphatic tissues.)

(*continued*)

Table 19–1
Water and Electrolyte Disturbances in Cirrhosis of the Liver With Ascites (continued)

Disturbance	Cause and Comments
Generalized edema	Reduced plasma protein level allows fluid to leave the plasma space and enter the tissue space. Excessive aldosterone and ADH levels cause fluid retention. Portal hypertension causes increased hydrostatic pressure and, thus, edema of the lower extremities.
Potassium deficit (A severe K^+ deficit is often a late manifestation of liver disease.)	Prolonged use of potassium-losing diuretics Hyperaldosteronism (recall that hyperaldosteronism causes K^+ loss) Poor dietary intake because of anorexia and nausea Vomiting and diarrhea (increased loss of K^+)
Respiratory alkalosis	May be related to hyperammonemia A high ammonia level acts as a respiratory stimulant and may induce hyperventilation.
Magnesium deficit (Occurs in some cases of alcoholic cirrhosis)	Poor dietary intake Impaired renal conservation of Mg^{2+} Intestinal malabsorption Intermittent diarrhea
Calcium deficit	Possibly a result of inadequate storage of vitamin D by the diseased liver
Renal tubular acidosis (Type I) (May occur in some patients; the mild metabolic acidosis is rarely symptomatic)	Apparently there is a relationship between liver disease and renal tubular acidosis—the reason for the relationship is obscure.
Metabolic alkalosis (Common in patients treated with thiazides, furosemide, and eth-acrynic acid)	Related to the hypokalemia induced by these agents

(continued)

Table 19–1
Water and Electrolyte Disturbances in Cirrhosis of the Liver With Ascites (continued)

Disturbance	Cause and Comments
Metabolic acidosis (May occur in patients treated with spironolactone alone)	Spironolactone interferes with sodium–hydrogen ion exchange in the distal tubules.
Hyperammonemia	Under normal conditions, the large amounts of ammonia formed in the intestines by bacterial action are absorbed into the bloodstream and carried to the liver to be converted to urea for renal excretion. However, in cirrhosis, the liver cannot convert the ammonia to urea, causing the blood ammonia level to rise. Bleeding into the gastrointestinal tract increases the production and absorption of ammonia.

Medical Therapy

Sodium Restriction

- Patients with ascites are first treated with low-sodium diets; accurate intake–output records, weight charts, and abdominal girth measurements are helpful to the physician in planning the degree of sodium restriction. (See section on nursing assessment of the patient with cirrhosis.)
- Dietary restriction varies according to need; usually, daily limitations of 500 mg to 1000 mg are necessary. Restriction to less than 200 mg daily prevents additional fluid retention in most patients, but is not usually well accepted. (Low-Na^+ diets are described in Chap. 23.)

Diuretic Therapy

- If dietary sodium restriction alone does not suffice, diuretic therapy is necessary.
- When potassium-losing diuretics (such as furosemide, ethacrynic acid, or the thiazides) are used, the patient may need potassium supplements to prevent hypokalemia. (Hypokalemia, potassium-losing diuretics, and potassium supplements are discussed in Chap. 8.)
- Sometimes the physician will prescribe a combination of potassium-losing and potassium-sparing diuretics (such as Aldactazide or Dyazide). These agents not only help prevent disruption in potassium balance, they provide two mechanisms for the elimination of sodium. (Furosemide, ethacrynic acid, and the thiazides block NaCl reabsorption in the proximal nephron or the ascending limb of Henle's loop; spironolactone acts in the distal nephron.) With the use of this combination of drugs, there is no need for potassium supplementation or the need is greatly reduced. Routine potassium replacement in this instance is contraindicated. Serum potassium levels should be monitored closely if potassium supplements are used.
- In many cases effective diuresis can be induced by the use of a potassium-sparing diuretic (such as spironolactone) alone; potassium supplements can be lethal in this situation and should be avoided. Use of potassium-sparing diuretics is extremely dangerous in patients with renal damage since hyperkalemia can be readily induced.
- Patients undergoing sustained diuresis should not lose more than 10 to 12 lb/week. If excessive diuresis is produced, hyponatremia, hepatic coma, or progressive renal insufficiency (hepatorenal syndrome) may develop.
- Removal of ascitic fluid by sodium restriction and diuretic therapy should be achieved slowly (at about a rate of 500–1000 ml/day) to allow for equilibration between peritoneal and intravascular spaces.

Water Restriction

- A low serum sodium level is an indication to moderately restrict the patient's water intake; restriction to less than 1000 ml/day may be indicated for patients with dilutional hyponatremia.
- According to one source, ascites production ceases when a 200-mg sodium diet and a 500-ml fluid restriction are utilized.[1] (Fluids should be evenly spaced over the waking hours to avoid excessive thirst and discomfort.)

Limitation of Activity

- Assumption of the supine position is often associated with a spontaneous diuresis (resulting in significant weight loss with mobilization of peripheral edema fluid and ascitic fluid).
- The diuresis induced by bedrest results from improvement in cardiac output and effective circulating blood volume.
- The patient with severe ascites is usually better able to tolerate the lateral recumbent position.

Paracentesis

- Eventually, patients with ascites may become refractory to diuretics and require paracentesis to relieve severe symptoms of intraabdominal pressure and respiratory distress.
- Since paracentesis can be associated with serious problems, it should be done only when respiratory distress or patient discomfort is severe.
- As much as 20 liters of ascitic fluid may accumulate in 1 week; unfortunately, its removal only causes more to form. Since ascitic fluid forms faster than it is reabsorbed, hypovolemia may occur if too much fluid is removed at once. The mechanism involves a reaccumulation of ascitic fluid in the abdominal cavity resulting from flow of water and sodium from the interstitial fluid and plasma, causing a diminished plasm volume. In addition, abdominal blood vessels that have been under heavy external pressure and have carried little blood, open suddenly with the removal of the fluid pressure. As a result, blood is also suddenly rerouted, and shocklike symptoms can develop.
- Paracentesis can disrupt fluid balance in several ways. It results in protein loss, decreased effective blood volume, and increased secretion of aldosterone and antidiuretic hormone (ADH).
  1. Existing hypoalbuminemia is made worse by paracentesis. Although salt-free albumin is sometimes administered intravenously to help correct hypoalbuminemia, it is expensive and the results are negligible. (The amount that can be given is often ineffective in restoring a normal plasma albumin level.)

2. Acute sodium depletion with shock or renal failure may also follow paracentesis, particularly when done repeatedly. In addition to sodium loss in ascites, body water is retained in excess of sodium and further dilutes the plasma. This dilutional hyponatremia appears to be due to stimulation of ADH secretion.

Peritoneovenous Shunt

- A peritoneovenous (LeVeen) shunt is an alternative in the treatment of cirrhosis that does not respond to sodium restriction and diuretic therapy. The shunt consists of a long tube (with openings along the side) inserted into the abdominal cavity, running under subcutaneous tissue into the jugular vein. With the shunt, reinfusion of ascitic fluid into the venous system is allowed (a one-way valve prevents backflow of blood and clotting of the shunt). Pressure changes of respiration permit the shunt to operate.
- Hemodilution by ascitic fluid is a potential complication of this procedure, sometimes requiring packed cell transfusions; the hematocrit level is used to monitor the extent of hemodilution. Other complications include disseminated intravascular coagulation, wound infection, and peritonitis.

Preventing Ammonia Buildup

- Bleeding in the gastrointestinal tract increases ammonia formation (caused by digestion of blood proteins) and may precipitate hepatic coma. (Approximately 15–20 Gm of protein are added to the system with each 100 ml of blood digested.)[2] Sometimes enemas or cathartics are used to rid the intestine of blood, thus decreasing ammonia formation.
- Ammonia formation caused by excessive bacterial growth in the small bowel can be decreased by the administration of bowel sterilizing antibiotics, such as neomycin. Recall that neomycin (given orally or by enema) kills urease-producing bacteria, causing less urease to be produced; hence less urea to be broken down into ammonia.
- Since constipation contributes to the accumulation of ammonia, it must be prevented. Bowel evacuation removes blood from the intestine and thus decreases ammonia production; therefore, laxatives (such as Milk of Magnesia) may be used to produce bowel movements. Enemas may also be used.
- Ammonium chloride or carbonic anhydrase inhibitors (such as Diamox) should not be used because they may precipitate hepatic coma.
- As stated earlier, hypokalemia and metabolic alkalosis may occur in cirrhotic patients treated with potassium-losing diuretics. These imbalances should be prevented by the administration of potassium

supplements. Among other effects, hypokalemia favors increased production of ammonia by the kidneys; metabolic alkalosis favors conversion of ammonium to ammonia.[3]

- Hypovolemia also predisposes to ammonia production by interfering with renal excretion of urea, causing more urea to enter the bowel to be converted into ammonia.

- Impending hepatic coma is an indication to temporarily eliminate protein from the diet. A decrease in protein intake results in decreased ammonia formation; as the patient improves, more protein can be added. High ammonia levels must be discouraged since they contribute to hepatic encephalopathy.

Nursing Assessment of the Ascitic Patient During Therapy

Measure Abdominal Girth at Regular Intervals to Monitor the Degree of Ascites

- Measure at the same place each time; anatomical landmarks (such as the iliac crests or umbilicus) should be utilized or ink marks made on the skin to indicate where measurements should be taken.

- Take measurements with the patient in the same position each time.

- Expect to see a gradual decrease in abdominal girth with appropriate treatment to the level it was before ascites developed.
  (Remember, however, that abdominal girth measurements give only an estimate of the degree of ascites since gaseous distention of the intestine can vary from measurement to measurement.) Percussion of the abdomen should help confirm that ascites is reduced.

Measure Daily Weights

- Use the same scales each time.

- Measure weight at the same time each day (preferably before breakfast and after voiding).

- Expect to see a 1-lb to 2-lb weight loss/day with appropriate sodium restriction and diuretic therapy (until stabilized). An even greater loss may occur if generalized edema was present.

Measure Intake-Output Daily

- Expect to see urinary output exceeding fluid intake during the diuresis of excess body fluids.

Monitor Vital Signs at Regular Intervals

- Look for hypotension and increased pulse rate with weak volume, when large doses of diuretics or severe dietary sodium restriction are employed.

- Be especially alert for symptoms of hypovolemia (such as hypotension, rapid, weak pulse, and pallor) following the removal of a large amount of fluid by paracentesis. (Removal of as little as 1000–1500 ml may precipitate hypotension in some patients.)

(continued)

- Expect to see improved respirations as ascitic fluid accumulation diminishes.

Look for Signs of Hypokalemia

Particularly in patients receiving potassium-losing diuretics look for

- Muscle weakness and flaccidity
- Sluggish bowel activity, gaseous intestinal obstruction
- Irregular pulse

Look for Signs of Hyperkalemia

In patients receiving potassium-sparing diuretics look for

- Intestinal colic
- Diarrhea
- Irregular pulse, bradycardia

Look for Signs of Hyponatremia

Particularly after paracentesis or large doses of diuretics or severe sodium restriction look for

- Nausea and vomiting
- Abdominal cramping
- Hypotension

Look for Signs of Hyperammonemia

Especially when the patient is bleeding into the gastrointestinal tract or is constipated look for

- Somnolence
- Lethargy
- Loss of memory
- Slow, slurred speech
- Hepatic flap (a spasmodic flexion and extension at the wrists and fingers, as well as lateral finger twitching; can be elicited in susceptible patients by elevating the arms, hyperextending the hands, and spreading the fingers)
- Personality change
- Confusion, disorientation
- Convulsive seizures
- Coma

Look for Signs of Hypocalcemia and Hypomagnesemia

(especially hypomagnesemia when the precipitating cause of cirrhosis is alcoholism)

- Increased neuromuscular irritability
- Paresthesias
- Positive Chvostek and Trousseau signs (see Chap. 2)

Monitor Bowel Movements

Constipation must be detected early and treated to help prevent hyperammonemia.

Monitor the Patient's Ability to Hear When Neomycin is Administered

Particularly monitor when neomycin is given in conjunction with furosemide (Lasix) or ethacrynic acid (Edecrin). Be alert for signs of hearing difficulties.

Acute Pancreatitis

Causes and Pathologic Changes

- Pancreatitis may be associated with gallstones, excessive alcohol consumption, direct trauma, and hyperlipoproteinemia, and with a number of drugs (such as thiazides, furosemide, oral contraceptives, and isoniazid).
- Pathologic changes include edema and cellular infiltration of the acinar cells, hemorrhage from necrotic blood vessels, and intrapancreatic and extrapancreatic fat necrosis.[4] Part of all of the pancreas may be involved.

Clinical Signs

Clinical signs of acute pancreatitis can include

- Abrupt onset of steady and severe epigastric pain (often with radiation to the back and flanks)
- Restlessness due to pain (patient often assumes a sitting position and leans forward to obtain relief)
- Nausea and vomiting
- Diaphoresis
- Abdominal tenderness (usually without guarding and rigidity) and distention (bowel sounds may be absent in association with paralytic ileus)
- Diarrhea
- Fever (38.4° C–39° C [101.1° F–102.2° F])
- Laboratory abnormalities
  1. Elevated serum amylase and urine amylase levels
  2. Elevated hematocrit (early) due to fluid volume deficit (FVD); may fall below normal after fluid replacement if hemorrhagic pancreatitis is present
  3. Elevated blood urea nitrogen (BUN) due to FVD
  4. Serum glucose may be normal, elevated, or decreased.

- Hyperglycemia and abnormal glucose tolerance curves occur in 50% of the cases.[5]
- If beta cells (which produce insulin) are affected by tissue destruction, hyperglycemia develops; if alpha cells (which produce glucagon) are affected, hypoglycemia will develop.[6]
- Chronic pancreatitis may produce destruction of both beta and alpha cells and thus cause the development of brittle diabetes mellitus.
  5. White blood cell (WBC) count is often elevated, with a shift to the left (reflecting pancreatic inflammation).
  6. Serum triglyceride levels may be elevated.
  7. Alkaline phosphatase, bilirubin, and serum glutamic-oxaloacetic transaminase (SGOT) are elevated due to biliary obstruction.
- Shock (particularly if severe necrotizing hemorrhagic pancreatitis is present)
  1. Tachycardia and hypotension
  2. Cool, clammy skin
  3. Decreased renal perfusion (predisposes patient to acute tubular necrosis and renal failure)
- Predisposition to pulmonary complications (such as pneumonia, pleural effusions, pulmonary edema, pulmonary emboli, and atelectasis)
  Contributing factors may include the following:[7]
  - Immobility and retained secretions
  - Release of pancreatic enzymes into the systemic circulation, which may cause damage to pulmonary tissues
  - Pleural effusions (a result of inflammation from pancreatic enzymes and extravasation of fluids)
  - Overzealous fluid replacement therapy

Associated Fluid and Electrolyte Disturbances

Major fluid and electrolyte problems associated with acute pancreatitis include the following:

- FVD
  1. FVD in acute pancreatitis is due to a loss of fluids through vomiting, nasogastric suction, diaphoresis, and the escape of pancreatic secretions into the peritoneal space.
  2. Hypovolemia, if untreated, may become severe enough to decrease perfusion of vital organs (particularly if the patient is losing blood from hemorrhagic pancreatitis). A most dreaded com-

plication of hypovolemia is acute tubular necrosis (ATN) and renal failure.

• Calcium deficit

1. Calcium deficit may be due to binding of calcium in areas of fat necrosis, caused by the action of pancreatic enzymes on peritoneal fat.

2. It is thought that parathyroid hormone (PTH) secretion is altered in patients with pancreatitis.[8] For reasons that are still unclear, it appears that the feedback mechanism that would stimulate the release of calcium from the bone to maintain serum levels is not operative in pancreatitis.

3. A serum calcium level less than 7 mg/100 ml is a poor prognostic sign.[9]

• Hyponatremia and hypokalemia may follow loss of gastric fluid from vomiting, nasogastric suction, and extravasation of fluid into the peritoneal cavity.

• Magnesium deficit may follow vomiting, gastric suction, and diarrhea in patients with pancreatitis, particularly if they are malnourished.

• Metabolic alkalosis may follow the loss of chloride from vomiting and gastric suction (hypochloremic alkalosis).

• Respiratory alkalosis may follow hyperventilation from anxiety and pain.

• Metabolic acidosis may follow ATN and renal failure (caused by severe hypovolemia).

• A decreased serum albumin level may result from the loss of protein into the peritoneal cavity (secondary to inflammation of the pancreas and surrounding tissues).

Medical Management

• Usually an attempt is made to treat pancreatitis by medical measures before surgical intervention is undertaken.

• Measures are instituted to prevent stimulation of pancreatic secretions and thus reduce inflammation, tissue destruction, and loss of secretions into the peritoneal space. These may include the following:

1. Maintenance of fasting (food substances in the stomach stimulate pancreatic secretions)

2. Insertion of a nasogastric tube and connection of the tube to suction to remove gastric secretions (hydrochloric acid in gastric fluid is a chemical trigger to pancreatic secretions)

3. Administration of antacids to alkalize gastric secretions and administration of cimetidine (Tagamet) to decrease gastric hydrochloric acid secretion

4. Maintenance of bed rest (keeps the metabolic rate at a minimum and thus decreases pancreatic stimulation)

- Parenteral fluid replacement is undertaken to correct fluid losses from vomiting, nasogastric suction, sweating, and pooling of fluid in the peritoneal space. Both crystalloids (electrolyte solutions) and colloids (plasma, albumin, and blood) may be used. Fluid replacement therapy is based on the following:

  1. Urinary output
  2. Serum electrolyte values
  3. BUN, hematocrit, and albumin levels
  4. Other clinical parameters such as blood pressure, pulse rate, skin turgor, and mucous membrane moisture
  4. Pressure readings from central venous, arterial, and pulmonary artery catheters (as necessary) to monitor vascular volume and cardiac status

- When necessary, hypocalcemia is treated with intravenous calcium salts (such as calcium gluconate, calcium gluceptate, or calcium chloride).

  1. Calcium preparations should be given carefully while electrocardiographic patterns are monitored, since calcium administration has been known to precipitate cardiac arrest.
  2. Intravenous calcium administration should not exceed 0.7 to 1.5 mEq/min.[10]
  3. Care must be taken to avoid extravasation of the drug, since calcium causes tissue sloughing.
  4. Great care must be taken when calcium is administered to a digitalized patient, since calcium potentiates digitalis and may precipitate toxicity.

- Hypomagnesemia may require the intravenous administration of magnesium sulfate. It should be remembered that it is difficult to correct a low serum calcium level when hypomagnesemia is present.

- Transient hyperglycemia may produce wide swings in serum glucose levels and thus must be cautiously managed by monitoring serum glucose levels. (Serum glucose levels are better indicators of insulin need than are urinary glucose levels.)

- Pain is managed by analgesics (usually meperidine) and measures to reduce pancreatic stimulation (such as initiation of fasting and gastric suction).

Nursing Considerations in the Management of Fluid and Electrolyte Disturbances in Patients With Acute Pancreatitis

- Carefully measure and record fluid intake and output from all routes.
- Be alert for signs of hypovolemia.

1. Urinary output less than 30 ml/hr
2. Urinary specific gravity greater than 1.020
3. Systolic blood pressure less than 80 mm Hg
4. Pulse rate greater than 100 beats/min
5. Respiratory rate greater than 24/min
6. Decreased skin turgor
7. Dry mucous membranes
6. Central venous pressure less than 4 cm water
7. Pulmonary artery capillary pressure less than 6 mm Hg
8. Cardiac output less than 4 liters/min

- Maintain correct position and patency of the nasogastric tube.

    1. It is desirable that the tip of the tube be well into the stomach (near the pylorus); if the tip rests high in the stomach, hydrochloric acid may escape through the duodenum and stimulate pancreatic secretions.

    2. Since most gastric tubes are radiopaque, their position can be confirmed by x-ray films. Once correct position is established, the tube should be marked and every effort made to retain its proper positioning.

    3. Patency of the tube should be maintained by the introduction of approximately 20 ml of isotonic saline every 2 hours (the irrigating fluid should be withdrawn). Viscous gastric secretions may necessitate more frequent irrigations. Between irrigations, the tube is usually hooked to low intermittent suction.

- Be aware that liquid antacids may be prescribed by the physician for administration through the nasogastric tube.

    It is necessary to follow the antacid administration with enough saline to clear the tube and thus maintain patency. The tube is usually clamped for 30 minutes and then reconnected to suction for 30 minutes. A test of gastric *p*H may then be indicated; the physician may attempt to keep the *p*H of gastric fluid above 3.5 (because reduction of hydrochloric acid content reduces pancreatic stimulation).

- Monitor serum levels of electrolytes; also monitor albumin, BUN, hemoglobin, and hematocrit levels. Be alert for serious abnormalities. (See Chap. 3 for normal values and the significance of variations.)

- Be alert for signs and symptoms of likely electrolyte disturbances.

    1. Calcium deficit

        - Note complaints of numbness or tingling in the extremities or circumoral region.

        - Observe for latent tetany by testing Trousseau's and Chvostek's signs (see Chap. 2).

        - Be alert for laryngospasm and convulsions (signs of severe hypocalcemia).

        - Look for a prolonged Q–T interval on electrocardiograms.

        - See Chapter 9 for a more thorough discussion of hypocalcemia.

(continued)

2. Magnesium deficit
   - Observe for tremors, confusion, hallucinations, hypotension, tachycardia, and positive Trousseau's and Chvostek's signs.
   - See Chapter 10 for a more thorough discussion of hypomagnesemia.
3. Potassium deficit
   - Observe for arrhythmias, weakness, gastrointestinal hypomotility, and paresthesia
   - See Chapter 8 for a more thorough discussion of hypokalemia.

- Assess pain by considering its location, severity, duration, and precipitating events. Also, determine whether other symptoms occur with the pain.

  Be aware that subtle changes in the nature and frequency of pain may be indicative of ensuing complications.

References

1. Krupp M, Chatton M: Current Medical Diagnosis and Treatment—1980, p 398. Los Altos, Lange Medical Publications, 1980
2. Leonard B, Redland A: Process in Clinical Nursing, p 177. Englewood Cliffs, Prentice-Hall, 1980
3. Ibid
4. Krupp and Chatton, p 409
5. Ibid
6. Evans M, Shinert J, Gross J: Acute Pancreatic and Hepatic Dysfunction, p 26. Bethany, CT, Fleschner Publishing, 1981
7. Ibid, p 81
8. Robertson G, Moore E, Switz D et al: Inadequate parathyroid hormone response in acute pancreatitis. N Engl J Med 294:512–516, 1976
9. Krupp and Chatton, p 410
10. Nursing '83 Drug Handbook, p 531. Springhouse, PA, Intermed Communications, 1983

chapter 20

Diabetic Ketoacidosis and Nonketotic Hyperosmolar Coma

Deficiency of insulin can lead to diabetic ketoacidosis (DKA) or nonketotic hyperosmolar coma (NKHC). These conditions are discussed next.

Diabetic Ketoacidosis

- DKA is a medical emergency that accounts for approximately 14% of all hospital admissions for diabetics; it is the most common cause for hospital admissions of diabetics under 20 years of age.[1] A 1000-bed community hospital may average one case of DKA every week.[2]

- Although improved treatment of DKA has significantly reduced the mortality associated with this disease, there is still at least a 5% to 10% mortality rate reported from most medical centers.[3]

- DKA may be precipitated by a number of factors, most of which are associated with poor understanding of the disease by the diabetic. For example, the patient may stop taking insulin during an illness when food intake is diminished because he fears a hypoglycemic reaction. Or, he may be unaware of the need for extra insulin during an acute infection or severe emotional stress.

- Health care personnel should remember that diabetics frequently need extra insulin during acute stressful situations such as surgical emergencies, severe trauma, or myocardial infarction.

Metabolic Consequences of Insulin Lack

Hyperglycemia and Hyperosmolality

- Lack of insulin leads to hyperglycemia primarily by promoting
    1. Under-utilization of glucose. (Recall that cells are relatively impermeable to glucose in the absence of insulin.)
    2. Excessive production of glucose from fats and amino acids by the liver (gluconeogenesis).

        Because of these two processes, the blood glucose level rises markedly and increases plasma osmolality. One becomes aware of the influence of glucose on plasma osmolality when considering the formula for calculating plasma osmolality

$$pOsm = (Na^+ + K^+) + \frac{G}{18} + \frac{BUN}{2.8} = 280\text{--}295 \text{ mOsm/kg}$$

Example: Patient with hyperglycemia

$$
\begin{aligned}
Na^+ &= 139 \text{ mEq/liter} \\
K^+ &= 4 \text{ mEq/liter} \\
\text{Serum glucose} &= 1800 \text{ mg/100 ml} \\
BUN &= 30 \text{ mg/100 ml}
\end{aligned}
$$

$$pOsm = 2(139 + 4) + \frac{1800}{18} + \frac{30}{2.8} = 397 \text{ mOsm/kg}$$

- The elevated osmolality of extracellular fluid (ECF) associated with hyperglycemia produces cellular dehydration, a result of a shift of water from the cells to the ECF.

Osmotic Diuresis and Fluid Volume Deficit

- When the blood glucose level exceeds the renal threshold, glucose is spilled out into the urine. As the glucose spills into the urine, it takes water and electrolytes with it, increasing the urine volume. Thus, one would initially expect to see polyuria and a high urinary specific gravity (SG) due to the high glucose content.
- The polyuria eventually leads to fluid volume deficit (FVD). As the FVD increases, glomerular filtration rate and renal blood flow progressively diminish, causing the patient to become oliguric or even anuric in spite of marked hyperglycemia. (A potential danger of profound FVD is renal tubular damage and resultant acute renal failure.)

Electrolytes

Sodium

- Despite losses of water in excess of solute in DKA, serum sodium is usually below normal. If vomiting is present, the magnitude of hyponatremia is increased. Also, sodium moves into the cells as they become depleted of potassium, further lowering the plasma sodium level.

Potassium

- Probably the most important electrolyte disturbance that occurs in DKA is the marked deficit in total body potassium. Causes of potassium depletion include
    1. Starvation effect with lean tissue breakdown
    2. Loss of intracellular potassium
    3. Potassium-losing effect of aldosterone (stimulation of aldosterone secretion is produced by FVD)
    4. Loss of potassium with osmotic diuresis
    5. The presence of severe anorexia (no intake) and vomiting (increased loss)
- In the untreated patient with DKA, the *serum* potassium level may be normal or elevated even though there is a marked deficit of *total* body potassium. Factors that tend to elevate the serum potassium level in the untreated patient include
    1. Plasma volume contraction with oliguria, interfering with renal excretion of potassium
    2. Metabolic acidosis (which causes K^+ to shift into the extracellular compartment as H^+ is buffered intracellularly)

    The hyperkalemia is usually quickly alleviated by fluid replacement therapy and reestablishment of urine output.
- After institution of therapy, the serum potassium falls rapidly and usually reaches its lowest point within 1 to 4 hours. Reasons for the decreased serum potassium level at this time include
    1. Dilution by the intravenous fluids
    2. Increased urinary potassium excretion due to plasma volume expansion
    3. Formation of glycogen within the cells, involving utilization of potassium, glucose, and water (which causes further withdrawal of K^+ from the ECF)
    4. Correction of the acidotic state with reentry of potassium into the cells

Phosphorus

- Hypophosphatemia almost invariably occurs in the patient with DKA during treatment. One potentially serious consequence of phosphorus deficiency is decreased erythrocyte 2,3 DPG (diphosphoglycerate); a low level of 2,3 DPG may result in decreased peripheral oxygen delivery. Also, depressed myocardial function has been observed when the serum phosphate level is less than 2 mg/100 ml.[4]
- When the serum phosphate concentration goes below 0.5 mg/100 ml, serious disturbances in metabolism can result; seizures, respira-

tory failure, impaired leukocyte and platelet function, and gastrointestinal bleeding have been reported.[5]

Ketosis

- Insulin lack allows greater release of free fatty acids from peripheral fat stores and also may activate ketogenic pathways in the liver; the excess fatty acids are converted by the liver to ketones, resulting in ketosis. Contributing to the excessive buildup of ketones in the bloodstream is the decreased utilization of ketones by peripheral tissues (again, due to insulin lack).

- Impaired ketone metabolism, coupled with ketone overproduction by the liver, overloads the body's buffers, and metabolic acidosis ensues. (Recall that ketones are strong acids.) The anionic charge of bicarbonate is replaced by the negatively charged ketones.

- The type of metabolic acidosis that occurs in DKA is manifested by a fall in bicarbonate with a reciprocal rise in the anion gap (AG). AG can be calculated by the following formula:

$$AG = Na^+ - (HCO_3^- + Cl^-) = 12\text{--}15 \text{ mEq/liter}$$

Example: Patient with ketoacidosis

$Na^+ = 131$ mEq/liter
$HCO_3^- = 5$ mEq/liter $AG = 131 - (5 + 95) =$
$Cl^- = 95$ mEq/liter 31 mEq/liter

- The elevated AG in ketoacidosis is due to flooding of the extracellular fluid with ketones. AG is very important in diagnosing acid–base disturbances. For example, the vomiting that frequently accompanies DKA can superimpose a metabolic alkalosis on the preexisting ketoacidosis, sometimes normalizing the pH. However, the ketone anions remain elevated, disrupting metabolism. Measurement of AG reveals the abnormal level of ketones despite the pH.

- Ketones are made up of acetoacetic acid, β-hydroxybutyric acid, and acetone. Excessive ketosis leads to ketonuria, and even excretion of volatile acetone from the lungs (resulting in the classic "fruity" odor of the breath associated with DKA).

- It is not unusual for the plasma pH to drop to 7.25 or below and for the bicarbonate level to drop to a level of 12 mEq/liter or less. Possibly the greatest risks of prolonged uncorrected acidosis are decreased cardiac function, arrhythmia, and impaired hepatic handling of lactate.[6]

- The body attempts to compensate for the metabolic acidosis associated with DKA by means of the kidneys and lungs. The kidneys eliminate hydrogen ions and conserve bicarbonate ions (resulting in decreased urinary pH). The lungs attempt to lighten the acid load by blowing off extra carbon dioxide by deep relatively rapid respirations (thus decreasing the PCO_2).

- The expected fall in PCO_2 for compensation of metabolic acidosis can be calculated by the following formula:

$$\text{Expected } PCO_2 \text{ (mm Hg)} = 1.5(HCO_3{}^-) + 8 \pm 2$$

Example: The expected PCO_2 in a patient with a bicarbonate level of 12 mEq/liter would be between 24 and 28 mm Hg.

$$PCO_2 = 1.5(12) + 8 \pm 2 = 24\text{--}28 \text{ mm Hg}$$

- A fall below the calculated amount indicates a superimposed respiratory alkalosis; failure of the PCO_2 to decrease to the expected level indicates a complicating respiratory acidosis (a dangerous combination).

Clinical Manifestations

It is important that the nurse be aware of common clinical manifestations of DKA; they are summarized in Table 20-1.

Differentiation Between Diabetic Coma and Hypoglycemic Coma

- It is imperative that the presence of hypoglycemia be ruled out before insulin is given. Hypoglycemia is characterized by anxiety, sweating, hunger, headache, dizziness, double vision, twitching, convulsions, nausea, pale wet skin, dilated pupils, normal breathing, and normal blood pressure.
- When in doubt, it is advisable to administer 50 ml of $D_{50}W$ intravenously. If hypoglycemia is the problem, the patient's condition will quickly improve; if ketoacidosis is the problem, this small amount of dextrose will do no harm.

(*Text continues on p. 260.*)

Table 20-1
Clinical Manifestations of DKA and Their Probable Causes

Clinical Manifestation	Probable Cause
Glycosuria	Blood glucose level exceeds renal threshold, causing glucose to spill out into the urine
Polyuria (initially) with high SG	Osmotic diuretic effect of hyperglycemia; high renal solute load
Polydipsia	Thirst due to cellular dehydration (cells become dehydrated when water is drawn from them by the hypertonic ECF)

(continued)

Table 20–1
Clinical Manifestations of DKA and Their Probable Causes (continued)

Clinical Manifestation	Probable Cause
Ketonemia and ketonuria	Excessive accumulation of ketones in the bloodstream causes them to spill out into the urine
Tiredness, muscular weakness	Lack of carbohydrate utilization; hypokalemia
Increased longitudinal furrows in tongue (poor tongue turgor)	FVD
Poor skin turgor, dry mucous membranes	FVD
Anorexia, nausea, and vomiting	Follows onset of DKA; interferes with fluid intake (hastening the development of FVD)
Gastric dilatation	Neuropathy (possibly) Water and electrolyte loss (possibly)
Abdominal pain (can simulate appendicitis, pancreatitis, or other acute abdominal problems)	Apparently is secondary to the DKA *Note*: Anorexia, nausea, and vomiting precede the abdominal pain when it is due to DKA—this is in contrast to most surgical emergencies, in which the pain usually occurs first
Deep "air hunger" respirations (Kussmaul)	Compensatory mechanism to increase plasma pH by the elimination of large amounts of CO_2 from the lungs
Cherry-red skin and mucous membranes	Marked peripheral vasodilatation associated with ketosis
Postural hypotension	Changing the patient's position from supine to sitting causes more than a 10 mm drop in the systolic blood pressure—eventually the blood pressure becomes low even in the supine position
No fever—in fact, hypothermia is often present (if fever is present, it is almost always associated with the precipitating factor of the DKA)	FVD
Acetone breath odor (similar to that of overripe apples)	Acetone content of body is increased; since acetone is volatile, some of it is vaporized in the expired air. *Note*: Vomitus odor to the breath frequently obscures this finding
Oliguria or anuria (late)	FVD causes decreased renal blood flow and decreased GFR *Note*: Before assuming that urine formation is scanty, check for a distended, atonic bladder

(*continued*)

Table 20–1
Clinical Manifestations of DKA and Their Probable Causes (continued)

Clinical Manifestation	Probable Cause
Acute weight loss	FVD (the amount of acute weight loss is a good approximation of the magnitude of the fluid loss)
Decreased level of consciousness (Patients with DKA are frequently stuporous, unresponsive, and, at times, in frank coma)	Level of consciousness correlates best with the level of hyperglycemia (and plasma osmolality) at the time of admission

Laboratory Findings

Blood sugar (Normal is 80–120 mg/100 ml) Elevated above normal, often 400–600 mg/100 ml; may be as high as 2000 mg/100 ml	Faulty glucose metabolism causes glucose to accumulate in the bloodstream (lack of insulin decreases glucose uptake by most cells and also increases gluconeogenesis in the liver)
Plasma osmolality (Normal is 280–295 mOsm/kg) Elevated in DKA	Due primarily to hyperglycemia
Bicarbonate (Normal plasma HCO_3^- is 24 mEq/liter) Usually decreased below 15 mEq/liter; may be as low as 5 mEq/liter, (or even unmeasurable) in severe DKA	Excessive ketonic anions in the bloodstream cause a compensatory drop in bicarbonate anions
Arterial *p*H (Normal is 7.38–7.42) Usually < 7.25 in DKA; may be as low as 6.8 in severe uncompensated states	Associated with the metabolic acidosis caused by ketosis
High anion gap (AG) acidosis (Normal AG is < 12–15 mEq/liter) Elevated in DKA	Due to excessive ketones in the bloodstream
Blood Urea Nitrogen (Normal is 20 mg/100 ml) Elevated in DKA to an average of 40 mg/100 ml	Partially due to increased protein metabolism with increased hepatic production of urea (because of insulin lack) Also related to FVD and diminished GFR
Polymorphonuclear leukocytosis Usually 15,000–20,000/mm^3 in DKA; may be as high as 90,000/mm^3 (with a shift to the left)	May be secondary to FVD, acidosis, and adrenocortical stimulation

Treatment

Fluid Replacement

- Adequate and prompt rehydration is vital, particularly if the patient is in shock. The initial fluid of choice in most instances is isotonic saline (0.9% NaCl) or lactated Ringer's solution. (Hypotonic fluids will not expand the extracellular fluid as quickly.)
- Isotonic saline is usually infused at the rate of approximately 1 liter/hr in adults without cardiac failure until the blood pressure is stabilized and the urine output is 1 ml to 2 ml/min.[7] If there is a question as to the adequacy of the patient's cardiovascular status, a central venous pressure (CVP) line may be necessary to help gauge the best rate of fluid replacement. (Nursing responsibilities associated with monitoring CVP are discussed in Chap. 4.)
- After the first 2 to 3 liters of isotonic saline have been administered, the intravenous fluid may be changed to ½-normal saline (0.45% NaCl) to dilute the hyperosmolar plasma and provide free water for renal excretion.
- When the blood pressure is stabilized and the urine output is adequate, the rate of fluid administration can be reduced to 1 liter every 2 hours and then to 1 liter every 3 to 4 hours. (Of course, the volume and rate of fluid administration must vary with the clinical status of each patient.) As soon as oral intake is adequate, intravenous fluids can be discontinued.
- If the patient is stuporous, an indwelling urinary catheter must be inserted until lucidity is regained; it is imperative that hourly urine volume be closely monitored in the early phase of fluid replacement therapy. (Of course, a catheter should not be used unless absolutely necessary because of the danger of urinary tract infection.)
- Glucose is not administered in initial intravenous fluids when the blood sugar is already markedly elevated; to do so would only potentiate the existing hyperglycemia and enhance glycosuria with renal salt and water loss.
- However, carbohydrate replacement is important in the overall treatment of DKA. It should be remembered that the total body stores of carbohydrates are greatly diminished in spite of hyperglycemia.
- Glucose solutions should be started when the blood sugar has fallen to approximately 250 mg/100 ml. A typical fluid might be 5% dextrose in 0.45% NaCl, or D_5W.
- It is important that the patient *not* be allowed to become hypoglycemic. A precipitous fall in blood sugar might produce cerebral edema because the glucose concentration in the brain decreases at a slower rate than that of the plasma, allowing fluid to be drawn into brain tissue. Not allowing the blood sugar to drop rapidly below 250 mg/100

ml should prevent the problem. (Most reported cases of brain edema associated with DKA treatment have been associated with a rapid decrease in blood sugar to levels close to 100 mg/100 ml.)

Insulin Administration

- Rapidly acting insulin may be given intravenously, intramuscularly, or subcutaneously in the treatment of DKA. The continuous intravenous (IV) route is best for the hypotensive patient because absorption from poorly perfused muscle and fat depots may be erratic.

- The aim of insulin therapy is to give enough insulin to correct the problem without subjecting the patient to the risk of hypoglycemia.

- Studies have demonstrated that large insulin doses are generally no more effective in correcting DKA than are small doses. The insulin dose for continuous IV infusion is usually 4 to 8 units/hr; for the hourly intramuscular (IM) or subcutaneous (SC) routes the dose of insulin is usually 5 to 10 units/hr.

- These low doses have been shown to lower blood sugar smoothly, improve ketosis, and repair acidosis at rates indistinguishable from those obtained by higher-dose regimens.[8] In addition, patients receiving low-dose insulin therapy are less likely to develop hypoglycemia and hypokalemia than are patients receiving large doses.

- Of course, all patients must be closely monitored for lack of response to insulin therapy; if necessary, more aggressive insulin therapy must be instituted. Some patients may be resistant to insulin and may require hundreds of units of insulin before a response can be demonstrated.

- There has been some concern that with the low-dose constant infusion of insulin there would be substantial loss of insulin by adsorption to containers or tubing. This should not significantly affect delivery, however, if sufficiently high concentrations of regular insulin are added to the container and the tubing is adequately rinsed with the solution. According to Peterson, the insulin-binding effect can be offset almost entirely by letting 50 ml to 100 ml of a solution containing 50 units of regular insulin/liter run through the tubing before administering it to the patient.[9] Apparently this high concentration and "washout procedure" allows for saturation of the insulin-binding sites. The exact amount of insulin adsorbed by intravenous infusion containers and tubing is dependent on (1) the type of IV infusion system, (2) the concentration of insulin in the infusion, (3) the rate of flow of the infusion, and (4) the type of solution containing the insulin.

- A dose sufficient for most patients can be achieved by administering 1 ml/min of a liter of isotonic saline containing 100 units of regular insulin (after 50 ml have been run through the tubing). More accurate delivery of the intravenous solution can be assured by infusion through an infusion pump; if one is not available, the solution should be administered through a volume-controlled administration set.

- The IM and SC routes are suited only for patients with normal blood pressure and tissue perfusion. Sometimes patients with DKA are initially given regular insulin (0.2 units/kg of body weight) by deep IM injection. Thereafter, a dose of 5 to 10 units may be given each hour until the blood sugar drops to 250 mg/100 ml and the DKA clears. At this time, insulin is administered SC as indicated by blood glucose levels or the glucose content of second voided specimens.

Electrolyte Replacement

Potassium

- Recall that hypokalemia is probably the most important electrolyte disturbance that occurs in DKA; therefore, potassium replacement during therapy is imperative. Paradoxically, however, *elevation* of serum potassium may be present before fluid therapy is instituted and adequate urine flow is established. (It is important to remember that serum potassium concentration does not accurately reflect total body potassium in the presence of acidosis; this does not hold as acidosis is corrected.)

- As stated earlier, the first 2 to 3 liters of fluid administered for the treatment of DKA usually consists of plain isotonic saline; potassium is not added until adequate urine flow has been established.

- Once good urine flow is established and the blood urea nitrogen (BUN) level is near normal, it is usually safe to add 20 mEq to 40 mEq of potassium to a liter of ½ normal saline (0.45% NaCl). In the presence of a low serum potassium level, potassium is usually administered at a rate of 20 mEq/hr IV (if the urinary flow is adequate). This amount of potassium will usually prevent clinical manifestations of hypokalemia. If the urine flow is questionable, potassium is infused at somewhat less than 20 mEq/hr (along with careful ECG monitoring). Since the intravenous administration of potassium is always associated with the risk of hyperkalemia, it is wise to monitor the serum potassium level at 1- or 2-hour intervals and utilize serial ECG tracings. (Rules for safe potassium administration are discussed in Chap. 4.)

- When signs of severe hypokalemia are present (such as acute respiratory paralysis, muscle weakness, cardiac arrhythmias, or ileus), it may be necessary to administer potassium at a rate of 40 mEq/hr.[10] Great care must be exercised in administering such large doses; clinical response must be continuously assessed by means of electrocardiography.

- Potassium may be administered as potassium chloride, potassium phosphate, or potassium acetate. Some authorities prefer the potassium phosphate salt because phosphate is also depleted in DKA; administration of the phosphate ion allows it to move into the cells with potassium to replace intracellular deficits of both electrolytes.

- IV Potassium supplementation is continued until the plasma potassium is within the normal range. Oral potassium-containing fluids (such as orange juice) may be administered once the patient is lucid, and free of nausea, vomiting, and gastric dilatation.

Sodium

- As mentioned earlier, the initial IV fluid frequently used for the treatment of the hypotensive patient with DKA is isotonic saline (0.9% NaCl). This solution not only expands the extracellular fluid, it supplies extra sodium to correct any sodium deficit.
- It should be remembered, however, that the patient with DKA has lost relatively greater amounts of water than of sodium. Thus, eventually, hypotonic fluids such as ½ normal saline (0.45% NaCl) must be administered. Hypotonic solutions provide free water to correct cellular dehydration.

Phosphorus

- As stated earlier, there are significant losses of phosphate with DKA. Some authorities favor replacing at least a part of the lost potassium as the buffered phosphate salt. To do so replaces both potassium and phosphate and allows for correction of intracellular deficits of both. It is imperative that significant renal failure be ruled out before phosphate is administered IV; serum phosphate levels should be monitored to prevent possible hyperphosphatemia. When oral intake is tolerated, skim milk is a good source of phosphorus. Occasionally phosphate is replaced orally in the form of Phospho-Soda, 5 ml, tid.
- It is thought that phosphate replacement accelerates the recovery of reduced red blood cell 2,3 DPG levels, thereby decreasing hemoglobin–oxygen affinity and improving tissue oxygenation.

Bicarbonate

- There is disagreement as to the necessity for alkali therapy in the treatment of DKA. Some feel that bicarbonate replacement is indicated in severe ketoacidosis; others have argued that bicarbonate replacement is unnecessary since insulin therapy reverses the biochemical abnormalities of DKA (including the bicarbonate deficit).
- Those who favor bicarbonate replacement feel it is necessary when severe acidosis is present (such as with an arterial pH < 7.15 and bicarbonate level < 8 mEq/liter).
- Profound acidosis may lead to life-threatening pulmonary edema, hypotension, and shock.[11] Therapy might start with the administration of 44 to 88 mEq of sodium bicarbonate in the adult and 0.5 to 1.0 mEq/kg in children.[12]

- Bicarbonate should not be given to completely restore the serum bicarbonate to normal levels since too rapid correction of acidosis can induce hypokalemia. Some authorities theorize that bicarbonate replacement is not only unnecessary, it may actually be detrimental in that rapid correction of acidosis might contribute to impaired tissue oxygenation. (Hemoglobin's affinity for oxygen is decreased in acidosis, allowing hemoglobin to unload more oxygen at the tissue level; conversely, hemoglobin's affinity for oxygen is increased in alkalosis, interfering with oxygen release.)

Nursing Assessment of Response to Therapy

- Monitor the patient with DKA closely during treatment. Both medical and nursing interventions must be based on the patient's response to therapy.
- Maintain vigilance over the clinical and the laboratory status of the patient during the acute phase; only constant observation will permit the critical judgments necessary to provide optimal therapy.
- Maintain a *diabetic flow sheet* during the acute phase to record sequentially the vital signs, laboratory data, and treatments. Data on the flow sheet is usually recorded hourly at first and may include the following:
  1. Temperature, pulse, and respirations
  2. Blood pressure(supine and sitting)
  3. Hourly urine volume
  4. Glycosuria and ketonuria
  5. Blood glucose level
  6. Plasma ketone level
  7. Plasma pH
  8. Electrolyte levels
  9. Fluid intake (amount and kind)
  10. Electrolyte intake
  11. Level of consciousness
  12. ECG findings
  13. Deep tendon reflexes

Degree of Fluid Volume Deficit

Vital Signs

- Monitor the degree of FVD by assessing the blood pressure, neck veins, hourly urine volume, skin and tongue turgor, and body weight.
- Measure the blood pressure in the supine position and then in the sitting position. A drop in systolic pressure by more than 10 mm Hg

on position change is indicative of FVD. Eventually, the volume-depleted patient is hypotensive in the supine position also. Elevation of the systolic pressure toward normal is indicative of fluid volume expansion and is, of course, desirable.

- Observe for collapse of the neck veins with raising the head off the bed (since this is also a sign of FVD).
- Measure pulse rate and pulse volume. Remember that fluid volume deficit is associated with tachycardia as the heart attempts to compensate for a diminished plasma volume by pumping faster. Thus, a decrease in pulse rate, associated with fluid volume replacement, is a favorable sign. Pulse volume should become less weak and thready and more normal with plasma volume expansion.

Urine Output

- Measure hourly urine volume because it is one of the best parameters for gauging adequacy of fluid volume replacement. (The desired urinary volume/hr for an adult is 30–50 ml.) Urine volume should be measured in a device calibrated for accurate readings of small amounts.

  Failure of the DKA patient to excrete a normal urine volume is related to the fall in glomerular filtration rate (GFR) and renal blood flow associated with FVD. It is important that fluid volume replacement be prompt and of sufficient magnitude to prevent renal tubular damage (a potential result of profound FVD).

- Look for a rise in hourly urine volume after institution of appropriate fluid replacement therapy.

  Failure of urine volume to increase can indicate either (1) inadequate rate and volume of fluid replacement, or (2) renal tubular damage. If the problem is the latter, the patient must be treated for acute renal failure. Urine of the patient with severe renal damage is usually of a fixed low SG regardless of the small urinary volume (as opposed to the high SG associated with scanty urine volume in the patient without renal damage).

- Be aware that an indwelling urinary catheter is indicated for stuporous patients to allow for accurate urine measurement. It is imperative to maintain meticulous aseptic technique to prevent secondary urinary tract infection.

Body Weight

- Weigh the patient with DKA on admission and try to ascertain his pre-illness weight in the history.

  It is important to know the patient's pre-illness weight in order to approximate the degree of FVD. For example, one might expect, at most, a 1 lb tissue weight loss per day in the patient who is not eating, and, at most, a ½-lb tissue weight loss per day in the patient who is eating poorly. Any acute weight loss larger than these amounts represents a fluid volume loss. (Recall that 2.2 lb is equivalent to 1 liter of fluid.) An acute weight loss of 5% of body weight constitutes a moderate FVD; an acute weight loss of 8% of more constitutes a severe FVD. It is not unusual for a patient with DKA to lose more than 10% of body weight during illness.

Plasma Osmolality

- Be aware that it is often difficult to determine the patient's pre-illness weight and thus other methods are required to more closely approximate the degree of FVD. The following formula, using the patient's plasma osmolality, is sometime used by the physician for this purpose:

$$\text{FVD (liters)} = \frac{\text{Patient's pOsm} - \text{Normal pOsm}}{\text{Normal pOsm}} \times \begin{array}{c}\text{liters of body fluid}\\ (0.6 \times \text{body} \\ \text{weight/kg})\end{array}$$

Example: Assume an adult patient has a plasma osmolality of 340 mOsm/kg and weighs 70 kg. (Use 280 mOsm/kg as the normal value.)

$$\text{Liters of body fluid} = 0.6 \times 70 = 42$$

$$\text{FVD (liters)} = \frac{340 - 280}{280} \times 42 = 9 \text{ liters}$$

Other Parameters

- Total and compare the 24-hour intake–output record. It is obvious that during treatment fluid intake must exceed fluid output until the deficit is corrected.
- Monitor tongue turgor since it is one of the best gauges of FVD (it is not influenced by age, as is skin turgor). Assess the tongue for excessive longitudinal furrows; as fluid volume replacement progresses, tongue turgor should improve. Skin turgor is usually a valid measure of hydration in all but the elderly and should be assessed periodically throughout the treatment phase.
- Assess the degree of firmness of the eyeballs. Recall that "soft eyeballs" are associated with severe FVD, indicating a deficit of intraocular fluid. This situation should improve with fluid volume replacement.

Degree of Hyperglycemia and Hyperosmolality

Blood and Urine Sugar Levels

- Be aware that the degree of hyperglycemia is best monitored by directly measuring the blood sugar level (either in venous or capillary blood) rather than by arriving at it indirectly by urine specimens.

A number of factors make urine glucose levels less reliable than blood sugar levels in monitoring the diabetic patient. For example, (1) renal threshold for glucose often increases with age, and (2) renal disease, which is common in diabetics, may alter the renal threshold for glucose. A number of other factors involving urinary glucose measurement must also be considered. A study by Feldman and Lebovitz using usual urine glucose testing methods reported a 23% incidence of falsely high results and a 33% incidence of falsely low results.[13]

- Record urine test results in percentages rather than in plusses because there is considerable variation among common urine tests. (Table 20-2.)

- Be aware that in addition to periodic laboratory measurements of blood sugar, it is highly desirable to use a product like Dextrostix (an enzyme test strip) to quickly monitor blood glucose levels from finger-stick samples. (Particularly helpful in the Eyetone reflectance colorimeter to register blood sugar levels from the test strip; Fig. 20-1.)

- Monitor blood sugar levels closely during insulin therapy to prevent a precipitous drop in the blood sugar, which can lead to cerebral edema. *The blood sugar should not be allowed to drop rapidly below 250 mg/100 ml.*

Sensorium

- DKA is associated with a variety of neurologic findings; depressed sensorium is the most frequent. According to one study, depression of sensorium occurred in 90% of the cases.[14]

- Whereas acidosis may undoubtedly affect the brain, diabetic coma seems to correlate best with the degree of hyperosmolality of body fluids.

- Depression of sensorium in DKA is usually readily responsive to corrective therapy and is one gauge of effectiveness of treatment.

Table 20–2
**Comparison of Degrees of Sugar Content
Indicated by Various Urine Tests**

Tes-Tape	Diastix	Clinitest (5-Drop Method)	Clinitest (2-Drop Method)
0% Negative	0% Negative	0% Negative	0% Negative
	$^1/_{10}$% Trace	¼ % Trace	¼ % Trace
$^1/_{10}$% (+)	¼ % (+)	½ % (+)	½ % (+)
¼ % (+ +)	½ % (+ +)	¾ % (+ +)	1% (+ +)
½ % (+ + +)	1% (+ + +)	1% (+ + +)	2% (+ + +)
2% (+ + + +)	2% (+ + + +)	2% (+ + + +)	3% (+ + + +)
			5% (+ + + + +)

Place the GLUCOMETER instrument on a flat surface, away from vibration or shock. Place switch in the ON position.

1. Remove DEXTROSTIX Reagent Strip from bottle and replace bottle lid.

2. Wipe the puncture area with an applicator saturated with alcohol. Allow alcohol to dry. After performing a puncture, allow a large drop of blood to form and wipe it away with a clean, dry cotton ball.

3. Allow another drop of blood to form, then press the "time" button on the GLUCOMETER instrument.

4. At the sound of the buzzer, freely apply a large drop of blood sufficient to cover the entire reagent area of a DEXTROSTIX Reagent Strip (cover the reagent area generously). Keep DEXTROSTIX in a horizontal (level) position to avoid spilling.

5. As the buzzer sounds a second time (signalling 60 elapsed seconds), wash blood off reagent area with a sharp stream of water from a wash bottle for two seconds. Direct the stream of water across the entire reagent area, and then blot on lint-free paper towel. Do not wash under faucet.

6. Lift Test Chamber Lid and insert reagent strip to the end of the Strip Guide with reagent area face down.

7. Close Test Chamber Lid and press "read" button. The blood glucose (sugar) value will appear within seconds on the digital display.

Figure 20–1. Procedure for blood glucose determination using Ames Dextrostix® and Glucometer® Reflectance Photometer. (Redrawn with permission of Ames Division, Miles Laboratories Inc, PO Box 70, Elkhart, IN)

- Besides assessing the patient's response to therapy, it is important to be aware of the level of consciousness concerning maintenance of an airway.

    An *adequate airway must be maintained* in the patient with a depressed sensorium, particularly when gastric dilatation is present (making aspiration of regurgitated stomach contents likely). A nasogastric tube should be inserted when gastric dilatation is present to minimize the danger of aspiration pneumonia. If the patient is comatose and there is evidence of airway obstruction, endotracheal intubation is indicated.

Degree of Ketoacidosis

- Monitor respirations, level of consciousness, skin color, acetone breath odor, and, of course, plasma pH, plasma ketone levels, and ketonuria to assess the degree of ketoacidosis.
- Look for labored, deep, costal and abdominal respirations because they reflect the lungs' attempt at compensation for metabolic acidosis by blowing off more carbon dioxide than normal. (More simply stated, respiration is stimulated by the acidemia, and the consequent decrease in $PACO_2$ helps bring the arterial pH back toward normal.)
- Be alert for a change from deep, rapid respiration to a rapid, shallow, gasping respiration; this may indicate a severe drop in blood pH (below 7) or impaired blood flow to the respiratory center because of FVD and circulatory collapse. When this occurs, the patient loses the compensatory action of Kussmaul breathing, and acidosis increases.
- Monitor the quality of respirations and the arterial pH at frequent intervals during the acute phase of DKA.
- Monitor the level of consciousness. Consciousness may be depressed with progressive acidosis and lethargy; disorientation and stupor may appear. These changes should be reversed with clearing of the ketoacidosis and hyperosmolality.
- Monitor skin color. A cherry-red color to the skin and mucous membranes is characteristic of DKA and is due to marked peripheral vasodilatation. An improvement in skin color is indicative of response to therapy.
- Monitor breath odor. The classic "fruity" breath odor ascribed to patients with DKA may be present, but it is unreliable in measuring the level of ketosis because it is frequently masked by the odor of vomitus.
- Monitor the presence of abdominal pain. Many feel that the abdominal pain commonly seen in DKA patients is secondary to the ketoacidosis itself. With clearing of the ketoacidosis, one would expect abdominal pain to subside. It is important to differentiate between the abdominal pain of DKA and that of an acute surgical

emergency. As stated earlier, under Clinical Manifestations, the abdominal pain in DKA is always preceded by anorexia, nausea, and vomiting. The patient with a surgical emergency frequently experiences pain first.

One must always keep in mind that an acute abdominal emergency (such as pancreatitis) can be present and, indeed, may have triggered the DKA. An intraabdominal emergency may be more difficult to detect if ileus or gastric distension, commonly seen in DKA, is present.

Electrolyte Derangements

Potassium Alterations

- Monitor serum potassium levels and ECG readings; be alert for hyperkalemia in the early phase. (Recall that an elevated serum K^+ level may be present before treatment.) Hyperkalemia usually responds rapidly to fluid replacement and insulin therapy.
- Look for signs of hypokalemia, especially in the first 2 to 4 hours into therapy because the potassium level falls rapidly with treatment.

1. Decreased serum potassium
2. Decreased deep tendon reflexes (be aware that deep tendon reflexes may also be due to peripheral neuropathy)
3. Abdominal distention
4. Cardiac arrhythmia
5. Muscle weakness
6. Ileus
7. Flat or inverted T waves; depressed ST segments; Q–T intervals often appear prolonged because of superimposed U waves on T waves.
8. Acute respiratory paralysis in severe hypokalemia

Phosphorus Deficiency

- Be alert for phosphorus deficiency; it usually becomes manifest within 4 to 12 hours after institution of therapy.

1. The clinical consequences of hypophosphatemia are not clear. Arthralgias, myopathies, seizures, acute respiratory failure, impaired leukocyte and platelet function, and gastrointestinal bleeding have been ascribed to hypophosphatemia.
2. When the serum phosphate concentration falls below 0.5 mg/100 ml, serious disturbances in metabolism and organ function may ensue. Depletion of phosphate has deleterious effects on the oxyhemoglobin dissociation curve and on cardiac output; therefore, phosphate depletion has the potential for interfering with tissue oxygenation and initiating lactic acidosis.

Patient Education

- Development of ketoacidosis in a previously diagnosed patient should not occur; when it does, it is frequently due to inadequate patient education.
- Ketoacidosis is not a sudden complication; it takes several hours to several days to develop. Because of this, there is sufficient time to take preventative measures provided the condition is recognized. Thus, the nurse should be sure that the patient and his family are familiar with the signs of ketoacidosis, and the physician should furnish specific directions for the patient to follow once early symptoms develop. Figure 20-2 is an example of a patient instruction sheet to prevent ketoacidosis during intercurrent illness.

Nonketotic Hyperosmolar Coma

- *Nonketotic hyperosmolar coma* (NKHC) primarily occurs in middle-aged to elderly diabetics, with the mean age reported in one study to be 62 years.
- Often the occurrence of NKHC is the initial harbinger of diabetes mellitus; in fact, as many as two-thirds of the patients who develop NKHC have not yet been diagnosed as having diabetes.
- The typical patient has a mild enough form of diabetes to be managed by oral hypoglycemic agents or diet alone.
- NKHC is found equally in males and females; as many as 85% have underlying renal or cardiovascular impairment, or both.
- The reported mortality rate ranges from 50% to 60%.
- While the literature reports that NKHC is only one-sixth as frequent as DKA, it is by no means uncommon. In fact, NKHC is becoming almost commonplace with the rising percentage of elderly patients receiving more aggressive forms of medical and surgical therapies.

Pathophysiology

- It has been postulated that in NKHC there is sufficient insulin to prevent ketosis but not enough to prevent hyperglycemia. The hyperglycemia results in hyperosmolality of the extracellular fluid (ECF), causing water to diffuse from the cellular to the extracellular space. The condition is made worse by the inability of the kidney to properly adjust water and electrolyte balance.
- Severe hyperglycemia causes osmotic diuresis with large losses of body fluids. Relatively more water than sodium is lost through the urine. Dehydration, hyperosmolality, and hyperglycemia are responsible for many of the clinical manifestations of NKHC.

(*Text continues on p. 274.*)

**This plan should be used *only during illness,*
not for day-to-day regulation of diabetes.**

1 **At the first sign of illness, check second-voided urine for glucose before each meal and at bedtime**

Frequent urine testing will give clear warning of diabetic ketoacidosis, which usually takes several days to develop. Clinitest[R] is the preferred method for testing urine glucose during illness.

If your urine test shows a glucose level of 2 percent or higher, then urinary ketones should be measured with Ketostix[R].

2 **Continue to take your usual insulin injection each day**

Your need for insulin usually is increased by another illness. If your usual insulin dose is not taken, your blood sugar level will rise quickly, even if you do not eat anything. So, even if you are nauseated and vomiting and cannot eat, *never omit your usual dose of insulin.*

---

3 **Take extra insulin if necessary**

If you feel ill *and* your urine test shows a 2 percent level of glucose or higher, then you need more insulin than usual.

When to take extra insulin

Before each meal and at bedtime:
- Check your second-voided urine test.
- Each time your glucose level is 2 percent or higher, you should take extra insulin at that time.

How much extra insulin to take

The amount of *extra* insulin you should take is calculated as follows:
- Add up the total number of units you usually take each day (be sure to include all mixtures or split dosages in the sum).
- Divide the total by 4, and administer this number of extra units of *regular* insulin, repeated before each meal and at bedtime, whenever your urine glucose level at that time is 2 percent or higher. (See example).

Figure 20–2. Patient instruction sheet: Prevention of ketoacidosis during intercurrent illness. (Redrawn from Benson R, Metz R: Diabetic ketoacidosis. Hospital Medicine, May 1979, p 34. ©1979. Reproduced from HOSPITAL MEDICINE, May 1979, with permission of Hospital Publications, Inc)

This is an example only. You will need to substitute your own numbers, according to your usual daily dose of insulin and your glucose levels when you test your urine.

If your usual daily injection is 28 units of NPH plus 4 units of regular insulin, your total daily dose is 32 units of insulin.

Divide 32 by 4:

$$4 \enclose{longdiv}{32} = 8 \text{ units}$$

SAMPLE daily administration of insulin when you are sick:
Before breakfast
   Always take your usual injection.   28 units NPH
                                          4 units regular
   Urine glucose level is 2%. Add   8 units regular
Before lunch
   Urine glucose level is 3%. Take   8 units regular
Before supper
   Urine glucose level is 2%. Take   8 units regular
At bedtime
   Urine glucose level is down to 1%.
   Do not take extra insulin this time.   0 units
        TOTAL FOR DAY   56 units

4   **If you are nauseated, substitute bland or liquid foods for your usual diet**

To avoid ketoacidosis or possible hypoglycemia, patients who take insulin must consume at least 150 grams of carbohydrate during each day.

What will satisfy this requirement?
   1½ quarts of a fluid containing sugar (such as gingerale, cola, or fruit juice) taken in small amounts over a 24-hour period.

If vomiting prevents even this intake, you will need hospital care

5   Notify your physician if:
- **Your condition gets worse**
- **Your illness lasts longer than 24 hours**
- **You continue vomiting**
- **Ketones persist in your urine, or**
- **You have any questions**

When you are ill, call your physician to let him know the results of your urine tests. It is far better to call early in the course of an illness than to wait until you become more seriously ill.

Your physician's telephone number: _____

Predisposing Factors

Many of the patients who develop NKHC have concurrent serious illness or are receiving medical treatments that can precipitate NKHC in susceptible persons. (See Contributing Factors to Development of NKHC in Susceptible Persons.)

Contributing Factors to Development of NKHC in Susceptible Persons

ILLNESSES

- Gram-negative infections (particularly pneumonia)
- Myocardial infarction
- Uremia
- Acute pyelonephritis
- Lactic acidosis
- Pancreatitis
- Heatstroke
- Gastrointestinal hemorrhage
- Severe burns
- Intracranial disorders
- Inadequate fluid intake for any reason (such as withholding fluids before surgery or for tests)

MEDICAL TREATMENTS

- Hyperosmolar tube feedings
- Hyperalimentation
- Peritoneal dialysis
- Medications
  1. Steroids
  2. Phenytoin (Dilantin)
  3. Diuretics, especially thiazides
  4. Cimetidine (Tagamet)
  5. Propranolol (Inderal)
  6. Chlorpromazine
  7. Immunosuppressive agents
  8. Mafenide

Clinical Manifestations

- The prodromal period is longer in NKHC (several days to several weeks) than in DKA (several hours to several days).
- The blood sugar tends to be higher than that seen in DKA; it is typically greater than 600 mg/100 ml, and possibly as high as 4800 mg/100 ml. Serum osmolality is high because of the hyperglycemia.

- Ketones in plasma and urine are absent or minimal. (Recall that patients with NKHC have sufficient insulin to prevent ketosis but not enough to prevent hyperglycemia.)
- The plasma bicarbonate level is normal or only slightly reduced; plasma pH is usually normal or only slightly low.
- Most striking in NKHC are symptoms of FVD.
    1. Dry skin and mucous membranes
    2. Poor tongue turgor (increased longitudinal furrows)
    3. Postural hypotension
    4. Acute weight loss
    5. Soft, sunken eyeballs
- Other symptoms of NKHC may be of a neurologic nature.
    1. Depressed sensorium
    2. Motor seizures occur in about 15% of patients (usually focal, but may be grand mal).
    3. Hemiparesis
    4. Aphasia
    5. Sensory defects
    6. Hyperreflexia
    7. Autonomic nervous system changes
- NKHC patients may often be misdiagnosed as having a cerebrovascular accident or having organic brain syndrome. Fortunately, most of the above NKHC changes revert to normal after correction of the plasma osmolality.
- Kussmaul respiration is not a feature of NKHC as it is in DKA; instead, patients with NKHC usually have rapid, shallow respirations.
- It has been reported that the average BUN is between 60 and 90 mg/100 ml in NKHC, as opposed to an average of 40 mg/100 ml in DKA, reflecting a more severe FVD in NKHC. (The elevated BUN is also indicative of the catabolic state associated with starvation.) In addition, serum creatinine is often elevated in NKHC, reflecting the common association of NKHC with underlying renal impairment.
- The serum sodium level in patients with NKHC is usually above normal although there is a deficit of total body sodium; the serum sodium level is higher than normal because of a relatively greater loss of water. Potassium and phosphate depletion syndromes are seen in NKHC, just as in DKA.

Treatment

Fluid Replacement

- Since the major problem in NKHC is severe FVD, a primary treatment concern is fluid replacement. In fact, it is impossible to develop NKHC if adequate hydration is maintained.

- A hypotonic electrolyte solution (such as 0.45% NaCl) is a logical fluid choice since the patient with NKHC has lost far more body water than electrolytes. If signs of hypovolemic shock are present, however, the initial fluid should be normal saline (0.9% NaCl) to expand plasma volume; after correction of shock, hypotonic fluids are administered.

Electrolytes

- Potassium replacement should be started as soon as adequate renal function is confirmed. As a rule, 20 mEq to 40 mEq of KCl can be added to each liter of fluid during the period of fluid replacement and insulin administration.
- Severe phosphate deficiency states should be treated by the administration of potassium phosphate; caution should be exercised because of the risk of causing hyperphosphatemia in patients with renal damage. (Recall that a substantial number of patients with NKHC have at least some degree of renal impairment.) (Hyperphosphatemia is described in Chap. 11.)

Insulin

Administration of insulin is of secondary importance to fluid replacement. Good responses have been demonstrated by the administration of low doses of intravenous regular insulin. (See the earlier discussion of IV insulin administration in the treatment of DKA.) As in DKA, care must be taken to avoid a precipitous fall in the blood sugar level below 250 mg/100 ml because of the danger of cerebral edema.

Prevention

- Vigorous efforts must be made to limit the increasing occurrence of NKHC by (1) proper patient education, and (2) awareness by health care professionals of common precipitating factors associated with NKHC.
- The diagnosis must be made quickly and appropriate and individualized treatment must be instituted to reduce the morbidity and mortality of NKHC.

References

1. Guyton A: Textbook of Medical Physiology, 6th ed, p 946. Philadelphia, WB Saunders, 1981
2. Ezrin C, Godden J, Volpe R, Wilson R (eds): Systematic Endocrinology, p 163. Hagerstown, Harper & Row, 1973
3. Maxwell M, Kleeman C (eds): Clinical Disorders of Fluid and Electrolyte Metabolism, 3rd ed, p 1307. New York, McGraw-Hill, 1980

4. Martin D, et al: Effect of hypophosphatemia on myocardial performance in man. N Engl J Med 297:901, 1977
5. Kyner J: Diabetic ketoacidosis. Crit Care Q :73, 1980
6. Park R, Arieff A: Lactic acidosis. Ann Intern Med 23:33, 1980
7. Kyner J, p 70
8. Kitabechi A, Ayyaguri V, Guerra S: The efficacy of low-dose versus conventional therapy of insulin for treatment of diabetic ketoacidosis. Ann Intern Med 84:633, 1976
9. Peterson L, Caldwell J, Hoffman J: Insulin adsorbence to polyvinylchloride surfaces with implications for constant infusion therapy. Diabetes 25:72, 1976
10. Hockaday T, Alberti K: Diabetic coma. Clin Endocrinol Metabol 1, No. 3:751, 1972
11. Wildenthal K: The effects of acid–base disturbances on cardiovascular and pulmonary function. Kidney Int 1:375, 1972
12. Maxwell M, Kleeman C (eds), p 1357
13. Feldman J, Lebovitz F: Tests for glycosuria: An analysis of factors that cause misleading results. Diabetes 22:115, 1973
14. Skillman T: Diabetic ketoacidosis. Heart and Lung. 7, No. 4:594, 1978

chapter 21

Head Injuries

Overview of Fluid Balance Problems

Abnormalities in water and sodium balance are likely occurrences in patients with head injuries.

- Disrupted water balance (and, thus, disrupted Na^+ balance) may be due to excessive or deficient secretion of antidiuretic hormone (ADH).

  1. Recall that ADH (vasopressin) plays a vital role in maintaining homeostasis. ADH promotes water reabsorption through its action on the kidney; it is sometimes referred to as the *water conserving hormone.*

  2. Excessive secretion of ADH (as occurs in the syndrome of inappropriate antidiuretic hormone [SIADH] secretion) causes abnormal water retention and hyponatremia.

  3. Conversely, deficient secretion of ADH (as occurs in central diabetes insipidus) causes abnormal water loss and hypernatremia (if water intake is inadequate).

- Hypernatremia can also be caused by excessive solute gain (as in hyperosmolar tube feedings) with inadequate provision for water. Other causes of hypernatremia in the neurologic patient can include inadequate water intake related to the patient's inability to perceive or respond to thirst, and to excessive water loss through hyperventilation.

- Hyperosmolality of the extracellular fluid (ECF) in the head-injured patient can be due to hypernatremia (as stated above), or hyperglycemia, as a result of steroid therapy (particularly in borderline diabetic, stressed, and aged individuals). It can also be due to excessive use of osmotic diuretics (such as mannitol and urea).

- For various reasons, cerebral edema is a frequent complication in head-injured patients.

Hyponatremia Due to SIADH

- An overview of the syndrome of inappropriate antidiuretic hormone secretion (SIADH) is presented in Chapter 7. The information presented below is specifically directed at SIADH in the head-injured patient.
- SIADH is a common problem in head-injured patients, particularly when facial trauma is present. Steinbok and Thompson reported the occurrence of SIADH in 6 of 33 severely head-injured patients; Fox, Falik, and Shalhour reported the occurrence of SIADH in 23 of 80 neurosurgical patients.[1,2]

Pathophysiology

- SIADH was once commonly referred to as cerebral salt wasting. It involves inappropriate urinary elimination of sodium in the presence of hyponatremia.
- In the head-injured patient with SIADH, there is evidently a sustained secretion of ADH by the pituitary mechanism despite the presence of decreased serum osmolality and increased circulatory volume. (Recall that in normal persons, ADH secretion decreases when hypoosmolality is present, or when circulating blood volume is increased.) Since these factors normally suppress ADH secretion, the continued release of the hormone in their presence is said to be *inappropriate.*
- Apparently there is suppression of aldosterone activity in patients with SIADH. It is thought that expansion of the ECF (due to excessive water retention) is responsible for aldosterone suppression. Because of decreased aldosterone activity, there is increased sodium loss in the urine; this is the salt-wasting aspect of SIADH.
- Swelling of cerebral cells results when water from the relatively hypoosmolar ECF moves into the area of greater solute concentration (in this case, the intracellular fluid). Irreversible brain damage can result if severe hyponatremia is allowed to persist.
- Excessive administration of fluids aggravates the hyponatremic state.

Clinical Manifestations

- Clinical signs of SIADH usually appear when the serum sodium level is below 120 mEq/liter (normal is 135–145 mEq/liter); however, the rate of fall is probably as important as the absolute serum sodium level in causing symptoms. As a general rule, the slower the rate of development, the lower the sodium level may become before symptoms appear.

- See Symptoms and Findings Associated with SIADH.
- Be aware that the clinical signs of SIADH are primarily neurologic; thus, the appearance of focal neurologic signs in the head-injured patient may be due to hyponatremia rather than to other causes

---

Symptoms and Findings Associated With SIADH

- Lethargy
- Headache
- Loss of appetite
- Nausea and vomiting
- Abdominal cramping
- Irritability
- Personality changes (uncooperative, hostile, and confused)
- Muscular twitching
- Deep tendon reflexes diminished or absent (particularly if serum Na^+ < 110 mEq/liter)
- Positive Babinski sign (particularly if serum Na^+ < 100 mEq/liter)
- Hemiparesis
- Aphasia
- Stupor
- Convulsions
- Absence of peripheral edema
- Cerebral edema may occur, leading to irreversible brain damage if not detected and treated early (the cerebral edema is due to intracellular volume expansion)
- Weight gain (usually 3–5 liters of water are abnormally retained in SIADH; recall that 1 liter of fluid weighs approximately 2.2 lb)
- Laboratory Findings
    - Serum Na^+ level < 130 mEq/liter (symptoms do not usually occur until the level is < 120 mEq/liter)
    - Serum osmolality < 275 mOsm/kg (normal is 280–295)
    - Urinary osmolality is higher than serum osmolality because the urine contains important amounts of Na^+ and the plasma is diluted with water.
    - Urine Na^+ content is > 20 mEq/liter.
    - Urinary SG may rise above 1.012 (usually in Na^+ deficit a low urinary SG is expected, perhaps 1.002–1.004).
    - Low BUN (result of overhydration)

(such as intracranial hematoma). *Frequent measurement of serum sodium levels is indicated to detect sodium balance problems early!* It is also important to monitor serum and urine osmolalities.

Treatment

- Most authorities recommend that fluid intake be at least mildly restricted in head-injured persons to prevent the occurrence of cerebral edema or hyponatremia.
- When mild hyponatremia secondary to SIADH occurs, it is often treated by water restriction alone. Authorities disagree as to the desired level of fluid restriction in this situation. Recommendations range from 600 to 1000 ml/day. Most agree that the degree of restriction should be regulated according to serum osmolality and sodium levels, and clinical status. (Of course, care must be taken to avoid excessive fluid restriction, particularly when diuretics are administered.)
- In addition to fluid restriction, administration of hypertonic saline (such as 5% NaCl) is generally recommended for patients with neurologic symptoms. It is often given in conjunction with furosemide (Lasix) to avoid circulatory overload; urinary losses of sodium and potassium are measured and replaced.
- Essential to effective therapy is the frequent measurement of serum and urine electrolytes and osmolality, and careful monitoring of fluid intake and output including daily body weights.
- Nursing responsibilities during therapy for SIADH are discussed in the section on SIADH in Chapter 7.

Hypernatremia Associated With Central Diabetes Insipidus

Overview of Diabetes Insipidus

- In review, ADH plays a major role in aiding the body to conserve water to prevent dehydration. This hormone acts by increasing renal reabsorption of water. Without ADH, the ingestion of approximately 10 to 20 liters of water daily would be necessary to match urinary losses. However, with ADH, urine output can be reduced to as little as 500 ml/24 hr.
- There are two types of diabetes insipidus (DI)—central and nephrogenic.
  1. *Central* (or neurogenic) *DI* is due to a relative lack of ADH; it is sometimes referred to as *vasopressin-sensitive DI*, since it responds

to vasopressin administration. This form of DI may occur after head trauma (particularly in the presence of fractures at the base of the skull), or surgical procedures near the pituitary or as a result of infection, primary tumor or metastatic tumor; or, it may be idiopathic.

2. *Nephrogenic DI* is not due to lack of ADH, but to failure of the kidneys to respond to the hormone; it is sometimes referred to as *vasopressin-resistant DI*, since the administration of vasopressin does not help. It may occur as an X-linked recessive trait; it may be acquired after primary renal disease or electrolyte disorders (such as hypokalemia or hypercalcemia), or may be pharmacologically induced by certain drugs (such as demeclocycline and lithium).

3. Central DI is seen most frequently in neurosurgical patients. It may be complete or partial, permanent or transient. The transient form is most common and usually subsides prior to discharge. Permanent DI is most likely to occur in patients with extensive damage to the hypothalamic area.

Clinical Manifestations of Central Diabetes Insipidus

- Classic symptoms of DI are presented as follows:
  1. There is excessive urinary output regardless of fluid intake.
  2. Urine output often exceeds 200 ml/hr.
  3. Urinary SG is less than 1.005, urine osmolality is less than 200 mOsm/liter.
  4. Serum osmolality is increased, as is the serum sodium level (due to water loss).
  5. Intense thirst is present in the alert patient. (The volume of fluid intake corresponds to the urinary volume.)
  6. There is an inability to concentrate urine by fluid restriction (a common test for central DI).
  7. Polyuria in central DI ranges from 3 to 10 liters or more daily (somewhat less in nephrogenic DI).
  8. Central DI is responsive to vasopressin administration.

Water and Electrolyte Disturbances

- As a rule, the patient with an intact thirst mechanism will drink sufficient fluids and will maintain balance (that is, an essentially normal serum osmolality and serum Na^+ level). Unfortunately, the frequency of urination and drinking interferes with other activities.
- If the patient is not able to perceive or respond to thirst, or if parenteral fluid replacement is inadequate, the polyuria will lead to se-

vere dehydration (hypernatremia and hyperosmolality of plasma) with weight loss, tachycardia, and even shock.

Treatment

- The standard treatment of central DI has been ADH replacement by means of vasopressin administration. Vasopressin may be administered in different forms, depending on the clinical situation.
- It may be given by injection in the form of aqueous vasopressin in acute care settings (duration of 1–4 hr). Aqueous vasopressin's short duration of action allows for greater control of fluid balance in rapidly fluctuating situations. General indications for its administration include urinary output exceeding 200 ml/hr for 2 consecutive hours, urinary SG less than 1.005, and an elevated serum sodium level.
- Vasopressin may also be given as Pitressin Tannate in Oil (duration 24–72 hr) or intranasally in the form of Diapid (lypressin) or DDAVP (desmopressin acetate).
- Until the polyuria is controlled by vasopressin therapy, careful attention must be paid to replacement of fluid (particularly if the patient has a decreased level of consciousness or other disturbances interfering with the perception of thirst or the ability to drink). It is important to keep in mind that a potential complication of the administration of vasopressin is water intoxication (excessive retention of water with a low serum Na^+ level).
- If the patient with central DI has some residual capacity to secrete ADH, drugs that increase the release of ADH or enhance its action on the kidney may be used instead of hormonal replacement therapy. Such drugs include chlorpropamide (Diabinese), an oral hypoglycemic agent; clofibrate (Atromid S), a hypolipidemic drug; and carbamazepine (Tegretol), an anticonvulsant.
- Paradoxically, administration of thiazide diuretics has been used in therapy for DI; it is the specific form of therapy for nephrogenic DI. However, it is at most an adjunct in the treatment of central DI. It is thought that the thiazides act by decreasing the amount of sodium ions that reach the distal tubules of the kidneys. Patients taking thiazides for treatment of DI should be instructed to avoid liberal use of salt since it decreases the effectiveness of the drug.

Nursing Measures to Aid in Detection of Diabetes Insipidus

- Maintain an accurate intake–output record.
  1. Be alert for polyuria in head-injured persons (an indwelling catheter is often indicated for accuracy and convenience)
  2. Calculate and record cumulative amounts for several shifts to obtain a more accurate account of the patient's fluid balance status.

(continued)

- Measure hourly urine output in postoperative neurosurgical patients (particularly those with known involvement of the pituitary–hypothalamus region).

   Be alert for and report large hourly urine outputs in the postoperative period (such as more than 200 ml in each of 2 consecutive hr or more than 500 ml in a 2-hr period).

   Be aware, however, that polyuria may occur postoperatively for reasons other than central DI. For example, polyuria may be due to excessive intraoperative fluid administration—in this case, the serum Na^+ level would not be elevated, and body weight would be greater than the preoperative level.

- Monitor urinary SG; a persistently low SG ($<$ 1.005) should be reported.
- Monitor serum Na^+ levels at least once a day (more often as indicated); report the presence of hypernatremia (serum Na^+ > 145 mEq/liter).
- Monitor body weight; be alert for weight loss paralleling polyuria. (A baseline preoperative weight is necessary for comparison.)

Other Causes of Hypernatremia

Tube Feedings

- Hypernatremia following brain injury may be secondary to hyperosmolar tube feedings with inadequate water intake.
- When inadequate water is supplied, the high-solute load in the tube feeding pulls water from ECF to allow for its renal excretion; the water loss is particularly severe in patients with poor renal concentrating ability. At first, urinary output is large; later, as the ECF becomes depleted, urine volume greatly diminishes.
- The literature reports numerous cases of hypernatremia in neurologic patients secondary to hyperosmolar tube feedings; some with sodium levels as high as 192 mEq/liter (normal is 135–145 mEq/liter). The blood urea nitrogen (BUN) level is also elevated (a sign of FVD with decreased blood flow to the kidneys).
- The reader is referred to Chapter 14 for a more thorough discussion of tube feedings (including nursing responsibilities).

Mechanical Ventilation

- The neurologic patient with respiratory depression often requires prolonged mechanical ventilation. Even with humidification there may be a tendency for the ventilator to remove water and produce hypernatremia. (One should remember, however, that mechanical ventilation with high humidification can also produce water overload.)

Inability to Experience or Respond to Thirst

- Unfortunately, many head-injured patients are not aware of thirst or are unable to communicate or alleviate their thirst—they must, then, rely on others to meet their fluid needs.
- Swallowing difficulties often cause the alert neurologic patient to avoid oral fluids because of a fear of choking. (The nursing staff may be reluctant to give oral fluids for the same reason.)

Fever and Hyperventilation

- Fever causes increased water loss through the lungs (secondary to hyperventilation) and through the kidneys (due to increased metabolic wastes associated with hypermetabolism). Fever may, thus, cause hypernatremia.
- It must be remembered, however, that fever may be a sign of hypernatremia. Fever after neurologic trauma calls for a check of the serum sodium level.
- Hyperventilation may occur without fever and, when prolonged, may be a significant cause of water vapor loss.

Clinical Signs of Hypernatremia

- Severe thirst (if the patient is conscious and the thirst mechanism is intact)
- Dry, sticky mucous membranes
- Flushed skin
- Fever
- Irritability (particularly in children)
- Urine SG > 1.030
- Serum Na^+ level above 145 mEq/liter
- Delirium and hallucinations
- Depressed level of consciousness

  The CNS signs of hypernatremia may be caused by altered cerebral electrolyte concentration and changed volume (shrinkage) of the brain cells. Hypernatremia may cause cerebrovascular damage (petechiae and subarachnoid hemorrhages).

Treatment of Hypernatremia

- Treatment consists of lowering the serum Na^+ level by giving extra water, preferably by mouth. If parenteral fluids are needed, a hypotonic electrolyte solution (such as 0.3% NaCl) may be used.
- The serum Na^+ level must be reduced gradually (no more than 15 mEq/liter in any 4- to 6-hr period) to decrease the risk of cerebral edema. It should be remembered that a rapid reduction in the serum Na^+ level renders the plasma temporarily hypoosmotic to fluid in the
  (continued)

brain (due to the blood–brain barrier); as a result, water can osmose into the brain tissue, causing cerebral edema.

- Correction of hypernatremia usually requires 24 to 48 hours to permit readjustment through diffusion between fluid compartments.
- Serum Na^+ levels should be monitored frequently (such as every 6 hr).

Nonketotic Hyperosmolar Coma

- This condition may result from massive osmotic diuresis in the diabetic neurosurgical patient (particularly in the older, mild or undiagnosed diabetic). Among the osmotic diuretics that can precipitate nonketotic hyperosmolar coma are mannitol and urea. Nonketotic hyperosmolar coma can also be precipitated by high dosages of steroids.
- Most striking in this condition are symptoms of FVD (caused by marked hyperglycemia, sometimes as high as 2000 mg/100 ml or more). Other symptoms are of a neurologic nature (such as depressed sensorium, seizures, hemiparesis, aphasia, and sensory defects).
- The hyperosmolar state may be overlooked because the symptoms may be mistaken for a worsening of the original neurologic disorder.
- Treatment is directed at intravenous replacement of fluid loss and insulin administration. The reader is referred to Chapter 20 for a more thorough discussion of nonketotic hyperosmolar coma.

Cerebral Edema and Elevated Intracranial Pressure

Pathophysiology

- In the neurosurgical patient the major cause of increased intracranial pressure is cerebral edema.[3]
- Cerebral edema can be classified into two categories: Cytotoxic (intracellular) cerebral edema and vasogenic (extracellular) cerebral edema.
    1. Causes of cytotoxic cerebral edema include hypoxia and the syndrome of inappropriate secretion of antidiuretic hormone (SIADH).[4]
    2. Causes of vasogenic cerebral edema, the most common type, include contusion, tumor, infarction, hemorrhage, and meningitis.[5]

- Since cerebral edema can lead to increased intracranial pressure, intensive measures are taken to prevent or minimize its occurrence in head-injured patients.
- To review, the intracranial compartment contains brain, blood, and cerebrospinal fluid (CSF); these compartments are normally in a state of pressure and volume equilibrium. The pressure is maintained by approximately 85% brain volume, 5% to 6% blood volume, and 9% to 10% cerebrospinal fluid volume.[6] When one component increases in volume, there must be a reciprocal change in one or both of the other components to maintain a constant total pressure. (For example, there may be a compensatory shunting of CSF into the spinal dural sac, or increased CSF absorption, or decreased cerebral blood volume.)

Clinical Signs of Increased Intracranial Pressure

Clinical signs of increased intracranial pressure (ICP) include
- Changes in level of responsiveness
    1. Lethargy
    2. Increasing drowsiness
    3. Confusion
    4. Shift from quietness to restlessness
    5. Less responsive to external stimuli
- ICP greater than 15 mm Hg (or 200 mm H_2O) as indicated by an ICP measuring device (such devices are highly desirable in the critically ill head-injured patient to provide valuable data for management of ICP)
- Pupil abnormalities
- Flexion or extension posture
- Vital sign changes (late signs indicating pressure on medullary structures)
    1. Slowed pulse rate
    2. Widened pulse pressure

Treatment and Nursing Considerations

- An effort is made to prevent cerebral edema in head-injured patients by limiting fluid intake and by other therapies as indicated (such as controlled ventilation and a variety of drugs).
- One must be aware that therapy aimed at preventing, diminishing, or reversing cerebral edema is often controversial (as indicated below).

Fluid Therapy

- Most physicians avoid the administration of free water (such as D_5W). Most prefer the administration of a hypotonic electrolyte solution, such as ½ normal saline (0.45% NaCl), with or without dextrose (2.5% or 5% solutions), in any condition likely to be associated with cerebral edema and increased intracranial pressure.
- Although most authorities agree that at least some fluid restriction is important in the control of elevated ICP, the degree of restriction is variable among medical centers (ranging from 800–2000 ml/day).
  1. In the past, severe fluid restriction was advocated in the management of head-injured patients at risk for elevated ICP. However, the present trend is moving toward a milder restriction, since more effective measures for controlling elevated ICP have been developed.
  2. It is generally agreed that the desired amount of fluid to be administered should be based on the patient's serum osmolality and electrolyte levels, and should consider abnormal routes of loss (such as gastric suction and ventricular drainage).
  3. Of course, fluid therapy must also provide for abnormal losses due to injuries other than head injuries. One must remember that the head-injured patient frequently has marked extremity fractures and thoracoabdominal injuries; such injuries predispose to severe hypovolemia.
  4. Failure to maintain an adequate vascular volume interferes with cerebral perfusion pressure. Also, patients with severe hypovolemia cannot be treated with mannitol and barbiturates because electrolyte imbalances, cardiac depression, and arrhythmias may develop.[7] In addition, hyperventilation in severely hypovolemic patients may have harmful effects by precipitously lowering the cardiac output, leading to irreversible ischemic damage.[8]
- Care must be taken not to excessively restrict fluids in patients receiving dehydrating agents (such as osmotic or loop diuretics).

Osmotic Diuretics

- Mannitol and urea are osmotic diuretics capable of temporarily relieving increased ICP when given by the intravenous route. By elevating the osmolality of plasma, these agents pull fluid from the interstitium of normal brain tissue into the bloodstream for subsequent excretion by the kidneys. The overall effect is a decrease in total intracranial volume. (It should be noted that mannitol has largely supplanted urea in the management of elevated ICP.)
- Serum osmolality should be maintained below 320 mOsm/kg in patients receiving osmotic diuretics; it is necessary to monitor serum osmolality at frequent intervals during such therapy (at least twice a day).

- Because of their potentially severe metabolic consequences, osmotic diuretics are not indicated for long-term use.
- Initially with the use of mannitol, there is expansion of the plasma volume caused by diffusion of water into the bloodstream; pulmonary edema or water intoxication (hyponatremia) are possible complications.
- Because of the diuretic effect of osmotic diuretics, an indwelling catheter should be inserted to avoid potential bladder overdistension. An indwelling catheter also facilitates measurement of urinary output, making it possible to quantify the amount of diuresis. *Urine output should be observed frequently, and if the urinary hourly output is less than 30 to 50 ml/hr, the drug should be stopped and the physician notified.*
- Among the contraindications to the administration of osmotic diuretics are FVD, renal disease of sufficient magnitude to produce anuria, and marked pulmonary congestion.
- A potential problem associated with the use of mannitol (and other osmotic diuretics) is a rebound increase in ICP (the rebound may result from retention of mannitol in the brain tissue as the blood mannitol level is dropping; this situation reverses the pressure gradient and allows water to diffuse back into the brain tissue, increasing ICP).
- Glycerol, an osmotic dehydrating agent, may also be used to lower ICP. Although it is slower acting than mannitol or urea, it can be used for longer periods. Glycerol can be given by mouth or by way of nasogastric tube. One should be alert for complications such as a rebound increase in ICP with its use also. (Rebound increases appear to be more frequent with the use of glycerol than with mannitol.)[9]

Nonosmotic Diuretics

- Furosemide (Lasix) has been recommended by some authorities to control elevated ICP, particularly for patients with congestive heart failure from chest injury or from previous cardiac disease. It may also be indicated in the treatment of patients with neurogenic pulmonary edema, when there is concern about administering the large fluid volume needed to give mannitol.[10]
- Nonosmotic diuretics are reported to decrease CSF production; this may be a valuable side effect for their usage.[11] Some authorities have found that nonosmotic diuretics are less reliable than mannitol and urea in producing sustained reduction in ICP.[12]
- The patient receiving loop diuretics should be monitored closely for signs of hypokalemia and FVD.

Controlled Ventilation

It is important to remember that cerebral blood flow is affected by carbon dioxide concentration and oxygen concentration in the blood.

- Cerebral blood flow increases as the $PaCO_2$ increases because carbon dioxide is a potent vasodilator. Dilatation of cerebral blood vessels and the resultant increased blood flow contributes to increased ICP.
- Hypoxia can also cause increased cerebral blood flow and thus contribute to elevated ICP. Danger of a high ICP is greatest when the PaO_2 is very low (such as < 50 mm Hg). A PaO_2 greater than 70 mm Hg should be maintained (by intubation if necessary).
- The patient with a significantly elevated ICP is usually hyperventilated to reduce the $PaCO_2$ to a level of 25 to 30 mm Hg. The resultant decreased cerebral blood flow decreases the ICP. Hyperventilation is an effective treatment for acute elevations of ICP; it is also useful in the treatment of chronically elevated ICP. Care should be taken to avoid dropping the $PaCO_2$ below 20 mm Hg because at this low level severe vasoconstriction and decreased cerebral blood flow occur. Secondary brain ischemia due to severe vasoconstriction has been observed.[13]

Position

- Unless contraindicated (as by shock or spinal cord injury), a body position with a 30-degree head elevation is recommended in head-injured patients to reduce ICP. Elevation of the head to 30 degrees maximizes venous drainage from the intracranial cavity. (Gravity can be utilized to enhance drainage because the venous vessels from the brain are valveless.)[14]
- Flexion of the neck should be avoided; sharp flexion of the neck can increase intracranial pressure by interfering with venous outflow (thus increasing the amount of blood in the cranial cavity).
- Although intermittent use of the Trendelenberg position is useful in draining the tracheobronchial tree, it should not be used when the patient has an elevated ICP.

Barbiturate Therapy

- The literature reports that long-term barbiturate administration may be helpful in reducing ICP in patients unresponsive to other therapies.
- The comatose state that is induced by a high serum barbiturate level renders the patient totally dependent; the protective gag, swallowing, and corneal reflexes are lost. Thus, full supportive care and extensive monitoring devices are essential. Intubation and total assisted ventilation is required due to loss of spontaneous respirations, and an ICP monitoring device must be used. In addition, arterial pressure, central venous pressure, pulmonary artery pressure, and cardiac function must be monitored. (Untoward cardiovascular responses may be elicited by barbiturate coma.)

- Goals of barbiturate coma are the preservation of viable ischemic brain tissue and the prevention of irreversible cellular damage.
- The mechanism of action is poorly understood. It has been hypothesized that a number of events may occur during barbiturate coma.
  1. Reduced cerebral blood volume due to vasoconstriction of cerebral vessels
  2. Decreased cerebral oxygen demands due to decreased rate of cerebral metabolism
  3. Reduced cerebral venous pressure due to total muscle relaxation
  4. Decreased metabolism due to secondary hypothermia
- Cerebral perfusion pressure (CPP) should ideally be maintained within a range of 60 to 70 mm Hg.[15]
  CPP equals mean arterial pressure (MAP) minus intracranial pressure (ICP). Dopamine may be administered when the MAP falls below 90 mm Hg to maintain adequate cerebral and renal perfusion. Mannitol and/or hyperventilation may be needed to keep the ICP in an acceptable range.
- Recall that the patient in barbiturate coma is prone to fluid and electrolyte imbalances secondary to fluid restriction, use of diuretics, and gastric fluid loss through suction.
- Arterial blood gas studies are required at frequent intervals to evaluate ventilation. The $PaCO_2$ is usually kept lower than normal (such as 25–30 mm Hg) to favor decreased ICP. The PaO_2 should be kept at a level greater than 70 mm Hg.
- Other parameters related to fluid and electrolyte balance requiring close monitoring include serum and urine osmolality, serum electrolytes, BUN, and creatinine.
- Fluid intake and output should be closely monitored; results should be totalled every 24 hours and cumulatively. Hourly urine output and SG readings may be required. Urine output should be maintained at 30 ml/hr or greater.
- In addition, central venous pressure and pulmonary artery pressures should be evaluated in terms of adequacy of fluid replacement.

Steroids

- Glucocorticoids are sometimes administered in the management of head-injured patients although research has not clearly supported their efficacy in this area. Some feel that massive doses in the early post-injury period may significantly improve mortality and morbidity from intracranial trauma.
- The nurse must be alert for complications associated with steroid therapy; these may include hyperglycemia with hyperosmolar reaction, gastrointestinal bleeding, and increased incidence of infection.

1. It is customary to monitor urine glucose levels during massive steroid therapy.
2. Antacids are usually administered to patients receiving large doses of steroids; they may be given by nasogastric tube to the comatose patient.
3. Cimetidine is also often used to decrease gastric acidity in patients receiving large doses of steroids.

Other Therapies

- A major effort is made to prevent seizures since the increased metabolic activity threatens cerebral perfusion.
- An attempt is made to maintain normothermia since hyperthermia increases the metabolic demand of the brain. Temperatures in excess of 38° C (100.4° F) should be avoided in patients with severe head injuries. Hypothermia blankets and medications to reduce body temperature are frequently used. Severe hypothermia must also be avoided since it is associated with cardiac arrhythmias and increased incidence of hypoxia.[16]

Summary of Nursing Considerations in the Care of Head-Injured Patients With a Potential for Elevated ICP

- Be aware of nursing activities that may increase ICP.
    1. Suctioning the patient frequently (particularly when rotating the head to reach the mainstem bronchi)
    2. Turning the patient frequently
    3. Discussing the patient's condition or prognosis at the bedside
    4. Placing the patient on the bedpan
    5. Rotating the patient's head
    6. Rapidly shifting the patient's position
- Use interventions to prevent or minimize increased ICP as indicated.
    1. Elevate the head of the bed to a 30° angle.
    2. Keep the patient's head in a neutral position; avoid neck flexion and extension.
    3. Use a turn sheet to move the patient.
    4. Encourage the patient to exhale while turning, or pushing self up in bed.
    5. Minimize noxious stimuli; use medications as indicated.
    6. Prevent constipation (straining for stools elevates ICP).
    7. Maintain a patent airway.
    8. Use intermittent brief suctioning if necessary. (Remember, hypoxemia predisposes to vasodilation and increased ICP.)
        - Mild manual hyperventilation prior to respiratory care helps mitigate the harmful effects of endotracheal suctioning, as does sedation.

- Some authorities recommend that suctioning and other respiratory care maneuvers be discontinued if ICP is > 30 mm Hg; they can be reinstituted when the ICP drops below 20 mm Hg.
9. Maintain a quiet environment. (Noise can precipitate elevated ICP in the patient with an unstable ICP.)

- Monitor ICP readings if a monitoring device is in place. Be aware that the normal range of ICP is 4 to 15 mm Hg (50–200 mm H_2O).
- Monitor neurologic signs at regular intervals to observe for complicating increased ICP.
- Be aware that the cerebral perfusion pressure (CPP) should ideally be maintained within a range of 60 to 70 mm Hg.

    CPP = MAP (mean arterial pressure) − ICP (intracranial pressure)

    (Of course, this cannot be done without the use of an ICP monitoring device and an arterial pressure monitoring device.)
- Administer fluids cautiously to avoid accidental fluid overload; be aware that most authorities advocate at least some degree of fluid restriction in head-injured persons.
- Monitor serum osmolality, electrolytes, and urine osmolality. Be aware that diuretics should not be used when the serum osmolality exceeds 320 mOm/kg. Be alert for Na^+ imbalances. (See early part of chapter.)

    Monitor blood gases. Be alert for changes in PaO_2 and $PaCO_2$.[18] Recall that the PaO_2 should be maintained at a level > 70 mm Hg; also, the $PaCO_2$ should not be allowed to elevate above normal. Indeed, the $PaCO_2$ is frequently kept below normal (such as 25–30 mm Hg) in patients with elevated ICP.

References

1. Steinbok P, Thompson G: Metabolic disturbances after head injury: Abnormalities of sodium and water balance with special reference to the effects of alcohol intoxication. Neurosurgery 3:9–15. 1978
2. Fox J, Falik A, Shalhour R: Neurosurgical hyponatremia: The role of inappropriate antidiuresis. Neurosurgery 34:506–514, 1971
3. Hickey J: The Clinical Practice of Neurological and Neurosurgical Nursing, p 184. Philadelphia, JB Lippincott, 1981
4. Nikas D: Fluid resuscitation of the head-injured patient. In Ellerbe S (ed): Fluid and Blood Component Therapy in the Critically Ill and Injured, p 80. New York, Churchill Livingstone, 1981
5. Ibid, p 80
6. Ricci M: Intracranial hypertension: Barbiturate therapy and the role of the nurse. Neurosurg Nurs 11, No. 4: 247–252, 1979
7. Bowers S, Marshall L: Severe head injury: Current treatment and research. Am J Neurosurg Nurses 14, No. 5: 214, 1982

8. Ibid
9. Ibid, p 216
10. Ibid
11. Ibid
12. Ibid
13. Ibid
14. Hickey J, p 184
15. Ricci M, p 249
16. Bowers S, Marshall L, p 214
17. Ibid, p 213
18. Ibid

chapter 22

Renal Failure

Acute Renal Failure

- Acute renal failure is of rapid onset (over days or weeks) and implies a pronounced reduction in urine flow in a previously healthy person. The condition may be caused by
  1. Severe and prolonged shock
  2. Severe FVD
  3. Hemolytic blood transfusion reaction (Lysis of red blood cells results in renal vasoconstriction and tubular blockage with hemoglobin.)
  4. Severe crushing injuries (Severely crushed muscles release large amounts of myoglobin into the bloodstream; myoglobin can block the tubules and might also produce vasoconstriction.)
  5. Nephrotoxic chemicals (such as lead, mercury, arsenic, and carbon tetrachloride)
  6. Drugs (such as phenacetin, sulfonamides, streptomycin, kanamycin, radiographic contrast agents, tetracycline, and amphotericin)
  7. Endotoxemia
  8. Renal vascular occlusion
- Acute renal failure occurs in a variety of clinical settings. For example, it can occur in obstetrical patients as a result of pre- and postpartum hemorrhages, toxemia, septic abortion, and the use of nephrotoxic abortifacients. Acute renal failure in postoperative patients has its highest incidence among those having aortic aneurysm, biliary, gastrointestinal, and cardiac surgical procedures. The use of contrast agents in diagnostic radiology (particularly in large doses)

has been incriminated as predisposing to acute renal failure; instances have been observed following gallbladder series, renal angiography, and intravenous pyelography.

- Indeed, clinical settings can be used to classify acute renal failure. An average breakdown is as follows: 40% surgery, 10% trauma, 10% nephrotoxins, 10% pregnancy, and 30% miscellaneous medical causes.[1]

- Acute renal failure may be classified as either reversible or irreversible. In irreversible renal failure, the kidney function does not return and uremia progresses. Reversible acute renal failure can be divided into two phases—(1) oliguria and (2) diuresis.

Oliguric Phase

- The first manifestation of acute renal failure is usually decreased urinary output, appearing within a few hours after the causative agent.

- It should be noted that although decreased urine volume (oliguria) is most characteristic of acute renal failure, urine output can vary from total anuria to polyuria. Categories have been defined as follows: Total anuria (no urine output), anuria (< 50–100 ml/24 hr), oliguria (< 400 ml/24 hr), and polyuria (> 1000–2000 ml/24 hr).[2] The incidence of nonoliguric acute renal failure, once thought to be rare, has been noted in 20% to 30% of the cases of acute renal failure.[3]

- Anuria is rare (except with urinary tract obstruction); usually, a 24-hour output of about 50 ml to 150 ml is the rule for the first few days. After this time, the urine output gradually increases.

- The oliguric phase may last 1 day or several weeks; the average duration is 10 to 12 days in severe cases.

- Major fluid and electrolyte problems in this phase include potassium excess, fluid volume excess, and metabolic acidosis. Others include calcium deficit, phosphorum excess, sodium deficit, and magnesium excess. Reasons for these imbalances along with common clinical signs and treatment measures are presented now.

Hyperkalemia (Potassium Excess)

- Hyperkalemia (Potassium Excess) usually occurs in the oliguric phase and can be life-threatening. Recall that the kidneys normally excrete 80% or more of the potassium lost daily from the body. When the kidneys are not functioning, potassium excretion is greatly reduced.

- This fact, in addition to the normal endogenous breakdown of muscle protein, causes an accumulation of potassium in the extracellular fluid.

Factors that can further increase protein catabolism include

1. Inadequate caloric intake
2. Acute stress or infection
3. Massive crushing injuries
4. The presence of large quantities of necrotic tissue (Recall that the catabolism of necrotic tissue releases potassium into the extracellular fluid.)

- The presence of metabolic acidosis augments the intracellular to extracellular shift of potassium, thereby hastening potassium buildup in the plasma.
- Potassium intoxication is one of the main causes of death in acute renal failure.

Clinical Signs

Clinical signs of hyperkalemia to be alert for include the following:

- ECG changes (tall-tented T waves, depressed ST segments, low amplitude P waves, prolonged P–R intervals, and widened QRS complexes; see Fig. 8-1)
- Anxiety and restlessness
- Muscular weakness progressing to flaccid paralysis (primarily affects limbs and respiratory muscles)
- Respiration decreased as a result of respiratory muscle weakness
- Paresthesias or numbness (particularly of the mouth, hands, and feet)
- Decreased pulse rate, finally resulting in bradycardia
- Cardiac arrhythmias
- Cardiac arrest and death

Management

Frequent electrolyte studies and ECGs are necessary to guide the physician and nurse in the management of patients.

Restricted Intake

- Dietary potassium intake should be sharply restricted and potassium-containing medications should be discontinued. The potassium content in certain medications is often overlooked. For example, Penicillin G contains 17 mEq of potassium per 100 million units.
- Salt substitutes should not be allowed since they contain potassium also.
- Potassium-conserving diuretics, while not adding potassium directly, cause retention of potassium and thus could prove lethal. (Potassium-conserving diuretics include spironolactone, triamterene, and amiloride.)
- Stored bank blood may contain as much as 30 mEq of potassium per liter and should not be used to transfuse patients with renal damage.

Cation Exchange Resins

- Cation exchange resins may be used to increase potassium excretion from the bowel. The resins may be taken by mouth, if tolerated, or may be instilled as a retention enema.
- While the cation exchange resins are given to remove potassium ions from the intestinal tract, they also remove other cations, such as sodium, calcium, and magnesium. It may be necessary to replace these ions if the resins are used for more than a few days. Such patients should have regular electrolyte studies.
- Sodium polystyrene sulfonate (Kayexalate) is a sodium–potassium exchange resin. During its use, sodium ions are partially released by the resin in the intestine and replaced by potassium ions from the body. The bound potassium is then excreted in the feces.
  1. Hypokalemia may occur if the effective dose is exceeded; therefore, it is necessary to determine the plasma potassium level daily.
  2. Sodium is released in the intestine during electrolyte exchange; therefore, the resin should be given with caution to patients with heart failure. Each gram of resin adds approximately 2 mEq (46 mg) of sodium. Signs of sodium retention include hypertension and edema.
  3. The average adult oral dose of sodium polystyrene sulfonate is 15 Gm, 1 to 4 times daily in a small quantity of water or syrup (3 ml to 4 ml/Gm of resin).
  4. Since obstruction has been reported with the use of exchange resins, the use of sorbitol, a liquid that promotes osmotic diarrhea, can be mixed with the resin to increase the excretion of the bound potassium. The patient should be told that mild diarrhea is desired when Kayexalate is used.
  5. The resin, mixed with sorbitol, may also be given as an enema; each adult dose, usually 30 Gm, is administered in 150 ml to 200 ml of water. Best results are achieved when the emulsion is warmed to body temperature before use. The enema should be retained for at least 1 to 2 hours (if leakage occurs, the patient's hips should be elevated on pillows, or a knee–chest position should be assumed temporarily). The enema vehicle administered with the cation exchange resin should be measured before instillation and after expulsion of the cation resin enema. (A Harris flush may be given to facilitate removal of the previously administered resin prior to instillation of another retention enema.)
  6. Oral administration of Kayexalate is preferred because it allows the resin to come into contact with a greater surface area of the gastrointestinal tract than is possible with rectal administration.
- Cation exchange resins actually remove potassium from the body (approximately 1 mEq of K^+ /Gm of resin). However, it is impor-

tant to remember that cation exchange resins are of the most value as a preventive rather than an emergency method to reduce severe potassium excess.

Emergency Measures

- Emergency measures to temporarily relieve dangerously high hyperkalemia (> 7 mEq/liter) include the administration of calcium gluconate to protect the myocardium from the effects of hyperkalemia; the administration of sodium bicarbonate to alkalinize the plasma and force potassium into the cells; and the administration of hypertonic dextrose and regular insulin intravenously to force potassium into the cells. These treatments are described in Chapter 8.
- It is important to remember that the emergency measures listed above for the treatment of severe hyperkalemia have only temporary effects. They serve to "buy time" for the institution of treatment measures to actually rid the body of excess potassium (such as cation exchange resins, hemodialysis, or peritoneal dialysis).

Measures to Reduce Catabolism

- Factors contributing to catabolism (and thus the release of K^+ from body tissues into the bloodstream) include immobilization, infection, and fever.

  Therefore, moderate activity should be encouraged as soon as possible and infections should be treated promptly with appropriate antibiotics usually in reduced doses owing to diminished renal function. If the patient has dirty wounds, undrained pus collections, or necrotic tissue, the plasma potassium concentration rises as a result of catabolism. To prevent severe potassium excess, it is important that necrotic tissue and pus be removed.

- Androgens have been demonstrated to decrease protein catabolism; however, their effect in patients with high rates of protein catabolism from trauma or infection seems negligible. The virilizing effects of testosterone may outweight its anabolic effect.

Extracellular Fluid Volume Excess

- Fluid volume excess (FVE) is a frequent problem during the oliguric phase when the body is unable to excrete excess fluid. Usually it is due to the excessive administration of fluids, either orally or intravenously.
- Clinically, FVE is manifested by
  1. Elevated central venous pressure
  2. Distention of neck veins and other peripheral veins
  3. Edema
  4. Bounding pulse
  5. Shortness of breath

6. Moist rales
7. Acute weight gain
8. Congestive heart failure and pulmonary edema if severe

Treatment and Nursing Implications

- The amount of fluid administered must be carefully planned to suit the patients's needs, with the body considered as a closed system. Excessive fluid intake should be avoided.
- Fluid intake should equal urinary output and other losses of water (such as occurs in diarrhea, vomiting, and fever), plus a basic fluid ration of 400 ml/day for the average adult. (Each degree of centigrade temperature elevation/24 hr may necessitate adding 100 ml to the total daily fluid intake.)
- The nurse must keep an accurate account of all routes of fluid gains and losses, because an inaccurate record could lead to a fluid overdose and FVE with its dangerous sequelae.
- If the nurse were ever to err in calculating fluid intake, it should be in the direction of too little rather than too much fluid. Some authorities recommend that the electrolyte content of all fluids excreted by the patient be measured. In general, patients with acute renal failure should not be given electrolytes in parenteral fluids except to replace documented losses.[4]
- Accurate body weight charts should be maintained to help guide fluid replacement therapy.
- Sodium intake usually is restricted to help avoid FVE. It is sharply restricted when hypertension and congestive heart failure are present.
- The patient should be allowed to help in deciding how his limited fluid intake will be distributed over the 24-hour period.

Metabolic Acidosis

- Metabolic bodily processes normally produce more acid wastes than alkaline wastes. Thus, when the kidneys fail to function, acid accumulation exceeds alkali accumulation. Metabolic acidosis is the result of the retention of these acid metabolites; their accumulation in the bloodstream causes the pH to drop. Metabolic acidosis is a high anion gap (AG) acidosis due to flooding of the plasma with the unmeasured anions of uremic poisoning and parallels uremia in severity.
- Decreased food intake causes increased utilization of body fats and the accumulation of ketones in the bloodstream; these acids further decrease the pH. Acidosis may develop more rapidly if marked tissue trauma or uncontrolled infection is present.
- Vomiting commonly occurs with the development of uremia. If vomiting occurs at the time metabolic acidosis is developing, it is possible that the metabolic alkalosis accompanying vomiting may help to counteract acidosis. However, because gastric hypoacidity frequently

occurs with uremia, vomiting may not have a significant effect on the *p*H. Extensive diarrhea contributes to the development of metabolic acidosis.

Clinical Signs

- Metabolic acidosis causes a compensatory increase in pulmonary ventilation, which causes the elimination of large amounts of carbon dioxide from the lungs with a resultant decrease in the carbonic acid content of the blood. The *p*H is partially corrected by this mechanism.
- Other signs may include anorexia, weakness, apathy, and coma.

Treatment

- The metabolic acidosis is tolerated fairly well and usually does not require treatment unless the plasma bicarbonate level is below 16 mEq/liter (normal is 22–26 mEq/liter) or frank Kussmaul breathing, stupor, or coma occurs.
- Alkalinizing agents such as sodium bicarbonate or sodium lactate may be given orally or intravenously, as indicated.
    1. Sodium salts must be used cautiously to prevent fluid overload, particularly in patients with congestive heart failure.
    2. Usually the depressed bicarbonate level is only partially restored to normal to avoid precipitating symptoms of hypocalcemia. (Recall that calcium ionization is decreased when alkalinity of the ECF increases.) Calcium salts may be indicated when alkalinizing agents are used to prevent this problem.
- Treatment of metabolic acidosis by dialysis may be preferable with many patients because the infusion of alkalinizing agents such as sodium bicarbonate or lactate may precipitate pulmonary edema or may cause clinical tetany in the presence of hypocalcemia.

Calcium and Phosphorus Derangements

- The serum calcium concentration is often below normal and may be related to the increased concentration of phosphorus in the bloodstream. (Recall that phosphorus is a constituent of one of the retained metabolic acids.) A reciprocal relationship exists between calcium and phosphorus, so that an increase in one causes a decrease in the other. However, the elevated serum phosphorus correlates poorly with the hypocalcemia of acute renal failure (which can drop as low as 6.5–8.5 mg/100 ml within 2 days of oliguria).[5]
- Decreased absorption of calcium from the gut occurs in uremia and contributes to the hypocalcemia.
- The cause of hypocalcemia in acute renal failure is not altogether clear; its severity is seldom so marked as occurs in chronic renal insufficiency.

- The major significance of hypocalcemia in acute renal failure is that it enhances the toxic effects of potassium on the heart. (Recall that Ca^{2+} and K^+ have antagonistic actions on heart muscle.)

Clinical Signs

- A serum calcium level below 2 mEq/liter usually causes tetany; however, these symptoms rarely occur in renal failure since the decreased blood pH of metabolic acidosis favors calcium ionization. (Recall that Ca^{2+} ionization increases in acidosis and decreases in alkalosis.)
  Chvostek's sign may be positive even though other symptoms of calcium deficit are not present.
- If metabolic alkalosis develops as a result of treatment with alkaline agents, calcium deficit becomes manifest with the development of muscle twitching and convulsions.
- The ECG is usually normal, but may show a prolonged Q–T interval.
- Sustained hypocalcemia resulting from hyperphosphatemia causes the parathyroid glands to hypertrophy in an attempt to compensate for the low calcium level by production of parathormone. Calcium is withdrawn from the osteoid tissue and eventually results in bone changes if the calcium–phosphorus derangement persists.

Treatment

- Calcium deficit can be treated with calcium gluconate by the oral or intravenous route—orally if nausea and vomiting are absent, intravenously otherwise.
- The slow administration of 30 ml of calcium gluconate intravenously will temporarily ameliorate severe symptoms of hypocalcemia.
- Calcium administration can cause pronounced improvement in ECG changes produced by hyperkalemia, even though the serum potassium concentration is not changed.
- Some physicians administer vitamin D in an attempt to increase calcium absorption from the gut.
- Measures to reduce the high phosphate level may cause an increased calcium level. Therefore, aluminum hydroxide gels (Amphojel and Basaljel) may be administered to combine with phosphate in the gastrointestinal tract. Phosphate-binding gels may not be well-accepted by patients because they can cause constipation, nausea, and vomiting. (Excessive use of these gels can cause phosphate depletion and hypercalcemia.)

Sodium Balance

- The majority of patients with acute renal failure exhibit some degree of hyponatremia. Contributing to sodium deficit is
  1. Administration of excessive amounts of water, which dilutes the serum sodium level

2. Shift of sodium into the cells, particularly if acidosis is present

3. Excessive loss of sodium in vomiting and diarrhea

- Moderate hyponatremia is usually asymptomatic. If the sodium level is markedly reduced (below 120 mEq/liter), enough water may be shifted into the cells to cause signs and symptoms of water intoxication (nausea, emesis, muscular twitching, grand mal seizures, and finally, coma).

- Sodium deficit usually is treated by limiting the water intake. Rarely, a hypertonic solution of sodium chloride (such as 3% or 5% NaCl) is administered slowly by the intravenous route in a small dose to correct a severe sodium deficit; sodium chloride should be administered with great caution because it can easily result in FVE with congestive heart failure and pulmonary edema.

- It is possible that failure of the kidneys to excrete sufficient sodium, caused by reduced glomerular filtration rate (GFR), may result in sodium retention. Sodium retention, in turn, leads to water retention with resultant FVE (described earlier). FVE is treated by limiting sodium and fluid intake and, when severe, by dialysis.

Magnesium Excess

- The serum magnesium level may increase when oliguria is present as a result of decreased renal excretion of this substance.

- Hypermagnesemia may be intensified to a serious level when magnesium-containing drugs are administered, such as magnesium citrate, magnesium sulfate, milk of magnesia, Mylanta, Gelusil, Maalox, and Creamalin. Great care must be taken to avoid the accidental administration of one of these substances to a patient in renal failure (for example, roentgenography of the bowel often requires magnesium citrate as a laxative).

- Clinical signs of hypermagnesemia include weakness, drowsiness, impaired respiration, orthostatic hypotension, and coma. (Magnesium imbalances and their treatments are discussed in Chap. 10.)

- Magnesium excess is primarily a disorder of acute and chronic renal disease and contributes to the central nervous features associated with uremia.

Uremia

Uremia (azotemia) is a toxic condition caused by failure of the kidneys to excrete urea, potassium, organic acids, and other metabolic waste products. It progresses most rapidly during the first few days of oliguria. Uremia may result from acute or chronic renal failure. Symptoms of uremia include the following:

- Chronic fatigue
- Nocturia
- Insomnia

- Anorexia (frequently caused by sight of food)
- Intractable vomiting (often occurs in early morning)
- Gastritis, hematemesis, and hiccoughs may occur
- Ammonia odor to breath and unpleasant metallic taste in mouth
- Stomatitis (salivary urea is hydrolyzed by urease into ammonia, causing mucous membrane irritation)
- Pruritus—skin is dry and scaly, with excoriations caused by scratching (probably related to calcium and phosphorus disturbances)
- Pale sallow skin resulting from anemia and urochrome deposition in skin—discoloration most prominent on face and other body parts exposed to light
- Increased bleeding tendency revealed by epistaxis, bleeding gums, easy bruising, petechiae, and conjunctival hemorrhage (probably caused by platelet defects)
- Coarse muscular twitching, first occuring during sleep
- Deep, rapid respirations (a respiratory attempt to compensate for metabolic acidosis)
- Chest discomfort (may be caused by pericarditis or pleurisy)
- Involuntary leg movements (restless leg syndrome)
- Peripheral neuropathy (numbness, pain, and burning sensations in legs and arms; these changes begin distally and are symmetrical—they may progress to motor weakness and paralysis)
- Decreased libido and impotency in males; suppressed ovulation, libido, and menstruation in females
- Hypertensive symptoms such as headache and visual difficulties (visual difficulties caused by retinal hemorrhages)
- Delayed wound healing (sutures need to remain in place longer than usual) and increased susceptibility to infection
- Hypothermia (patients with severe uremia may have body temperatures of 95° F [35° C] or less, even when infection is present)
- Decreased tolerance for all drugs that are excreted by the kidneys
- Gradual diminution of mental acuity over a period of time, leading to coma (resulting from toxic substances in bloodstream and acidosis)
- Generalized convulsions
- Transient episodes of agitated psychotic behavior (hallucinations and paranoid delusions) interspersed with periods of lucidity
- Uremic frost (urea crystals excreted through sweat glands, heaviest on nose, forehead, and neck) is rarely seen today because of improved management of uremic patients
- Increased propensity of congestive heart failure (contributed to by the combination of hypertension, anemia, and acidosis)
- Muscle wasting and weakness
- Anemia (may be profound due to a combination of decreased erythropoietin, decreased red-cell survival time, and blood loss)

Anemia

- Anemia may develop within 48 hours after the onset of acute renal failure. The red blood cells are normal in color (normochromic) and shape (normocytic), but are decreased in number.
- The hematocrit usually stabilizes at a level of 20% to 25%.
- The inadequate production of a renal enzyme that stimulates release of erythropoietin is the probable cause of anemia. (Recall that erythropoietin normally stimulates the bone marrow to produce red blood cells—a deficit of this substance, thus, results in anemia.)
- Other factors that may contribute to the occurrence of anemia include high serum levels of urea, potassium, and hydrogen ions. It has been found that the life span of red blood cells is decreased when uremia is present, further contributing to the development of anemia.
- The low hemoglobin level predisposes to acidosis since the buffering action of hemoglobin is diminished.
- The anemia of renal failure usually is well tolerated unless the hemoglobin level falls below 7 Gm/100 ml of blood.
- Severe anemia may be manifested by fatigue, dyspnea on exertion, palpitations, tachycardia, and angina.
- Unless symptoms are present, anemia usually is left untreated. Sometimes transfusion is absolutely necessary owing to active bleeding, angina, or hypovolemia. If so, it is best to give fresh, frozen packed red cells, because the risk of hyperkalemia is less when fresh rather than bank blood is used; also, the risk of fluid volume overload is less.
- Unfortunately, the transfused cells have a shortened life span (probably the result of abnormal constituents in the patient's blood). Transfusions depress normal bone marrow reticulocytosis, may cause hepatitis, and increase antibody levels that can later cause rejection of renal transplants. Washed cells keep the leukocyte antigenic exposure to a minimum, making rejection of a future transplant less likely to occur.

Infection

- Patients with acute or chronic renal failure are highly susceptible to infection; in fact, infection is a major cause of mortality in such patients.
- Not only is the acutely uremic patient more susceptible to infections, but once started, the spread of bacteria is difficult to control.
- Exposure to others who are ill must be avoided; also, the patient should receive adequate rest periods and avoid chilling.
- Indwelling urinary catheters frequently cause urinary tract infections and should not be used in patients with oliguric acute renal failure once the diagnosis is established. If a catheter is required, it should

be connected to a closed drainage system to minimize contamination. Scrupulous perineal and catheter care is indicated along with the application of povidone-iodine (Betadine) or a topical antibiotic at the catheter's point of entry into the urethra.

- Recall that fever, one of the cardinal symptoms of infection, may be absent in the acutely uremic patient with an infection. (Hypothermia is a well-recognized complication of renal failure.)
- The dosage of potentially nephrotoxic antibiotics (such as kanamycin, polymyxin, and gentamicin) must be adjusted to each patient's renal function and metabolic status. Sulfonamides, nitrofurantoin, and tetracycline should be avoided altogether. Chloramphenicol, erythromycin, penicillin, and penicillin derivatives can be used with only slight dosage reduction.[6]

Urine Changes

- The urine usually is bloody for the first few days of the oliguric phase, becoming clear about the end of the oliguric phase.
- If renal failure is due to hemolytic blood transfusion reaction, the urine has a port-wine color.
- As renal failure advances, there is a tendency for the urinary SG to become fixed at a low level (isosthenuria), similar to that of the glomerular filtrate (approximately 1.010). A low SG indicates that the renal tubules have lost their ability to concentrate urine.

Diuretic Phase

- The early diuretic phase begins when the 24-hour urine volume exceeds 400 ml/24 hr, usually around the 10th day after onset; in some instances, it may not occur for 14 to 21 days.
- The amount of urinary output depends on the treatment the patient received during the oliguric phase. If fluid overloading was allowed, the urine volume may be more than 5000 ml daily. If the patient was well-managed, the urine volume is not excessive.
- Urinary output usually increases in a stepwise manner, but occasionally increases rapidly.
- During the early part of the diuretic phase, the partially regenerated tubules are unable to concentrate urine, and the glomerular filtrate is excreted virtually unchanged. Thus, the patient's condition does not improve in the first few days of the diuretic phase; indeed, uremia may be more severe during this period than at any other time.
  1. Convulsions, stupor, nausea, vomiting, hematemesis, bloody diarrhea, or hemorrhage may occur. The reason for the severity of the uremia lies in the rapid contraction of the total body fluid; early and prompt replacement of fluid is needed until the BUN level begins to fall.

2. Dialysis is indicated if uremia is severe.

- Hypokalemia and hyponatremia should be watched for during this phase since potassium and sodium losses are increased when urine output is great.

  Urinary losses of these substances should be measured and replaced quantitatively, unless hyperkalemia is present.

- Body weight should be measured twice daily; if it decreases too rapidly, water and electrolytes must be supplied in sufficient amounts to prevent hypovolemic shock.

- As a rule, fluid intake is restricted to less than that excreted in the urine (since quantitative replacement may serve to perpetuate the obligatory water and salt loss).

- Urine and serum electrolyte levels should be measured daily in addition to daily hematocrits. Electrocardiograms are of ancillary value and should be obtained periodically.

- Alkalinizing agents may be needed if acidosis is severe; however, they should be used cautiously to avoid precipitating symptoms of hypocalcemia.

- Prognosis is favorable for patients with little underlying disease. On the average, the BUN becomes normal within approximately 15 to 20 days after the onset of the diuretic phase; the clinical signs and symptoms of uremia usually subside rapidly, although it may be several months before the patient is well enough to resume full activities. The associated anemia improves gradually over a period of months.

- Prognosis is unfavorable for patients with severe underlying renal damage. Significant permanent renal damage (following acute tubular necrosis) can occur from a variety of insults such as heatstroke, severe or prolonged septic shock, severe FVD, especially with the administration of diagnostic contrast media, and nephrotoxins, such as methoxyflurane and ethylene glycol.

- Despite advances in treatment, the mortality rate remains at approximately 50%. Two-thirds of the deaths occur during the oliguric phase and one-third during the diuretic phase.[7] The mortality rate is related to the etiology, being highest in patients with marked tissue destruction (as in crush injuries, severe trauma, or major surgery).

Chronic Renal Failure

- Man is normally born with 2 million nephrons; he can survive, although with difficulty, with as few as 20,000 nephrons.
- Chronic renal failure is the result of progressive irreversible loss of functional nephrons and occurs gradually over years or months.

Causes may include glomerulonephritis, pyelonephritis, polycystic kidneys, essential hypertension, or urinary obstruction.

- Because of the advent of long-term dialysis and renal transplantation, death is no longer the inevitable result when the nephron population falls below the minimal number compatible with unassisted life.

- Chronic renal failure is sometimes classified into 3 stages.

   1. Diminished renal reserve—renal function is somewhat diminished but the blood urea nitrogen (BUN) level remains normal; symptoms may include nocturia and polyuria. One of the earliest regulatory functions of the kidney to fail is the maximal urine concentrating ability. It becomes necessary for the kidneys to excrete a large urinary volume (up to 3000 ml a day) to rid the body of wastes.

   2. Renal insufficiency—renal function is impaired to the point that metabolic wastes begin to accumulate in the bloodstream; homeostasis is usually maintained sluggishly.

   3. Uremia—renal function is so markedly impaired that homeostasis can no longer be maintained; the BUN level rises sharply and serious fluid and electrolyte disturbances occur.

- Prognosis is variable, depending on the degree of irreversible renal damage. Serious abnormalities in the body fluids may not appear until more than 75% of the nephron population has been destroyed, indicating the remarkable homeostatic ability of the kidneys.

- When the glomerular filtration rate (GFR) is as low as 15% to 20% of normal, the untoward consequences of nephron reduction begin to appear. The greater the fall in GFR, the more serious and extensive are the symptoms; virtually no organ system escapes damage from advanced renal insufficiency. Death is imminent when the GFR has reached 1% to 3%, unless the patient can be maintained on dialysis.

Potassium Imbalances

- Serum potassium concentrations vary widely in patients with chronic renal failure. Some have normal levels until oliguria and starvation occur, causing hyperkalemia. Other causes of hyperkalemia include severe metabolic acidosis, catabolic illnesses, excessive protein intake, administration of stored blood, and gastrointestinal hemorrhage. Use of potassium-sparing diuretics (such as triamterene, spironolactone, or amiloride) may result in fatal hyperkalemia in renal failure. As a rule, hyperkalemia doesn't usually occur until late in chronic renal failure.

- In some cases of chronic nephritis with polyuria, the serum potassium level may be low. Hypokalemia may also result from anorexia, vomiting, diarrhea, and excessive aldosterone production.

- Since potassium metabolism varies from patient to patient, each patient must be evaluated individually to ascertain his need for potassium restriction or replacement.

Metabolic Acidosis

- In late chronic renal failure, metabolic acidosis occurs as a result of the impaired renal excretion of acid metabolites and the inability of the tubules to form ammonia.
- The serum bicarbonate usually stabilizes at approximately 18 to 20 mEq/liter.
- Although the respiratory mechanism partially corrects the decreased blood *p*H, it is sometimes necessary to treat the acidosis medically; sodium bicarbonate tablets may be given, in dosages of several grams a day, depending on the severity of the acidosis and on the sodium intake limitations.
- Acidosis is usually not treated unless the serum bicarbonate is less than 16 mEq/liter or symptoms are present (such as anorexia, weakness, apathy, and coma).
- Drugs that impose an exogenous acid load (such as aspirin, ammonium chloride, or methionine) worsen acidosis and are contraindicated.
- Calcium administration may be necessary when alkalinizing agents are given to correct acidosis, since tetany may be induced by an increase in the serum *p*H.

Sodium Balance

- Most patients with chronic renal failure maintain sodium balance until relatively late in the course of their disease, provided they are maintained on an average salt intake.
- Because the need for sodium varies with each patient, the desired level of intake is best gauged by observing the 24-hour urinary sodium output when the patient is on a known salt intake.
- Excessive loss of sodium causes contraction of the extracellular fluid volume and constitutes one of the most common causes of an acute exacerbation of the uremic state.

  Excessive sodium loss can occur from the kidneys, vomiting, diarrhea, or profuse sweating. Salt depletion is treated by the oral or intravenous administration of sodium chloride, taking care to avoid overexpansion of the extracellular fluid; such treatment often results in a rise in the GFR and striking clinical improvement.
- If a patient exhibits sodium retention, the dietary intake of sodium must be reduced. (Sodium restricted diets are discussed in Chap. 23.) Diuretics may be used as an alternative to dietary sodium re-

striction since they promote sodium excretion; however, high dosage levels are often required at low GFRs and the potential for toxicity is increased.

- A delicate balance must be struck between (1) allowing enough salt to prevent sodium depletion with its associated FVD and (2) limiting salt intake to avoid FVE with its associated hypertension, congestive heart failure, and pulmonary edema.

- When the uremic patient presents with hyponatremia, the physician must determine if the problem is (1) sodium loss, or if it is (2) water overload (as may occur when the patient has received excessive amounts of water or hypotonic fluids).

    1. Usually it is the former; if it is the latter, water restriction will usually serve to correct the hyponatremia.

    2. If dangerous water intoxication is present, it may be necessary to partially correct the low serum sodium level by the cautious IV administration of 3% or 5% sodium chloride solution. (Dialysis, however, is frequently the treatment of choice for this complication.)

    3. If a low serum sodium level is due to actual sodium loss, it may be necessary to cautiously administer sodium-containing solutions.

Calcium and Phosphorus Derangements and Bone Changes

- Hypocalcemia and hyperphosphatemia occur in chronic renal failure when the GFR falls below 25 to 30 ml/min (normal 100–120 ml/min).

- Phosphates are retained in excess when renal function is impaired. Since a reciprocal relationship exists between phosphorus and calcium, hyperphosphatemia causes hypocalcemia.

- Decreased absorption of calcium from the gut occurs in uremia and contributes to the hypocalcemia.

- Efforts should be made to reduce the phosphate level in the bloodstream (to 5 mg/100 ml or less); dietary intake of phosphorus is restricted, and phosphate-binding gels are given orally to bind phosphorus in the intestine and thereby decrease the entry of phosphorus into the extracellular fluid (ECF). (Hyperphosphatemia is thus minimized and its contribution to a drop in the ionized Ca^{2+} level is minimized.)

- Symptoms of hypocalcemia rarely occur in renal failure patients because acidosis increases ionization of calcium. However, rapid correction (or overcorrection) of metabolic acidosis in a uremic, hypocalcemia patient can precipitate tetany and even grand mal seizures.

- Should hypocalcemic symptoms appear, calcium compounds may be administered orally or intravenously; severe symptoms require treatment by the intravenous route with either calcium gluconate or calcium chloride. (Caution should be exercised if the patient is receiving a digitalis preparation because Ca^{2+} ions enhance the action of digitalis.)

- Vitamin D may also be administered to increase calcium absorption from the intestines.
- The low serum calcium level stimulates the parathyroids, causing hyperplasia and increased secretion of parathyroid hormone. Excess of this hormone eventually causes dissolution of the bone and adds both calcium and phosphorus to the ECF; decreased bone density and strength occurs.
- Pruritis is often associated with calcium–phosphorus derangement and may disappear after subtotal parathyroidectomy, although this procedure is performed less frequently now than in the past.
- Metastatic calcification is said to occur when calcium precipitates in such areas as the joints, cornea of the eyes, heart, or lungs. An attempt is made to keep the serum calcium and phosphorus levels normal to prevent their precipitation into soft tissues.
- Chronic negative calcium balance produces renal osteodystrophy in patients with chronic renal disease.

  Usually the bone disease is not severe enough to cause symptoms, although pain and stiffness of the limbs and joints may occur. Demineralization of bone, most often of the hands, may be revealed in x-ray films; sometimes it isn't discovered until autopsy. In childhood, chronic renal failure may cause bone lesions similar to those of vitamin D deficiency rickets; efforts are directed toward early renal transplantation.

- One type of osteodystrophy (osteomalacia) is due to failure of calcium salts to be deposited in newly formed osteoid tissue.
- Another type of osteodystrophy (osteitis fibrosa) is due to reabsorption of calcium from the bone and replacement with fibrous tissue.
- Treatment of renal osteodystrophy may include phosphate-binding gels, supplemental calcium, vitamin D, and, in extreme cases, parathyroidectomy.

Anemia

Normocytic, normochromic anemia occurs in almost all patients with chronic renal failure. (See the discussion of anemia in the section on acute renal failure.)

Uremia

- Clinical manifestations of uremia are discussed in the section on acute renal failure.

Management

- Pruritis may be relieved by starch or vinegar baths. Nails should be kept short and clean to avoid skin trauma and infection caused by scratching.
- Application of lotions or Aquaphor ointment helps to relieve dry, cracked skin. Antipruritic medications should be given as necessary.
- In some patients with intractable itching, subtotal parathyroidectomy seems to provide relief (although this is done infrequently).
- Good oral hygiene is essential to prevent infections and alleviate oral irritation caused by ammonium hydroxide; a 0.25% acetic acid solution is beneficial as a mouthwash to help neutralize this substance. (Recall that ammonium hydroxide is formed in the mouth as a result of hydrolysis or oral urea.)
- Chewing gum, hard candy balls, and cold liquids are helpful in alleviating the unpleasant taste caused by uremia.
- Gastrointestinal bleeding may necessitate the use of enemas to remove blood from the intestine and, thus, prevent elevation of potassium and the BUN.
- If diarrhea is a problem, administration of Lomotil may be indicated.
- If laxatives are necessary for the treatment of constipation, they should be free of magnesium (since patients with renal damage do not excrete magnesium normally and may thus develop serious hypermagnesemia).
- Since the uremic patient is subject to disorientation and convulsions, his bed should remain in the low position with the siderails up. A padded tongue blade, suction apparatus, and oxygen equipment should be immediately at hand. (Seizures should be treated with IV diazepam [Valium] or short-acting barbiturates, or phenytoin [Dilantin].) The patient should be assessed frequently for changes in the level of consciousness and neuromuscular irritability.
- It is necessary to explain to the family that the patient's altered behavior is related to his toxic condition and not necessarily to anything they have done. The patient's family should be included in all explanations because the uremic patient frequently has some degree of impaired mental status.
- Seizures and other manifestations of uremic encephalopathy usually respond to control of uremia by dialysis.

Fluid and Dietary Implications

- The level of renal impairment at which dietary protein restriction should begin is controversial.
  1. Some authorities recommend maintaining chronic renal failure patients on an unrestricted diet as long as the GFR is above 15

to 20 ml/min and there are no special problems (such as severe acidosis, persistent hyperkalemia, or hypertension).[8]

2. Once the GFR falls below 10 ml/min, some degree of dietary protein restriction is indicated to reduce potassium and phosphorus intake and inhibit the development of acidosis.

- Substitution of essential amino acids may have the positive antiuremic effects of a low-protein diet without the disadvantage (malnutrition) of a severely restricted protein diet.
- Sodium requirements must be individualized for each patient; sodium restriction may be necessary for some and harmful for others.

    1. Excessive sodium intake can result in edema, hypertension, and congestive heart failure; hence, patients retaining excessive sodium and water require a low-sodium diet.
    2. On the other hand, patients who excrete large quantities of sodium in the urine require a normal diet to prevent sodium deficit.
    3. The physician has to determine sodium needs according to the response to test diets and clinical observations such as excessive weight gain.
    4. Abnormal losses of sodium (as in vomiting, diarrhea, or excessive sweating) increase the need for sodium. (Low-sodium diets are discussed in Chap. 23.)

- As a rule, potassium retention is not seen until the GFR falls below 5 ml/min.[9] However, patients should be instructed to avoid high-potassium foods (such as bananas, oranges and dried fruits).
- Vitamin D and calcium supplements are given to treat the renal bone dystrophy of chronic renal failure.
- Chronic renal failure patients who do not have oliguria or anuria should be urged to drink from 2000 ml to 3000 ml of water daily to aid in the elimination of urinary waste products. If the patient is unable to excrete large volumes of water, however, fluid intakes should be limited to his needs.

    1. Fluid intake should be increased during periods of abnormal extrarenal fluid loss (such as occurs in vomiting, diarrhea, fever, and excessive sweating).
    2. In the final stages of chronic renal failure, patients may experience thirst, probably a result of cellular dehydration produced by the increased osmotic effect on the high urea level.

References

1. Maxwell M, Kleeman C (eds): Clinical Disorders of Fluid and Electrolyte Metabolism, 3rd ed, p 768. New York, McGraw-Hill, 1980

2. Ibid, p 745
3. Anderson R, Linas S, Berns A et al: Nonoliguric acute renal failure. N Engl J Med, 296, No. 20:1134, 1977
4. Sanders J, Gordon L: Handbook or Medical Emergencies, 2nd ed, p 267. Medical Examination Publishing Co, 1978
5. Maxwell M, Kleeman C, p 776
6. Ibid, p 785
7. Ibid, p 778
8. Ibid, p 821
9. Hodges R: Nutrition in Medical Practice, p 244. Philadelphia, WB Saunders, 1980

chapter 23

Congestive
Heart Failure

- *Heart failure* refers to an inability of the heart to pump enough blood to meet the metabolic demands of the body.
- Causes of heart failure include factors within or affecting the heart itself, such as cardiomyopathy, infarction, arrhythmias, valvular disease, cardiac tamponade, electrolyte imbalances, and circulatory problems, including inadequate venous return and circulatory overload.
- Ventricular failure can cause symptoms primarily associated with pulmonary congestion (left-sided heart failure), venous congestion (right-sided heart failure), and decreased cardiac output (cardiogenic shock).
- Failure of one side of the heart eventually leads to failure of the other. Heart failure can be acute or chronic.

Left-Sided Heart Failure

- Left-sided heart failure is primarily a backward failure, which causes damming of blood back from the left side of the heart to the pulmonary vessels.
- Symptoms of pulmonary vascular congestion include dyspnea, orthopnea, dry cough, and pulmonary edema. Pulmonary capillary pressure above 30 mm Hg causes transudation of fluid into the alveoli and diminished oxygen–carbon dioxide exchange.[1]

This chapter was prepared with the assistance of Martha A. Spies, RN, MSN, Assistant Professor of Nursing, Saint Louis University, St. Louis, Missouri.

- Acute pulmonary edema is characterized by severe dyspnea, profound anxiety, cyanosis, noisy respirations, and pink, frothy sputum.
- Physical signs associated with left-sided heart failure include 3rd and 4th heart sounds (ventricular gallop and atrial gallop, respectively) and fine moist rales in the lungs. Ventricular gallop in adults is almost never present in the absence of significant heart disease.
- Rhythms associated with left-sided heart failure include sinus tachycardia, atrial premature contractions, paroxysmal atrial tachycardia, and ventricular premature beats.
- Other signs of left ventricular failure may include pulsus alternans (an alternating greater and lesser arterial pulse volume), expiratory wheezing breath sounds, and Cheyne-Stokes respirations.

Right-Sided Heart Failure

- In unilateral right-sided heart failure, blood is not pumped adequately from the systemic circulation into the lungs; back pressure of the right heart causes systemic edema.
- Right-sided heart failure may result from stenosis of the pulmonary tricuspid valve, cor pulmonale (chronic bronchitis, emphysema, bronchiectasis), or massive pulmonary embolism.
- Acute right-sided failure rarely occurs alone, when it does, symptoms are related to a low cardiac output. In rare conditions in which the right heart fails acutely, cardiac output may be so low that death occurs rapidly.
- Chronic right-sided heart failure leads to progressive development of peripheral congestion, hepatosplenomegaly, and ascites.
- Failure of the right ventricle causes a rise in the right atrial and vena caval pressures.
- Neck veins will appear distended when the patient is lying in bed with the head of the bed elevated between 30 and 60 degrees. (See Fig. 2-3).
- As with left-sided failure, sinus tachycardia and other rhythms associated with pump failure may be present. Right ventricular 3rd and 4th heart sounds may be heard.
- Usually, right-sided heart failure is the result of left-sided failure; in this case, symptoms of both right- and left-sided failure are present.

Adaptive Mechanisms in Decreased Left Ventricular Function

Nursing care of the patient in heart failure should be based on an understanding of the adaptive mechanisms involved.

Hemodynamic Mechanisms

- When cardiac output falls, the first response is stimulation of the sympathetic nervous system and inhibition of the parasympathetic nervous system. This tends to support arterial pressure even when the stroke volume is decreased.
- Blood is preferentially shunted to the brain and heart while being directed away from the skin, kidneys, and skeletal muscles; this shunting increases cellular oxygen extraction from the blood from 25% to 60%. If the shunting is unsuccessful in preventing cellular hypoxia, the cells shift to anaerobic metabolism, which can lead to metabolic acidosis.
- Venous tone is increased to maintain venous return and ventricular filling. There is also an increased heart rate (probably caused by increased release of norepinephrine).
- In chronic heart failure, the ventricular myocardium will hypertrophy (increasing the heart's ability to maintain adequate stroke volume when tissue needs increase).

Salt and Water Retention by Kidneys

- Decreased cardiac output causes a decrease in glomerular filtration rate (GFR). This, along with the shunting of blood away from the kidney, can result in almost complete anuria when the cardiac output decreases by one-third.[2] The kidneys respond by retaining sodium and water.
- If the cardiac output remains low, the renin–angiotensin system will be activated to increase secretion of aldosterone from the adrenal cortrex. The effect is further sodium and water retention while increasing potassium excretion.
- When cardiac output remains low, the release of antidiuretic hormone (ADH) from the posterior pituitary is stimulated. ADH acts on the distal tubules to cause water retention; patients may retain water in excess of sodium.
- The effects of aldosterone and ADH may be further increased if inactivation of these hormones by the liver is inhibited because of liver congestion or cirrhosis.
- Circulating blood volume may also be increased by an increase in the production of red blood cells. This is accomplished as hypoxia stimulates the release of erythropoietin, which then causes the bone marrow to increase the production of erythrocytes.
- All of these mechanisms increase blood volume 5% to 15% in mild heart failure and 30% to 50% in severe heart failure. These adaptive mechanisms aim at preserving stroke volume in compensated heart failure. But increased blood volume and venous return can increase preload beyond the ability of the heart to adapt. This leads to decompensated heart failure, resulting in even further fluid retention.

Summary of Water and Electrolyte Disturbances Accompanying Heart Failure

The adaptive mechanisms described above, in addition to some of the therapies for heart failure, can lead to a number of fluid and electrolyte alterations. These imbalances are summarized in Table 23-1.

Nursing Assessment of Patients With Heart Failure

Clinical Manifestations of Heart Failure

Clinical manifestations of chronic ventricular failure are presented in Table 23-2. These can be utilized to provide an understanding of the factors to assess patients at risk for developing heart failure. Knowledge of the underlying pathophysiology also allows the nurse to provide effective nursing therapy in this complex situation.

Hemodynamic Monitoring

- The use of hemodynamic monitoring in heart failure has made it possible to measure pressures on a continuing basis directly within the heart chambers and great vessels and to monitor cardiac output.
- Changes in these pressure measurements can indicate early left ventricular failure before other signs and symptoms develop, thus they have become an important part of nursing assessment for left ventricular failure.
- Left ventricular function is monitored indirectly by measuring pulmonary artery and pulmonary capillary wedge pressures with a flow-directed catheter. These pressures provide an indication of the left ventricular end-diastolic pressure (LVEDP).
  1. During left ventricular diastole, the mitral valve is open and the pressures become equalized in the left ventricle and left atrium. This pressure is reflected in retrograde fashion into the pulmonary veins, pulmonary capillary bed, and pulmonary artery. Thus, the pulmonary artery end-diastolic pressure (PAEDP) is a reflection of LVEDP.
  2. In addition to assessing left ventricular function, the PAEDP is useful in monitoring the effect of therapies aimed at restoring left ventricular function. These therapies include inotropic agents, vasodilator medications, diuretics, and dietary/fluid restrictions.
- Cardiac output can be obtained with the use of a Swan-Ganz catheter. Cardiac output refers to the quantity of blood pumped each

(*Text continues on p. 322.*)

Table 23–1
Causes of Water and Electrolyte Disturbances in Congestive Heart Failure

Cause	Water and Electrolyte Disturbance

Excessive aldosterone secretion
Decreased renal blood flow secondary to cardiac failure and vasoconstriction
	Increased retention of Na^+ and water by the kidneys resulting in
	  Increase in total Na^+ content of body
	  Increase in total extracellular water volume

Excessive secretion of ADH causes increased retention of water	Relatively greater retention of water than Na^+ May depress serum Na^+ to abnormally low levels, even though the total body sodium is above normal
Hydrostatic pressure is increased by the excessive venous blood volume	Shift of fluid from the intravascular compartment to the interstitial compartment with edema
Excessive aldosterone secretion promotes K^+ excretion Excessive use of K^+-losing diuretics or prolonged loss of K^+ by vomiting or diarrhea represent typical causes of K^+ deficit	K^+ deficit
Slowing of the circulation interferes with the excretion of metabolic acids Increased liberation of lactic acid from anoxic tissues and failure of the body to metabolize it rapidly	Mild metabolic acidosis
Pulmonary congestion interferes with the elimination of CO_2 from the lungs	Respiratory acidosis
Mercurial and thiazide diuretics cause a greater excretion of Cl^- ions than Na^+ ions; loss of Cl^- ions causes a compensatory increase in HCO_3^- ions, hence alkalosis	Metabolic alkalosis if mercurial or thiazide diuretics are used extensively
Extensive use of potent diuretics plus severely restricted Na^+ intake Excessive loss of Na^+ from other routes, such as repeated paracentesis, vomiting, or diarrhea	Na^+ deficit

Table 23–2
Clinical Manifestations of Congestive Heart Failure and Their Causes

Symptom or Sign	Cause
Fatigue with little exertion or at rest	Tissue anoxia due to decreased cardiac output
Dyspnea on exertion	Cardiac output is inadequate to provide for the increased oxygen required by exertion
Increased respiratory difficulty	Increased tissue hypoxia caused by progressive failure of the heart as a pump. Increased interstitial edema decreases the lung compliance and increases the work of breathing
Dyspnea at rest	
Orthopnea	
Paroxysmal nocturnal dyspnea (with left heart failure)	When recumbent, edema fluid from the dependent parts returns to the bloodstream and increases preload and causes decompensation
Troublesome cough producing noncharacteristic sputum, although it may at times be brownish or blood-tinged	Transudation of serum with hemosiderin-filled macrophages into the alveoli causes pulmonary congestion
Pulmonary rales and wheezing	
Tachycardia	Effort to compensate for decreased cardiac output
Abnormal heart sound — S_3	Associated with rapid ventricular filling in noncompliant ventricle
Elevated pulmonary capillary wedge pressure	The left ventricle cannot maintain stroke volume in the face of increased venous return
Elevated venous pressure	Increase in total blood volume
	Accumulation of blood in the venous system results from incomplete emptying of the heart
Cardiomegaly	Hypertrophy of myocardium helps to maintain stroke volume
	Venous distintion
Cyanosis, particularly of lips and nail beds	Inadequate oxygenation of blood

(continued)

Table 23–2
Clinical Manifestations of Congestive Heart Failure and Their Causes (continued)

Symptom or Sign	Cause
Visible distention of peripheral veins, most noticeable on face, neck, and hands	Elevated venous pressure
Decreased urinary output	Decreased cardiac output and renal blood flow
	Na^+ and water retention caused by excessive aldosterone secretion
	Increased water retention caused by excessive ADH secretion
Edema first appears in dependent parts	Hydrostatic pressure is greatest in dependent parts of the body
Edema later becomes generalized	Progressive cardiac failure causes substantial increase in hydrostatic pressure in all parts of the body
Engorgement of the liver and other organs	Decreased cardiac output causes damming of venous blood
	Increase in total blood volume and interstitial fluid volume
Nausea and vomiting	Edema of the liver and intestines
	Impulses arising from the dilated myocardium in acute CHF
	Digitalis toxicity
Anorexia	K^+ deficit
	Digitalis toxicity
Constipation	Poor nourishment and inadequate bulk in diet
	Lack of activity
	Depression of motor activity by hypoxia
Pulmonary edema with severe dyspnea, coughing and pink, frothy fluid, cyanosis, shock, and death	Increased venous pressure may cause serum and blood cells to transudate into the alveoli

minute. When compared to body surface area, this is called the cardiac index and the normal is 2.8 to 3.2 liters/min/m².

- The relationship between pulmonary artery wedge pressure readings and cardiac index can be used to identify the presence and type of left ventricular failure; it also indicates the type of medical therapy needed and assists in evaluating the therapy's effectiveness.

Treatment of Congestive Heart Failure: Nursing Implications

When possible, treatment involves elimination of the underlying disease producing the heart failure. For example, surgical correction of a valvular disorder or removal of a calcified pericardium may restore cardiac function to normal and produce a spontaneous diuresis. Unfortunately, most persons have irreversible cardiac damage, such as that caused by myocardial infarction. When the primary disease cannot be eliminated, the only alternative is to make the most efficient use possible of remaining cardiac function.

Decreasing Myocardial Workload

- Rest causes a reduction in the tissue's oxygen need and decreases the workload of the heart. It also produces a physiologic diuresis.
- Sometimes rest alone is sufficient in alleviating the symptoms of congestive heart failure. The amount of rest required varies with the individual and may range from complete bedrest to only slight restriction of activity.
- The prescription of physical rest by the physician must be specific enough to have meaning to the patient. Too often, patients are given ambigious directions "to rest" or "to take it easy." Such vague statements are not only useless; they may actually be harmful, because each person interprets rest differently. The nurse can help by encouraging the patient to ask the physician specific questions regarding the activities permitted.
- Another important nursing responsibility consists in observing patient responses to exercise such as pulse and respiratory rate changes. An increase in heart rate greater than 10% over the baseline rate indicates that an activity may exceed the capacity of the failing heart to respond. Careful reporting of these observations helps the physician determine the desired amount of activity.
- Emotional rest is also important in the management of congestive heart failure. Periods of tension are associated with increased sodium and water retention, while periods of emotional relaxation are associated with diuresis. Nursing efforts should, therefore, be directed toward avoiding emotional problems and achieving a relaxing environment.

- Other interventions to ensure decreased myocardial workload include
  1. Relief of any existing pain
  2. Reduction of fever, if present
  3. Oxygen therapy to maintain a PaO$_2$ of at least 80 mm Hg
  4. Placing the patient in a semi-Fowler's position (facilitates ventilation by allowing full chest expansion and diaphragmatic excursion. The upright position may also decrease venous return)

Decreasing Venous Return

- There are three aspects involved in the long-term management of venous return.
  1. Low-sodium diet
  2. Diuretics
  3. Fluid restriction

Sodium-Restricted Diets

- Restriction of sodium ions in the diet is a valuable aid in the management of congestive heart failure. In general, the fewer sodium ions in the body, the less water is retained.
- An average daily diet not restricted in sodium contains 6 Gm to 15 Gm of salt, whereas, low sodium diets can range from a mild restriction to as low as 250 mg of sodium a day, depending on the patient's needs.
- It should be noted that the sodium content of diets is often expressed in Gm of Na$^+$ rather than in Gm of NaCl. One gram of Na$^+$ equals 43 mEq, whereas 1 Gm of NaCl equals 17 mEq.[3] (Thus, a 4-Gm Na$^+$ diet is approximately equivalent to a 10-Gm NaCl diet in sodium content.)
- The American Heart Association has prepared booklets describing 500-mg, 1000-mg, and mildly restricted sodium diets.
  1. The 500-mg sodium diet can be changed to a 250-mg diet by substituting low-sodium milk for regular milk.
  2. Food exchange lists have been devised to simplify sodium-restricted diets; they include, list 1—milk and milk products, list 2—vegetables, list 3—fruits, list 4—breads, list 5—meats, list 6—fats, and list 7—free choices. Also listed are seasonings and miscellaneous items allowed and not recommended.
  3. These booklets are available to patients by a physician's prescription.
- A mild sodium-restricted diet requires (1) only light salting of food (about half the amount as usual) in cooking and at the table, (2) no addition of salt to foods that are already seasoned (such as canned

foods and foods ready to cook or eat) and (3) avoidance of foods that are very high in sodium.

- Examples of foods high in sodium content include
  1. Sauerkraut and other vegetables prepared in brine
  2. Bacon, luncheon meats, frankfurters, ham, kosher meats, sausages, salt pork, and sardines
  3. Relishes, horseradish, catsup, mustard, and worcestershire sauce
  4. Processed cheese
  5. Olives and pickles
  6. Potato chips, pretzels, and other salty snack foods
  7. Peanut butter
- A definite relationship exists between the protein content of a diet and the degree of sodium limitation that it permits; thus, foods such as eggs, cheese, ordinary milk, meat, fish, poultry, and seafood must be used in measured amounts.
- The patient should be aware that most canned foods and ready-to-eat foods already have salt added, and, thus, should be used only as their specific diets allow.
- Foods that may be used freely include most fresh vegetables and fruits and unprocessed cereals. Low-sodium milk, milk products, and bakery goods are available in most large cities.
- Numerous compounds containing sodium are used by manufacturers of processed foods or in home preparation of foods to improve texture of flavor or to maintain freshness. Among these are
  1. Sodium chloride (table salt; 1 teaspoon contains about 1955 mg of Na^+)[4]
  2. Sodium bicarbonate (baking soda; 1 teaspoon contains about 821 mg of Na^+)[5]
  3. Baking powder (1 teaspoon of commercial pyrophosphate baking powder contains about 486 mg of Na^+)[6]
  4. Monosodium glutamate (MSG; 1 teaspoon contains about 765 mg of Na^+)[7]
  5. Meat tenderizer (1 teaspoon of Adolph's meat tenderizer contains about 1745 mg of Na^+)[8]
  6. Soy sauce (1 teaspoon contains approximately 365 mg of Na^+)[9]
  7. Sodium benzoate (used as a preservative in many condiments)
  8. Sodium cyclamate (an artificial sweetener)
  9. Sodium propionate (used in pasteurized cheeses and some breads and cakes to inhibit mold growth)
  10. Sodium alginate (used in many chocolate milks and ice cream for smooth texture)
  11. Disodium phosphate (present in some quick-cooking cereals and processed cheeses)

- Patients should be instructed to look for the words sodium, salt, or soda on food labels. The degree of limitation of commercially prepared foods varies with the severity of sodium restriction indicated for each patient.
- Patients should be made aware that some foods do not list sodium as a content even though it is present; examples are standardized foods such as mayonnaise and catsup.
- Medicines not prescribed by the physician should be avoided, since some contain enough sodium to interfere with the desired intake. Among these are
  1. Alkalinating agents (such as Alka-Seltzer)
  2. Certain laxatives
  3. Salicylates
  4. Certain cough syrups
  5. Certain antibiotics
- It may be necessary for patients to use distilled water when the local water supply is very high in sodium content.
- In certain communities, the drinking water may contain too much sodium for a sodium-restricted diet. Depending on its source, water may contain as little as 1 mg or more than 1500 mg of sodium per quart.[10]
- Salt substitutes are sometimes used to make low-sodium diets more palatable.
  1. These preparations contains potassium and should be used cautiously in patients taking potassium-sparing diuretics (such as spironolactone, dyrenium, and amiloride), potassium supplements, or particularly in those who have renal impairment and thus have an altered ability to excrete potassium.
  2. Potassium-containing salt substitutes may be useful in preventing hypokalemia in patients taking potassium-losing diuretics (such as furosemide or the thiazides).
  3. Salt substitutes containing ammonium chloride can be harmful to patients with liver damage.
  4. Salt substitutes and their sodium and potassium contents are listed in Table 8-2.
- Occasional monitoring of the 24-hour urinary sodium excretion can be a useful compliance check on dietary sodium restriction.

Diuretics

- Diuretics are a valuable aid in the symptomatic treatment of congestive heart failure; their primary purpose is to promote the excretion of sodium and water from the body. If hemodynamic monitoring is being used, it may be noted that diuretics produce a decrease in pulmonary artery wedge pressure (preload). Most diuretics produce, in varying degrees, disruptions in potassium balance. Diuretics are discussed in Chapter 8.

- Primary nursing responsibilities in the care of heart failure patients receiving diuretics and restricted sodium diets include keeping an accurate account of fluid intake and output and measuring the weight daily. The data obtained from these measurements are of inestimable use to the physician in regulating the dose of diuretics and the degree of dietary sodium restriction for each patient.

Fluid Restriction

- Water intake is usually not restricted in the long-term management of congestive heart failure unless there is a body sodium deficit or dilution of the serum sodium by the excessive retention of water.
- The intravenous route for fluid administration may be necessary in critically ill patients with congestive heart failure. Many physicians are hesitant to administer fluids to such patients for fear of causing circulatory overload and pulmonary edema.
- While there is little doubt that intravenous administration of fluids to a cardiac patient carries some risk of causing circulatory overload, the fear of this complication has been exaggerated to the point that many cardiac patients receive inadequate fluid therapy. The recent increase in the use of hemodynamic monitoring devices has done much to alleviate this problem. Frequent checks of venous pressure and pulmonary artery wedge pressure during fluid administration give early warning of circulatory overload and serve as guides to the safe administration of needed water and electrolytes.
- It is important to pay careful attention to the volume, speed, and composition of fluids administered to patients with congestive heart failure. The response to fluids should be observed frequently and the flow rate adjusted accordingly.

Increasing Myocardial Contractility and Cardiac Output

Digitalis

- Digitalis preparations are given to patients with decreased left ventricular function because of their inotropic action (*i.e.*, digitalis increases the force of myocardial contraction). The mechanism of action for this inotropic effect is not completely understood; it may act by increasing the availability of free calcium ions to the contracting sites of the heart muscle.[11]
- As the heart contracts more forcefully, tissue perfusion increases and the compensatory responses caused by hypoxia decrease, allowing a corresponding increase in renal function and diuresis. Inotropic agents in general are given to increase cardiac output and

decrease pulmonary artery wedge pressure by increasing the strength of myocardial contraction.

- Excessive doses of digitalis may result in the following toxic symptoms:
    1. Aversion to food (which usually precedes other symptoms by 1 or 2 days)
    2. Nausea
    3. Excessive salivation
    4. Vomiting
    5. Abdominal pain
    6. Diarrhea
    7. Urticaria
    8. Gynecomastia
    9. Headache, fatigue
    10. Confusion (particularly in elderly patients with arteriosclerosis; CNS symptoms are often late effects)
    11. Blurred vision, diplopia, optic neuritis
    12. Yellowish-green "halo" vision or presence of white dots (frost) on objects
    13. Bradycardia (caused by A-V block)
    14. Variety of arrhythmias, including ventricular tachycardia, in which the heart beats rapidly and irregularly (arrhythmias may precede extracardiac manifestations)
    15. Bigeminal pulse
- It is important to differentiate between the combined anorexia and nausea of heart failure and that produced by digitalis toxicity. Patients receiving digitalis should have periodic electrocardiograms to detect the development of digitalis toxicity early. Serum digitalis levels also provide important information.
- Before administering the drug, it is important to check the apical–radial pulse for a full minute, noting rate, rhythm, volume, and pulse deficit. In order to evaluate significant changes, one must be familiar with the patient's baseline heart rate and rhythm.
- Unless otherwise ordered, withold the drug and notify the physician when
    1. The apical is below 60
    2. A marked change in rate or regularity occurs (premature ventricular contractions), bigeminy, atrial fibrillation, sudden marked change from a tachycardia (*e.g.*, a change in heart rate from 160 to 80 may indicate that a 2:1 heart block has developed), sudden spurts of a rapid pulse (*e.g.*, paroxysmal atrial tachycardia with varying block)
- An overdose of digitalis may have a depressant action, causing con-

duction disturbance and excessive slowing of the heart. It may also cause increased myocardial irritability, producing extrasystoles or tachycardias.

- One should be alert for a coupled pulse beat (bigeminy) in which the regular beat is followed almost immediately by a weak beat and a pause. A bigeminal pulse is a common sign of digitalis toxicity in adults; bigeminal pulse, and other irregularities should be noted and reported to the physician.
- Calcium ions enhance the action of digitalis; thus, decreasing the serum calcium concentration is helpful in counteracting the cardiotoxic effects produced by digitalis. Calcium should never be administered intravenously to a digitalized patient.
- Symptoms of digitalis toxicity may be induced by potassium deficit, since this deficit sensitizes the heart to digitalis. Patients prone to developing potassium deficit (such as those receiving furosemide or thiazide diuretics, or those with vomiting, diarrhea, or poor food intake) should be observed especially carefully for signs of digitalis toxicity.

Pulmonary Edema

- Pulmonary edema is a manifestation of severe left ventricular failure. The signs and symptoms of pulmonary edema include
  1. Restlessness
  2. Severe dyspnea
  3. Gurgling respirations
  4. Cyanosis
  5. Coughing up of frothy fluid
  6. Elevated pulmonary artery wedge pressure
- It is important to identify patients at risk for pulmonary edema and provide nursing care that decreases the heart's workload. This includes monitoring activity levels, careful patient teaching, and emotional support so that anxiety does not stimulate the sympathetic nervous system, causing increased cardiac output. Preventive interventions also include careful monitoring of intravenous fluids (especially when NaCl must be given to correct a profound Na^+ deficit).
- The occurrence of pulmonary edema is an emergency situation. The patient should be quickly placed in a high-Fowler's position, and oxygen started.
- Often, positive pressure breathing devices are used for assisting ventilation because this increases intraalveolar pressure so that fluid does not continue to move into the alveoli. Positive pressure ventilation also increases the intrathoracic pressure so that venous return and, therefore, preload is reduced.

- Intravenous administration of morphine sulfate may be utilized to achieve multiple effects in the patient with pulmonary edema.
  1. Decreases preload through peripheral venous vasodilatation (which decreases venous return)
  2. Decreases afterload by decreasing arterial blood pressure
  3. Decreases the myocardial workload
  4. Decreases anxiety (thus decreasing sympathetic nervous system stimulation)

  Morphine must be given cautiously to patients with increased intracranial pressure, with severe pulmonary disease, or with decreased level of consciousness.
- Alternating tourniquets may be used to obstruct venous return to the heart, and can remove up to 700 ml of blood from the circulating volume.
- The removal of 100 ml to 500 ml of blood by phlebotomy may be tried to relieve the workload on the heart and to reduce venous pressure.
- Diuretics may be given intravenously to produce a rapid diuresis. All of these measures help to decrease preload, which will be reflected in a decreased pulmonary artery wedge pressure.
- An intravenous vasodilator such as sodium nitroprusside may be given to decrease afterload. This decreases the workload of the heart by lowering the peripheral resistance to left ventricular output. This can cause an increased cardiac output and decreased pulmonary artery wedge pressure. Pulmonary venous pressure usually decreases significantly because of the peripheral venous pooling effect caused by vasodilators. Myocardial oxygen consumption is decreased.
- Careful monitoring of arterial pressure, cardiac output, and pulmonary artery wedge pressure should be done when parenteral vasodilator therapy is employed in the management of such patients. A major possible complication of vasodilator therapy is a pronounced drop in arterial pressure, which could increase myocardial ischemia by restricting coronary blood flow.
- An inotropic agent such as digitalis may be given. In the undigitalized patient, 0.5 mg to 1.0 mg of digoxin may be given intravenously. This can be followed by additional doses as indicated.
- One should remember that acidosis, hypoxia, and electrolyte imbalance can precipitate or prolong cardiac failure. Because of this, arterial blood gases and electrolyte levels should be monitored closely in the acutely ill patient. Any chronic abnormality must not be corrected too abruptly.
  1. Fluid volume deficit, if present, is corrected first.
  2. Then, pH alterations are treated, along with imbalances of potassium and calcium.

3. It should be noted that myocardial contractility may be decreased when the pH is greater than 7.55 or less than 7.20, the PCO_2 is less than 25 mm Hg, the serum potassium is elevated; or the serum calcium level is decreased.

4. Adequate ventilation is mandatory to prevent hypoxia and respiratory acidosis; if necessary, intubation and assisted ventilation should be employed.

References

1. Guyton A: Textbook of Medical Physiology, 6th ed, p 295. Philadelphia, WB Saunders, 1981
2. Wilson R (ed): Principles and Technique of Critical Care, Section D, p 8. Kalamazoo, Upjohn, 1976
3. Freitag J, Miller L (eds): Manual of Medical Therapeutics, 23rd ed, p 23. Boston, Little, Brown & Co, 1980
4. Pennington J, Church H: Bowes and Church's Food Values of Portions Commonly Used, 13th ed, p 49. Philadelphia, JB Lippincott, 1980
5. Ibid
6. Ibid
7. American Heart Association: Your Mild Sodium Restricted Diet (revised), p 10
8. Pennington J, Church H: Bowes and Church's Food Values of Portions Commonly Used, 13th ed, p 149. Philadelphia, JB Lippincott, 1980
9. American Heart Association: Your Mild Sodium Restricted Diet (revised), p 10
10. Ibid, p 9
11. Horvath P, DePew C: Toward preventing digitalis toxicity. Nurs Drug Alert 4, No. 4: 25, 1980

chapter 24

Burns

Burns cause a series of major water and electrolyte changes. The purpose of this chapter is to explore these changes and their implications for nursing care. A background discussion of physiologic changes accompanying burns precedes the discussion of treatment and nursing care.

Evaluation of Burn Severity

The severity of water and electrolyte changes is largely dependent on the burn depth and the percentage of body surface involved.

Burn Depth

- Burns may be classified as first, second, or third degree burns, according to the depth of skin damage. (A more recent classification of burns is partial-thickness and full-thickness burns; deep, partial-thickness burns extend into the dermis and full-thickness burns extend into the subcutaneous tissue or deeper.)
- Factors considered in determining burn depth (Table 24-1) include the following:
    1. Nature of burning agent plus length of exposure to it
    2. Appearance of the burned surface
    3. "Feel" of the burned surface with the gloved hand
    4. Amount of sensation remaining
- *First-degree burns* are characterized by simple erythema, with only microscopic destruction of superficial layers of the epidermis. This type of injury is of little clinical significance since the skin's water barrier is not disturbed. First-degree burns are not considered in the planning of fluid replacement.

Table 24-1
Diagnosis of Burn Depth

Degree	Nature of Burn	Symptoms	Characteristics	Course
Epidermal Burn				
First	Sunburn	Tingling Hyperesthesia Painful Soothed by cooling	Reddened; blanches with pressure	Complete recovery within 1 wk
Intradermal Burn (Partial-Thickness)				
Second	Scalds (spills) Flash flame	Painful Hyperesthesia Sensitive to cold air	Blistered; mottled red base; broken epidermis; weeping surface Feels moist; soft and pliable to touch Edema Hair stubble visible	Recovery in 2–3 wk Some scarring and depigmentation
Subdermal Burn (Full-Thickness)				
Third	Fire Electrical Scalds (immersion) Chemical contact	Painless Shock	Dry; pale white or charred Feels hard and dry to touch Edema Hair absent	Eschar (nonviable skin) sloughs Grafting necessary Scarring and loss of contour function

- *Second-degree burns* extend through the epidermis into the dermis; this type of injury is characterized by damaged capillaries and the appearance of blebs containing fluid. Some skin elements remain viable, from which epithelial regeneration can occur.
- In *third-degree burns* (full-thickness) burns, there is destruction of all skin elements; spontaneous regeneration of epithelium is not possible. Clinically the burn appears dry, hard, inelastic, and translucent, with thrombosed veins.

Percentage of Surface Involved

- The "rule of nines" is commonly used to estimate the severity of burns in adults. It divides the body surface into areas of 9% or its multiples:

Head	=	9%
Each arm	=	9%
Each leg	=	18%
Front of torso	=	18%
Back of torso	=	18%
Genitalia	=	1%

  1. The proportion of each of these areas with either second- or third-degree burns is estimated; summations of these estimates represents the percentage of total body surface area burned.
  2. Unless used cautiously, this method can result in dangerously high estimates.
- Factors other than the percentage of burned body surface area can influence the severity of the burn. For example, burns are made more serious by the presence of prior renal, cardiac, or metabolic disorders; concurrent injuries; burns of the face, hands, or genitalia; respiratory burns; and extreme age variation (very young children and the elderly).
- Mortality associated with burns has been markedly reduced in the past 25 years. In several series, 50% of the patients between the ages of 8 and 45 have been reported to survive burns of 75% of the total body surface. However, individuals over the age of 65 with burns of more than 25% total body surface have a guarded prognosis.[1] (Recall that chronic disease states in the aged interfere with appropriate physiologic responses to major burns.)

Water and Electrolyte Changes in Burns

Loss of Body Fluids in Burns

- Body fluids are lost in severe burns as
  1. Plasma leakage from intravascular space into interstitial space (edema)

2. Water vapor (lost from denuded burn site)
3. Blood leakage from damaged capillaries

Plasma-to-Interstitial Fluid Shift

• Intravascular water, electrolytes, and protein are lost through damaged capillaries at the burn site. Clinically, the shift results in edema; the magnitude of this shift depends upon the burn depth and the percentage of surface area involved.

Consider a specific example: An adult with a body surface of 1.75 m² has about 10.5 liters of extracellular fluid. If he sustains a 50% burn, the volume of edema fluid formed during the first day or two would approximate 5.25 liters, a quantity exceeding the total plasma volume of the patient. Obviously, all of the edema fluid is not derived from plasma; some of it comes from the body cells and some from administered fluids.

• Proportionately greater amounts of water and electrolytes than of protein are lost from the plasma. (Protein molecules are larger and, thus, fewer escape through the damaged capillaries.) As a result, the circulating plasma protein becomes more concentrated; the increased osmotic pressure draws fluid from undamaged tissues in all parts of the body, and generalized tissue dehydration results.

• The increased capillary premeability lasts for approximately 24 hours; at this time, the capillary walls seal and compensatory changes set in.[2]

Water Vapor and Heat Loss

• The intact skin serves as a barrier against the loss of water and heat. When skin is destroyed by a burn, increased water and heat loss result, and the larger the burned surface, the greater the loss of water vapor and heat. Burned infants are particularly vulnerable to heat loss.

• Maintenance of an environment saturated with water vapor decreases the transcutaneous water loss; for example, wounds covered with dressings of aqueous topical agents and dry covers lose only half the amounts of water vapor lost through exposed wounds.

• Some physicians feel that early coverage of burn wounds with a temporary biological dressing (such as pig skin) decreases evaporative water and heat loss.

Water and Electrolyte Changes in Major Burn Phases

• The nurse should be aware of the water and electrolyte changes occurring in burns so that significant changes can be observed for and recognized.

• An outline of expected water and electrolyte changes is presented in

Table 24-2. (The reader is referred to earlier chapters for detailed descriptions of these imbalances.)

Physiologic Basis for Treatment and Nursing Care During the Fluid Accumulation Phase

The adequacy of early burn treatment largely depends upon the physician's and the nurse's understanding of physiologic derangements caused by burns, the organization of equipment, and the ability to act quickly and skillfully.

Need for Early Treatment

- The shift of fluid from plasma to the interstitial space is rapid and well underway by the end of the first hour. The maximal speed of edema formation is reached by the end of the first 8 to 10 hours; the shift continues until the 36th to 48th hour. (By this time, the capillaries have healed sufficiently to prevent further fluid loss.)
- Oliguria or anuria are particularly threatening during this phase because of the excessive amounts of potassium flooding the extracellular fluid. (Remember that K^+ is mainly excreated in urine; decreased urinary output causes a dangerous excess to build up in the bloodstream.)
- The sodium deficit requires prompt attention, as does the acidosis so frequently present.
- Acute tubular necrosis may result from hypovolemia, hemoglobinuria, or myoglobinuria and is an entirely preventable complication. Prevention consists of adequate fluid replacement and, when necessary, the administration of mannitol to flush out the tubules.

Initial Patient Evaluation

- Pertinent questions to consider when the patient is first seen include
  1. When did the burn occur? (The degree of fluid shift is related to the length of time the burn has been present.)
  2. What was the nature of the burning agent? (Notice in Table 24-1 that burns are often classified in relation to burning agents.)
  3. What was the length of exposure to the burning agent? (Questions listed above are intended to help establish the burn depth —appearance of the burns on admission is often misleading.)
  4. Were any medications given prior to hospital admission? (Sometimes narcotics are given at the scene of the accident; it is

(Text continues on p. 340.)

Table 24–2
Water and Electrolyte Changes in Various Burn Phases

Phase	Water and Electrolyte Changes	Comments
Fluid accumulation phase (shock phase) First 48–72 hr in most patients; may last longer in the elderly	Plasma-to-interstitial fluid shift (edema at burn site) Generalized FVD	Plasma leaks out through the damaged capillaries at the burn site — edema forms Undamaged tissues give up fluids to help increase plasma volume — part of it leaks through the damaged capillaries and helps form edema
	Contraction of blood volume	Loss of plasma causes a decreased circulatory volume; a rise in pulse rate ensues
	Decreased urinary output	Secondary to Decreased renal blood flow Increased secretion of ADH Sodium and water retention caused by stress (increased adrenocortical activity) Severe burns may cause hemolysis of red blood cells; the ruptured cells release Hb, and it is excreted by the kidneys (hemoglobinuria can cause severe renal damage)
	Cellular uptake of Na^+ and water	The so-called "sick-cell" syndrome (an alteration that burn injuries share with other major, nonthermal forms of trauma)[3]

(continued)

Table 24-2
Water and Electrolyte Changes in Various Burn Phases (continued)

Phase	Water and Electrolyte Changes	Comments
	Hemoconcentration (elevated hematocrit)	Relatively greater loss of liquid blood components in relation to blood cell loss
Fluid remobilization phase (stage of diuresis) Starts 48–72 hours postburn	Interstitial fluid-to-plasma shift	Edema fluid shifts back into the intravascular compartments
	Hemodilution (hematocrit decreased)	The blood cell concentration is diluted as fluid enters the vascular compartment and, in addition, at this time a decrease in the number of cells becomes evident (destruction of red blood cells at the burn site causes anemia — as much as 10% of the total number of RBCs may be destroyed)
	Increased urinary output	Fluid shift into the intravascular compartment increases renal blood flow and causes increased urine formation
	Na$^+$ deficit K$^+$ deficit (may occasionally occur in this phase)	Na$^+$ is lost with water when diuresis occurs Beginning about the fourth or fifth postburn day, K$^+$ shifts from the extracellular fluid into the cells
Convalescent phase	Metabolic acidosis Ca^{2+} deficit	Since Ca^{2+} may be immobilized at the burn site in the slough and early granulation phase of burns, symptoms of Ca^{2+} deficit

(continued)

Table 24–2
Water and Electrolyte Changes in Various Burn Phases (continued)

Phase	Water and Electrolyte Changes	Comments
	Hemoconcentration (elevated hematocrit)	Relatively greater loss of liquid blood components in relation to blood cell loss
Fluid remobilization phase (stage of diuresis) Starts 48–72 hours postburn	Interstitial fluid-to-plasma shift	Edema fluid shifts back into the intravascular compartments
	Hemodilution (hematocrit decreased)	The blood cell concentration is diluted as fluid enters the vascular compartment and, in addition, at this time a decrease in the number of cells becomes evident (destruction of red blood cells at the burn site causes anemia—as much as 10% of the total number of RBCs may be destroyed)
	Increased urinary output	Fluid shift into the intravascular compartment increases renal blood flow and causes increased urine formation
	Na+ deficit	Na+ is lost with water when diuresis occurs
	K+ deficit (may occasionally occur in this phase)	Beginning about the fourth or fifth postburn day, K+ shifts from the extracellular fluid into the cells
Convalescent phase	Metabolic acidosis Ca²+ deficit	Since Ca²+ may be immobilized at the burn site in the slough and early granulation phase of burns, symptoms of Ca²+ deficit

(*continued*)

Table 24-2
Water and Electrolyte Changes in Various Burn Phases (continued)

Phase	Water and Electrolyte Changes	Comments
	K^+ deficit	may occur (recall that for some unknown reason Ca^{2+} rushes to damaged tissues)
		Extracellular K^+ moves into the cells, leaving a deficit of K^+ in the extracellular fluid
	Negative nitrogen balance (present for several weeks following burns)	Secondary to Stress reaction Immobilization Inadequate protein intake Protein losses in exudate Direct destruction of protein at the burn site
	Na^+ deficit	

important to avoid repeating drugs too soon, especially if respiratory tract burns or shock are present.)

5. Was the burn sustained in an enclosed area where heat and fumes were inhaled? (This question is highly significant in establishing the likelihood of respiratory burns.)

6. Are there any preexisting illnesses, such as cardiac or renal damage or diabetes, that will require therapy in addition to burn treatment? (Failure to ascertain the presence of such illnesses is not uncommon in the initial rush.)

7. What is the normal pre-burn weight? (The pre-burn weight is a baseline for comparison for later weight changes; the weight is also instrumental in determining drug dosages.)

8. Is pain present? If so, how severe? (It should be remembered that severe pain can cause a drop in blood pressure and further complicate the patient's condition.)

9. Is the patient known to have any drug allergies?

10. What is the status of tetanus immunization? (All patients with severe burns should receive prophylaxis against tetanus.)

11. Are there any associated injuries requiring treatment? (For example, head injuries or fractures)

- While asking these questions, the nurse can be busy with other activities, such as readying fluid equipment, removing loose clothing not stuck to the burns, and removing constrictive jewelry before edema becomes severe.

- Necessary laboratory work should be initiated. The physician will usually request the determination of electrolytes, hemoglobin, hematocrit, glucose, and blood urea nitrogen. Arterial blood gas levels should be determined as indicated.

Observing for Burn Shock

- The nurse should be particularly alert for symptoms of burn shock.

1. Extreme thirst (caused by generalized cellular dehydration).

2. Increasing restlessness. (However, be aware that persistent slight to moderate restlessness may be due to apprehension and discomfort produced by the burn rather than to inadequate fluid replacement.)

3. Sudden high pulse rate (heart beats faster to compensate for decreased blood volume).

4. Respiratory rate is often increased, but the character of breathing should be normal unless there are complicating factors.

5. Blood pressure, when measureable, is either normal or low (the body may be too extensively burned to permit application of a blood pressure cuff).

- The supine, burned patient can often tolerate a large fluid volume deficit with little or no change in blood pressure; however, upon sitting or standing, hypotension and even syncope may occur—thus, the badly burned patient should be left supine for at least the first 48 to 72 hours.
6. Cool pale skin in unburned areas (indicative of cutaneous vasoconstriction—a compensatory mechanism that helps preserve normal blood flow).
7. Oliguria (owing to decreased renal blood flow and to increased levels of ADH and aldosterone).
8. Delirium or coma (presumably these symptoms are due to inadequate cerebral blood flow and are serious signs).
9. Seizures sometimes occur (may be the result of cerebral ischemia).
- Fortunately, burn shock is slow in onset and can be prevented or corrected by the intravenous administration of fluids in sufficient volume to maintain an adequate urinary output.

Initial Urine Observations

- An indwelling urinary catheter is usually inserted when burns involve 20% or more of the body surface; the catheter should be connected to a device for hourly measurement of the urine flow.
- The urine should be observed for discoloration resulting from blood; if present, the patient, probably has severe third-degree burns. Hemolysis of red blood cells at the burn site (caused by trauma) causes the release of hemoglobin and, consequently, *hemoglobinuria.*
- Thermal trauma (particularly electrical burns) may also destroy muscle tissue, releasing myoglobin into the bloodstream to be filtered out by the kidneys.

Initial Observations and Treatment of Respiratory Burns

- The nurse should observe the patient closely for symptoms of respiratory burns. These include
  1. Increased respiratory rate
  2. Singed nasal hair
  3. Burns around mouth
  4. Hoarseness
  5. Red painful throat
  6. Dry cough
  7. Moist rales and dyspnea

8. Stridor

9. Bronchospasm with prolonged wheezing

- Pulmonary complications may be noted immediately in the burned patient, or detection may be delayed until swelling of upper respiratory structures exposed to hot gases occurs.

- Burns about the face and neck may produce sufficient edema to embarrass respiration; soft tissue edema is maximal between 24 to 48 hours.

- Upper airway obstruction is usually heralded by tachypnea, progressive hoarseness, and increased difficulty in clearing bronchial secretions (due to vocal cord swelling). In these instances, an airway *must* be established. If possible, an endotracheal tube is inserted; if not, a tracheostomy is performed.

- The head of the bed should be elevated (unless contraindicated) and the patient should be encouraged to deep breathe and cough every 20 minutes.

- It also is important that the patient be turned at least hourly.

- Fluid should be suctioned frequently from the respiratory tract to prevent its accumulation. In addition, humidifiers and bronchodilators are used to loosen secretions and improve ventilation; oxygen should be administered to decrease anoxia.

- Use of mechanical ventilation may be necessary.

- Prophylactic antibiotics are given to prevent pulmonary infection and consequent increased edema.

- Some physicians prescribe intravenous corticosteroids to decrease the pulmonary inflammatory reaction.

- Lastly, parenteral fluids are administered cautiously to avoid overloading the circulatory system and causing pulmonary edema. (The desired urinary volume is slightly less when respiratory tract burns are present.)

- If the patient has sustained excessive exposure to smoke, carbon monoxide poisoning may result. Carbon monoxide poisoning is manifested by hypoxia (which may range from pronounced tachypnea to respiratory arrest). Diagnosis is confirmed by analyzing the carboxyhemoglobin concentration in the blood. Treatment consists of the administration of 100% oxygen and ventilatory support if needed.

- It should also be noted that a burn of the circumference of the thorax may interfere with respiratory excursions.

Intravenous Fluid Replacement

- Intravenous fluids are lifesaving in the treatment of moderate and severe burns.

- Burns involving 20% or more of the body surface (less in the very young or the aged) usually require intravenous fluid therapy.
- The aim of early fluid therapy is to give the least amount of fluids necessary to maintain the desired urinary output and keep the patient relatively free of burn shock symptoms.

Kinds of Intravenous Fluids Used in Burn Treatment

- Physicians are of widely varied opinions about the kinds of intravenous fluids to be used in burn treatment. It is generally agreed, however, that the keystones of therapy are sodium and water.[4]
  1. Many physicians favor beginning treatment with a balanced salt solution, such as lactated Ringer's solution. It not only supplies sodium, it helps correct the metabolic acidosis associated with burns; a liter of lactated Ringer's contains 130 mEq of sodium and 28 mEq of lactate (a bicarbonate precursor).
  2. Isotonic saline (0.9% sodium chloride) can also be used to supply sodium, because a liter of isotonic saline contains 154 mEq of sodium and 154 mEq of chloride. It has the disadvantage of supplying an excessive amount of chloride ions, which can contribute to the acidosis.
- The use of colloids (plasma or plasma substitutes) is advocated by some physicians who reason that colloids must be given to replace the plasma leaked through injured capillaries into the burn site. They reason that excessive administration of salt solutions (such as lactated Ringer's) further contributes to dilution of the colloidal content of plasma; the decrease in colloidal pressure contributes to edema formation.
  1. Although many physicians usually incorporate colloids into the fluid regimen, the exact timing of its administration cannot be set rigidly. It is generally prescribed after the first 24 hours of fluid therapy.[5] Burke states that colloids should be administered before the plasma albumin drops to, or below, 2 Gm/100 ml.[6]
  2. Plasma contains sodium (sometimes as much as 200 mEq/liter) and thus helps correct sodium deficit. However, it should be remembered that the use of plasma is associated with the risk of serum hepatitis.
  3. A frequently used colloid, Plasmanate, is a plasma protein fraction that has good expansion properties without the disadvantage of possible viral hepatitis transmission (heat treated to destroy hepatitis virus).
  4. Albumin is a colloid sometimes used in burn treatment.
- Blood is usually not included in burn-fluid therapy unless the patient has been bleeding from an associated injury; it should be given if the hematocrit is less than 30%.[7]

- Hypertonic albuminated fluid-demand resuscitation (HALFD) is advocated by Jelenko.[8] This fluid regimen consists of the administration of a hypertonic albuminated salt solution. The fluid contains 240 mEq of sodium, 120 mEq of chloride and 120 mEq of lactate, to which is added 1.5 Gm of fresh albumin/liter.

- The use of a hypertonic lactated solution (HLS) is advocated by some physicians for the treatment of burned patients. One liter of HLS contains 250 mEq of sodium, 100 mEq of lactate, and 150 mEq of chloride; it supplies sodium in a minimal fluid load, causing a sustained hypernatremia and increase in serum osmolality. Proponents of this therapy feel that it is associated with less tissue edema and fewer pulmonary complications than seen with other forms of treatment.[9]

- Some physicians favor the use of dextran solutions as plasma volume expanders in the treatment of burns.

- A nonelectrolyte solution, usually D_5W, may be used to replace the normal insensible water loss and the increased transcutaneous water vapor loss. Although free water is lost from the burn site, it is generally not replaced in the first 24 hours of treatment, because during this early phase, water contributes little to the maintenance of normal cardiovascular function. (Note that the formulas listed in Table 24-3 allow for its replacement in the second 24 hr of treatment.)

Burn Formulas for Planning Intravenous Fluid Therapy

Whenever intravenous fluids are given there is a danger of giving too much or not enough; both hazards are always present in burn therapy. To serve as a guide for the amount to be given to burned patients, numerous formulas have been devised. Several of the more commonly used formulas are presented in Table 24-3.

Monitoring Fluid Replacement Therapy

- It is essential to remember that any formula serves only as a guide because there are many variations among individual patients; frequent clinical assessment is mandatory to adequately tailor fluid replacement to individual needs.

- Successful fluid resuscitation of the burned patient is signalled by
  1. Adequate urine volume
  2. Normal sensorium
  3. Near normal vital signs
  4. Normal central venous pressure (These criteria are discussed below.)

Table 24–3
Formulas for Use in Planning Intravenous Fluid Replacement in Burned Patients

Formula	Type and Volume of Fluid		Special Monitoring Recommendations
	First 24 Hours	Second 24 Hours	
Parkland	Ringer's lactate—4 ml/kg/% burn One-half of prescribed volume in first 8 hrs One-half of prescribed volume in next 16 hr	Colloid—0.5 ml/kg/% burn D_5W 2000 One-half of first day's Ringer's lactate*	Urine output 30–50 ml (adults) 1 ml/kg/hr (children < 3 yr) 15–25 ml/hr (children > 3 yr)
Hypertonic resuscitation	Na^+ 250 mEq/liter Cl^- 150 mEq/liter Lactate 100 mEq/liter Average rate 300 ml/hr	D_5W	Adjust formula to urine output 30–40 ml/hr (adults), mental state, peripheral capillary filling, vital signs
New Brooke	Ringer's lactate—2–3 ml/kg/% burn One-half of prescribed volume first 8 hr One-half of prescribed volume next 16 hr	D_5W—one-half of first 24 hr total + colloid 0.3–0.5 ml/kg/% burn	Urine output 30–50 ml/hr
Hypertonic albuminated (Jelenko)	1000 ml D_5W and 120 mEq NaCl 120 mEq Na^+ lactate 12.5 Gm albumin	Most patients complete resuscitation in 24 hr No recommendations for second 24 hr	Infuse so that mean arterial BP is 60–110 mm Hg, urine output 30–50 ml/hr

*Dr. Baxter no longer recommends the use of Ringer's lactate on the second day.
(Modified from Munster A: Burn Care for the House Officer, p 19. Baltimore, Williams & Wilkins, 1980)

Urine Volume

- Hourly urine volume collected through a Foley catheter is generally considered to be the best clinical sign of adequacy of fluid resuscitation.
- The ideal range is thought to be 30 to 50 ml/hr in the adult and 1 ml/kg/hr for a child weighing less than 30 kg.[10]
- Commercial devices are available for measurement of small urine volumes. When dealing with a small urine volume, any error can be significant.
- Absent or decreased urinary output can be due to
    1. Inadequate fluid replacement
    2. Gastric dilatation (trapped fluid and, thus, unavailable for use)
    3. Renal failure
    4. Clogged catheter (falsely indicating oliguria)
- If the oliguria is due to inadequate fluid replacement, the urine volume will increase when the fluid flow rate is increased. If it is due to renal failure, the output will remain small. In addition to oliguria, acute renal failure will manifest itself as a low urine osmolality.
- It should be noted that patients who have suffered a hypovolemic insult from burns (not sufficient to produce acute tubular necrosis) may develop a transient polyuric renal failure.
- The physician should indicate the desired urinary volume plus the variations in either direction to be reported. The desired urinary volume should be realistic and approximate the minimum, not the maximum. Attempts to increase fluid input sufficiently to cause large urine volume in the aged or the very young are dangerous.
- Lastly, *accurate* recording of intake–output is a necessity for assessing the patient's fluid balance status during burn treatment. (Fluid instilled as a catheter irrigant should be charted as intake; fluid withdrawn during the irrigation should be charted as output.)

Sensorium

- It is important to assess the patient's sensorium frequently since this is one measure of the adequacy of cardiovascular functioning. Is the patient alert and lucid? Does he respond in a normal fashion?
- With adequate fluid replacement, sensorium should remain normal unless other known factors (such as head injury, drug intoxication, carbon monoxide poisoning, or arterial hypoxia) preclude this.

Vital Signs

- Vital signs should be taken at least hourly in the newly burned patient.
- Blood pressure should be near normal; as a general guideline, a systolic pressure less than 100 mm Hg or greater than 200 mm Hg

should be reported to the physician.[11] If the blood pressure cannot be easily measured with a cuff (due to peripheral edema), an arterial line should be considered. One should, of course, remember that the patient's baseline, "normal" blood pressure must be considered when evaluating his current status.

- Temperature is usually slightly elevated in the burned patient; as a general guideline, a reading greater than 101.5° F (38.6° C), should be reported to the physician.[12]
- Tachycardia is common in the burned patient but should diminish with fluid replacement. As a general guideline, a pulse greater than 160/min or less than 60/min should be reported to the physician.[13]
- Peripheral pulses in the burned extremities should be checked every few hours; it should be remembered, however, that overlying soft tissue edema may prevent palpation of underlying arteries.

Capillary Filling of Unburned Skin

- Capillary refill time of unburned skin should be carefully observed.
- Warm, pink skin that displays a normal capillary filling time after blanching is a sign of a physiologically intact circulation.[14] (One must remember that vasoconstriction of unburned skin is a normal compensatory response to help preserve normal blood flow during the early hours of a severe burn, causing unburned skin to be cool and pale during this period; if it persists, it becomes a cause for concern.)
- The nail beds are good sites for observing capillary refill time.

Central Venous Pressure

- The central venous pressure (CVP) should remain within normal limits in the adequately treated burned patient.
- CVP measurement is frequently used to monitor the effects of intravenous fluid replacement therapy, particularly in infants, the aged, and patients with cardiac and renal disease.
- CVP measurements are generally indicated when the burned surface is greater than 50% or 60%. The margin for therapeutic error is small when the patient has deep extensive burns.
- Frequent CVP checks allow more aggressive fluid replacement therapy without the risk of circulatory overload.
- If the CVP becomes elevated above the reading designated by the physician, the fluid infusion rate should be curtailed.
- If the patient is responding poorly to fluid replacement therapy, a Swan-Ganz catheter should be considered to measure pulmonary artery pressure, wedge pressure, and cardiac output.
- In summary, the nurse must be alert for symptoms of inadequate or excessive fluid administration. Inadequate fluid therapy in burned patients is indicated by

1. Decreased urinary output (below the desired level)
2. Thirst
3. Restlessness and disorientation
4. Hypotension and increased pulse rate (Burn shock is described earlier in the chapter.)

- Circulatory overload is indicated by
  1. Elevated CVP
  2. Shortness of breath
  3. Moist rales
  4. Blood pressure increased above normal limits

Oral Electrolyte Solutions

- Oral electrolyte solutions may be the sole source of fluids for the patient with minor burns, and may be used in conjunction with intravenous fluids in moderately burned patients.
- Oral fluids should not, of course, be administered if the gastrointestinal tract is not functional. (Recall that paralytic ileus, gastric dilatation, and nausea are frequent complications of severe burns.)
- Some physicians prefer not to give fluids orally for the first day or two (or until bowel sounds are heard); fluids are offered slowly at first and increased as tolerated.
- Thirst is an early symptom following burns. The patient permitted unlimited quantities of plain water is in danger of developing water intoxication (sodium deficit) because of simple dilution.
- The nurse must explain to the seriously burned patient that plain water (usually in the form of ice chips) should be taken in *limited* amounts, if at all.
- Physicians favoring the use of oral electrolyte fluids differ as to the exact contents and proportions of the solutions to use. One solution consists of 1 tsp of sodium chloride and ⅛ tsp of sodium bicarbonate in 1 liter of chilled water. Occasionally isotonic saline or $\frac{1}{6}$ molar sodium lactate is used. (The solution should be prepared carefully; errors between teaspoons and tablespoons can be serious.)
- Oral electrolyte solutions have a definite taste and some patients find them difficult to accept. Measures that help to make the solution more palatable include chilling the solution, making ice chips from it, flavoring it with lemon, or disguising it in juices (when allowed).
- Orange juice and other potassium-containing fluids should be withheld until renal function is established and the physician approves their use.
- Oral solutions are not given in the presence of
  1. Mental confusion (danger of aspiration of fluid into the lungs)
  2. Acute gastric dilatation

- When gastric dilatation is present, fluids taken orally become trapped in the distended stomach. Thus, even though oral fluids are swallowed, they are not available for body use and urine formation.

Observations for Gastrointestinal Complications

- *Reflex paralytic ileus* will develop some time during the first 24 postburn hours in most patients with more than 20% total body surface area burns. Usual symptoms include nausea, effortless vomiting, hiccoughing, and abdominal distension. Vomiting carries a high risk of tracheal aspiration and must be prevented.
- Some physicians favor placing a nasogastric tube in all patients with major burns in order to decompress the stomach until normal bowel sounds have returned. Another purpose for insertion of a nasogastric tube is to allow for the observation of gastric secretions.
- Recall that patients with major burns are at risk of *hemorrhagic gastritis* due to stress. Bleeding from the stomach may manifest itself in gastric secretions or (if a slow ooze) by guaiac-positive stools and a gradual decrease in the hematocrit. Gastric aspirates should thus be observed frequently for signs of bleeding; antacids may be added through the nasogastric tube at hourly intervals to deter superficial erosions of the gastric mucosa. A common practice is to add 1 ounce of antacid through the tube every hour and clamp the tube for 30-minute intervals. The incidence of Curling's ulcers, a once-common burn complication, has been markedly diminished over recent years due to the use of antacids, enteral feedings, and when indicated, cimetidine.
- The presence of abdominal distention after the first few postburn days may indicate the presence of invasive *wound sepsis*, since ileus commonly occurs with this condition. Ileus due to wound sepsis may be associated with gastroduodenal perforation. In the past, 85% of the cases of upper gastrointestinal hemorrhage were associated with bacteremia.[15]
- The decreased incidence of gastrointestinal ulcers in burned patients has been attributed to reduced frequency of major sepsis, prophylactic use of antacids through nasogastric tubes, and improved provision of nutritional supplements (allowing for more rapid healing of small, eroded areas).

Control of Pain

- The amount of pain present varies with the depth of the burn, the extent of surface area involved, and the patient's pain threshold.
- Third-degree (full-thickness) burns are painless because the nerve endings are destroyed; however, pain is experienced around the periphery of third-degree burns where first-degree and second-degree

burns are present. As a rule, the requirement for analgesics is inversely proportional to the depth of the initial injury.

- After 48 hours, the requirements for analgesics are greatly diminished, except when wounds are being actively debrided during the waking state.[16]
- Burned patients complaining of severe pain may be given small doses of meperidine or morphine through the intravenous cannula put in place for fluid replacement. Fluid resuscitation should be underway before analgesics are administered.
- The intravenous route assures rapid and dependable concentrations of the drug in the central nervous system.
- The subcutaneous or intramuscular route should *never* be used to administer narcotics to a burned patient with circulatory collapse.
  Peripheral tissue perfusion is erratic when shock is present; thus, absence of the desired effect may prompt repeated doses by the same route. When peripheral circulation is improved after fluid replacement, there may be a rapid absorption and cumulative overdosage of the narcotic, resulting in respiratory narcosis and depression
- The administration of a large dose of an analgesic to an inadequately resuscitated patient in or near burn shock can be *lethal.*
- It is important not to confuse the restlessness of burn shock with pain. A patient thrashing about in bed, without complaints of pain, may well be in burn shock. In this case, a narcotic is contraindicated; the physician usually orders an increased rate of parenteral fluid administration.

Physiologic Basis for Treatment and Nursing Care During the Fluid Remobilization Phase

Remobilization of edema fluid represents an interstitial-to-plasma fluid shift, which begins on the second or third day after the patient has been burned. Its usual duration is 24 to 72 hours.

Observing Urinary Output

- Reabsorption of edema fluid takes place about the second to third postburn day. The blood volume is greatly increased and large amounts of urine are excreted. The nurse should be alert for increasing urine volume and should report its presence to the physician; parenteral fluid orders must be decreased during this phase to avoid overloading the circulatory system.

- If the expected diuresis does not occur, the possibility of renal damage must be considered.

Observing for Pulmonary Edema

- Fatal pulmonary edema may occur if the renocardiovascular system is not capable of handling the volume of water and electrolytes shifting from the interstitial fluid into the plasma. (Recall that the volume of edema fluid in a burn may equal the total normal plasma volume.)
- The nurse should be alert for signs of circulatory overload and pulmonary edema:
  1. Venous distention
  2. Shortness of breath
  3. Moist rales
  4. Cyanosis
  5. Coughing of frothy fluid

Parenteral Fluid Therapy

- Once the fluid remobilization phase is reached, parenteral fluids should be sharply curtailed or discontinued; infusion of large volumes of fluids could easily cause circulatory overload with pulmonary edema.
- Oral fluids and foods may supply adequate fluid and nutrition during this phase if tolerated; if not, moderate quantities of intravenous fluids may be necessary to meet daily needs. If possible, a high protein and high caloric diet is started by the second or third day.

References

1. Curreri W: Burns. In Schwartz S, et al (eds): Principles of Surgery, 3rd ed, New York, p 299. McGraw-Hill, 1979
2. Munster A: Burn Care for the House Officer, p 5. Baltimore, Williams & Wilkins, 1980
3. Ibid
4. Schwartz S: Consensus summary on fluid resuscitation. Trauma (Suppl) 19, No. 11: 876, 1979
5. Ibid
6. Burke J: Fluid therapy to reduce mortality. Trauma (Suppl) 19, No. 11: 866, 1979
7. Schwartz S, p 876
8. Jelenko C: Fluid therapy and the HALFD method. Trauma (Suppl) 19, No. 11: 866, 1979

9. Munster A, p 16
10. Schwartz S, p 876
11. Munster A, p 23
12. Ibid
13. Ibid
14. Monafo W: The Treatment of Burns—Principles & Practice, p 66. St Louis, Warren H. Green, 1971
15. Curreri W, p 298
16. Ibid, p 289

chapter 25

Heat Disorders

- Although heat disorders occur most often in tropical zones, the temperate climate of North America can cause heat stress. Many persons living in a temperate climate withstand heat stress poorly, hence the increased number of deaths during heat waves.
- The literature reports numerous deaths in the elderly during the hot summer months. In fact, two-thirds of the persons who suffer from heatstroke are over the age of 50.[1]
- Heatstroke is reported to be the second leading cause of death among American athletes.[2] Heat disorders are particularly common in athletes training vigorously before they have become acclimated to heat stress.
- The recent increase in the number of persons who have taken up recreational jogging has markedly increased the potential for serious heat disorders; the danger is greatest in novice runners unaccustomed to vigorous exercise in the heat.
- New military recruits undergoing basic training in southern states also are at risk for heat disorders.
- Industry is often associated with artificially created hot climates, resulting from deliberately designed environments for some industrial purpose.
- Because so many persons in our society may be subject to heat disorders, the nurse should become familiar with their prevention, recognition, and treatment.
- Continuation of heat stress without relief can eventually lead to heat cramps, heat exhaustion, or heatstroke. Some patients, after heat exposure, have a clinical picture that includes elements of more than one of these syndromes; usually, however, the heat disorders are seen in their "pure" form.

Heat Cramps

The heat cramps syndrome is the mildest of the three heat disorders.

Characteristics

- Characteristically occurs in well-conditioned athletes following intensive exercise.[3]
- Occurs only in individuals who sweat profusely—the exact mechanism of its production is not known.
- It is believed to result from an acute deficiency of sodium in the muscles.[4,5]
- Rarely occurs in the elderly since they are seldom able to perform the degree of physical activity necessary to produce such large sweating losses.
- Characterized by brief, intermittent, painful spasms of the large skeletal muscles, which have been subjected to intense activity. Pain may be mild to severe.
- Onset is usually sudden.
- Body temperature is normal.
- Thirst mechanism is intact; individual often replaces fluid loss with water only.
- Patient is rational and alert.

Treatment

- Ingestion of fluid containing salt (for example, a solution of 1 tsp of NaCl in 1 quart of water).[6]
- Some authorities favor adding extra salt to food in preference to the use of salt tablets.[7]
- Salt tablets are not as widely used as before. If they are used, 1 tablet of salt, enteric-coated, for every pound of weight lost by perspiration during the day has been suggested as the ideal amount to avoid gastrointestinal irritation.[8]
- Be aware that salt ingestion without adequate water intake, or vice versa, may make matters worse.
- Rest at least 12 hours.
- In the event of severe, repeated, unrelenting cramps, oral or intravenous salt solutions rapidly relieve all symptoms.[9]

Heat Exhaustion

Next on the continuum of heat disorders is the heat exhaustion syndrome.

Characteristics

- Results from inadequate cardiovascular responsiveness to the circulatory changes brought about by heat (the diversion of blood flow to the skin is not adequately compensated by vasoconstriction in other parts of the body or by volume expansion).[10]
- Frequently seen among elderly who are at special risk because of a high incidence of cardiovascular disease, sluggish vascular autonomic responses, and frequent use of diuretic agents.
- Signaled by cool, pale, moist skin (*note*: patient continues to sweat).
- Extreme weakness and copious sweating, giddiness, and postural syncope; sometimes accompanied by nausea, vomiting, and the urge to defecate.
- Body temperature may be normal or elevated.
- Patient usually conscious but may be unconscious.
- Tachycardia and hypotension may be present.
- In young athletes, signs of heat exhaustion may mimic a viral upper respiratory infection.[11]
- Can present with, or without, heat cramps.

Treatment

- Cool patient by swabbing (may or may not need tub cooling, depending on initial body temperature).
- Patient should lie down and rest in a cool place.
- Oral replacement of water and electrolytes is indicated if the patient is able to tolerate oral fluids.
- Intravenous fluid replacement may enhance recovery but is usually unnecessary.[12]
  (The type of parenteral fluid to be administered must be delineated by clinical laboratory findings, as patients may present with low, normal, or high sodium levels. For example, if the patient has predominantly a water deficit, D_5W may be used; if the primary problem is salt deficit, isotonic saline may be used.)
- Hospitalization is indicated for patients over 65 years of age and for those with predisposing disease, severe vomiting, diarrhea, muscle necrosis, or unstable blood pressure.
- Rest for 2 or 3 days.

Heatstroke

Heatstroke is the most serious of the three heat disorders; it constitutes a medical emergency and has a mortality rate of up to 80%.

Types of Heatstroke

- There are two major types of heatstroke: (1) exercise induced and (2) nonexercise induced. The former tends to occur in the young athlete, military recruit, or worker; the latter tends to occur in the sedentary elderly person.
- Strenuous muscular work contributes to heatstroke by enhancing production of body heat. The hyperpyrexia can be associated with extensive muscle damage characterized by myoglobinuria, oliguria, and severe hyperkalemia. These complications are not often seen in the elderly, in whom severe muscular work is not ordinarily a contributory factor.

Risk Factors for Heat Exhaustion and Heatstroke

The similarities between heat exhaustion and heatstroke suggest that the pathophysiology of the two disorders may be similar and that they may represent closely related points along a continuum of responses to heat stress.[13] At risk for heat exhaustion and heatstroke are

- Those exposed to high ambient temperature and humidity without adequate acclimatization
- The elderly (particularly those with known cardiovascular disease)
- Those with myocardial ischemia, arteriosclerosis, and hypertension
- Those performing heavy physical work in a hot environment (since they undergo the double stress of two sources of excessive heat)
- Those with recent illnesses, particularly when accompanied by fever
- Those with a water deficiency
- Diabetics (since they frequently have autonomic dysfunction and heart disease and, thus, may be unable to make the important hemodynamic alterations necessary to dissipate a heat load imposed by increased environmental temperatures)[14]
- The obese (an obese person has less body surface in proportion to body weight than does a person of slight build; thus, there is greater difficulty in dissipating heat)
- Those with impaired ability to sweat (as in dermatologic conditions affecting large areas of skin)
- Those with potassium deficiency
- Those taking medications that interfere with thermoregulation
    1. Diuretics (because they promote fluid loss)
    2. Antiparkinsonian drugs
    3. Antihistamines   } because they suppress sweating
    4. Anticholinergics

5. Phenothiazines—because they suppress sweating and possibly disturb hypothalamic temperature regulation
6. Haloperidol—because it decreases thirst recognition and might disturb hypothalamic temperature regulation
7. Amphetamines ⎫
8. Vasodilators ⎬ because they increase metabolic heat
9. Thyroid extract ⎭ production
10. LSD
11. Alcohol—because it causes an increased metabolic load, promotes diuresis, and impairs judgment and critical thinking
12. Propranalol—may place patients at risk because it impairs sweating and decreases cardiac output[15]
13. Tricyclics—because they increase motor activity and heat production

- Unalert bedridden patients who are unable to throw off bedclothes and to drink at will
- Those wearing impermeable clothing and exercising strenuously to achieve weight loss
- Short, stocky, heavily muscled athletes appear to be at higher risk than tall, lanky athletes.
- Those who falsely believe that water deprivation accelerates physical conditioning (and thus avoid drinking adequate amounts of water during heavy exercise)
- Those who overzealously take salt supplements (remember that ingestion of excessive Na^+ and insufficient water predisposes to severe water-depletion heat exhaustion that can culminate in frank heatstroke)

Characteristics

- Sudden onset in most of the cases
- May occur with premonitory symptoms (headache, fatigue, and disorientation)
- Rectal temperatures is usually 41° C (105.8° F) or higher
- Signaled by very hot, dry skin (*note*: patient has usually ceased to sweat)
- Presents with varying degrees of impairment of consciousness, such as lethargy, stupor, or coma (due to shunting of blood away from CNS)
- The classic picture is a comatose patient with hot, dry skin (this type occurs most usually in the elderly, alcoholic, or diabetic person).
- In young athletes, signs of heatstroke may resemble psychiatric illness (euphoria, confusion, belligerence, and assaultiveness).[16]

- Sinus tachycardia with heart rates of 140 to 150 is the rule (may approach 170 in severe heatstroke).
- Salt balance may be near normal unless heat exhaustion has preceded heatstroke (which is unusual).
- Can occur with or without heat exhaustion or heat cramps preceding it
- In the elderly, exogenous heat can cause a quietude resembling a senile dementia.
- Some patients report having felt chilly or "having goose flesh" just before collapsing (piloerection of upper chest and arms).
- Hypotension may be present (likely due to pooling of blood in the skin; should diminish as core temperature cools).
- Dehydration has been reported to be present in 10% to 20% of elderly patients.[17]
- Hyperpnea often present
- Electrolyte and acid–base disturbances may include
  1. Hypernatremia
  2. Hypokalemia
  3. Hyperkalemia
  4. Hypocalcemia
  5. Hypophosphatemia
  6. Respiratory alkalosis
  7. Lactic acidosis

Treatment

- Establish an intravenous line, intubate the trachea if patient is comatose, insert a thermistor into rectum to a depth of 20 cm (8 inches); then immerse patient in a tub of ice water and briskly massage the skin to stimulate exchange of heat, or cool patient by rubbing body surface with plastic bags filled with ice while the body is wet down with water (requires a fan in an air-conditioned room).[18]
- Monitor rectal temperature every few minutes; monitor vital signs, and be alert for arrhythmias.
- When core temperature drops to 39° C (102.2° F), discontinue hypothermic measures and continue to monitor vital signs.
- Some authorities recommend the use of chlorpromazine (Thorazine), 50 mg, when core temperature drops (otherwise, at about 40° C (104° F), violent shaking may occur).
- Monitor urinary output closely (remember, oliguria doesn't necessarily mean dehydration is present, can be due to renal failure).
- Administer intravenous fluids cautiously—avoid overloading the cir-

culatory system and causing congestive heart failure and pulmonary edema. Remember, most patients with heatstroke are not markedly dehydrated (500–1000 ml of fluid may be all that is needed).

The type of parenteral fluid to be administered must be delineated by clinical laboratory findings (as patients may present with low, normal, or high Na^+ levels)

- Avoid antipyretic drugs, such as aspirin or acetaminophen. (They are not helpful because their action requires the presence of intact heat-losing mechanisms, and may be harmful because of their tendency to produce bleeding.)[19]
- Obtain an electrocardiogram reading, particularly if patient is over 50 years of age.
- Obtain laboratory work such as a complete blood count CBC, platelet count, prothrombin and fibrinogen times, blood urea nitrogen, creatinine, glucose, electrolytes, and liver function tests.
- Observe for possible complications:
  1. Spontaneous recurrence of hyperthermia
  2. Acute tubular necrosis
  3. Myocardial infarction
  4. Clotting abnormalities
  5. Hepatic necrosis
- Avoid heparin and dextran solutions since heatstroke patients harbor a risk for bleeding disorders that these agents only compound (coagulation characteristics usually revert to normal as cooling is accomplished).[20]
- Preventive measures are listed at the end of the chapter.

Reduction of Body Temperature

- Quick and effective reduction of the high body temperature is essential. Even a few hours of delay may leave the patient with severe neurologic deficits. The longer the temperature remains high, the greater the possibility of irreversible brain damage.
- When the body temperature is above 106° F (41.1° C) damage to the cell parenchyma occurs throughout the entire body. Especially devastating is the loss of neurons, since neuronal cells, once destroyed, cannot be replaced.
- The body temperature should be reduced to 102° F (38.9° C) within the first hour of treatment. Recovery from heatstroke depends largely on reducing the degree and duration of fever.
- Electrically controlled hypothermia mattresses have been used successfully to reduce body temperature in heatstroke patients.
- A frequently recommended method for cooling consists in immersing the patient in a tub of water to which ice is slowly added. (The

use of a bathtub filled with ice can usually lower the body temperature to 102° F [38.9° C] within 1 hour.)

1. Since the patient may be comatose, or at least disoriented, he must be protected from drowning.

2. The trunk and extremities should be massaged during the bath to bring blood to the periphery for cooling.

3. The body temperature should be checked every 3 to 5 minutes. When it reaches approximately 102° F (38.9° C), the patient should be removed from the ice bath since, as a rule, the body temperature continues to fall another 3° F to 4° F (1.6° C to 2.2° C).

4. After removal from the tub, body temperature should be measured at frequent intervals so that any rise can be noted early.

5. A less effective method of cooling includes sponging the patient with alcohol or cool water; this method is made more effective when good air movement is ensured, as with an electric fan.

6. The patient should be placed in a cool room with low humidity.

Fluid Administration

- The initial status of body water and electrolytes depends upon the state of hydration prior to the onset of heatstroke. Some patients have no water or salt loss; others have a severe depletion of salt or water, or both.

- The type of fluid to be administered must be delineated by clinical laboratory findings (as patients may present with low, normal, or high Na^+ levels).

- Intravenous fluids should be administered cautiously to avoid overloading the circulatory system and causing congestive heart failure and pulmonary edema.

- Hypernatremia is present in a number of cases and requires treatment with dextrose in water, or hypotonic electrolyte solutions, or both.

- Metabolic acidosis, if present, may be treated with a bicarbonate solution.

Diuretics

- If the hourly urine output is consistently less than 20 ml/hr, mannitol or another diuretic, such as furosemide or ethacrynic acid, may be used in an effort to prevent the development of acute tubular necrosis.

- Acute renal failure is reported to occur in 5% of the patients with nonexercise-induced heatstroke and in 40% of those with exercise-induced heatstroke.

Rest

- After 1 to 3 days of intensive treatment, the sweat glands should become functional again, although it may be as long as 6 months before they begin to secrete normally. The patient should continue to rest in bed and avoid exposure to sunlight for 1 to 2 weeks after temperature reduction.

Emergency Measures Before Medical Aid is Available

Because time is so vital in preventing fatalities from heatstroke, one should take every measure to reduce body temperature as soon as possible. Unfortunately, heatstroke may occur in an area some distance from medical aid; furthermore, facilities for ice water baths or even sponging may not be available. The following points should be kept in mind to care for the heatstroke victim before medical aid is available:

- Move the patient out of the sun to the coolest, best-ventilated spot available.
- Remove most of the patient's clothing.
- Summon medical aid; if necessary, move the patient to a place for medical aid. The transporting vehicle should be air conditioned, or have all of its windows opened so that a draft can blow on the patient to promote cooling. If moving the patient entails further exposure to high heat stress, it is better to wait until a more suitable means of transportation is available. Additional heat stress could cause death.
- Investigate surroundings for any immediate means of reducing the patient's temperature until more effective measures can be made available. For example, if heatstroke occurs during an outing near a body of water, the patient may be partially immersed to promote cooling. Or, if a water hose is available, the patient can be sprayed continuously with water. If nothing but a drinking water supply is available, the patient can be sponged with it.
- Massage the patient's skin vigorously; this maintains circulation, aids in accelerating heat loss, and stimulates the return of cool peripheral blood to the overheated brain and viscera. Body heat may be lost rapidly in this manner.

Summary of Measures to Prevent Heat Disorders

Heat disorders are usually both predictable and preventable by simple measures. The nurse should become familiar with the preventive measures listed below so that she can offer sound advice about heat disorders. The following points should be remembered:

- Strenuous activity should be curtailed as much as possible on hot days. Prolonged exercise such as training programs or athletic events should be held in the cooler parts of the day (such as before 8 AM or after 6 PM). Care should be taken to avoid the hours of 11 AM to 2 PM because these are the hours of greatest solar heat.
- Persons moving from a temperate climate to a hot climate, or those subjected to heat stress in their work, should gradually build up a tolerance to heat through planned acclimatization. Sudden exposure of an unacclimated person to high heat stress invites heat disorders.
- Persons exposed to heat stress should maintain good physical condition. (Sufficient rest and proper food and fluid intake help prevent heat disorders.)
- Persons exposed to heat should wear loose, porous clothing. Under no circumstances should heavy exercise be performed in the heat while wearing plastic or rubberized clothing.
- Persons confined to bed should be protected from excessive bedclothing and hot, poorly ventilated rooms.
- *The use of salt tablets without adequate water replacement increases risks.* Salt supplements are mandatory in some form during the first week or so of conditioning in hot environments; however, *provision of water is much more important.* After acclimatization is achieved, salt loss in perspiration declines. The salt consumed with meals is generally adequate; thirst is the best guide to water regulation. The use of commercial drinks containing glucose, water, salt, and other minerals is unnecessary but probably harmless.[21] One kind of commercially available salt tablets (Thermotabs, Beecham Products) contains NaCl, 7 gr (0.45 Gm), and KCl, ½ gr (30 mg). Manufacturer's directions state that a full glass of water should be consumed with each tablet.
- Potassium deficit has been shown to help cause some heat disorders. Hence, persons prone to develop potassium deficits—those taking diuretics, for example, should guard against potassium deficit in hot weather. Food intake (*i.e.*, K^+ intake) usually decreases in hot weather; at the same time, more potassium than normal is lost in heavy sweating. Thus, potassium supplements are indicated. A decrease in the requirement for digitalis, caused by hypokalemia, may render the management of cardiac patients more difficult during their stay in a hot environment. (Recall that hypokalemia potentiates the action of digitalis on the heart.)
- Medications predisposing to heat stress should be stopped temporarily or cautiously diminished during prolonged periods of extreme heat (under the direction of a physician). These medications are discussed earlier.
- Offer regular periods of rest during work or heavy exercise when the individual can cool off and drink fluids.

- A cool place should be available for workers exposed to constant heat stress in their work. It has been pointed out that the predisposition to heat exhaustion or heatstroke may be related to a lack of a recovery period (such as the availability of a cool place to rest periodically).
- Water or cool drink dispensers should be placed in convenient locations in high-heat-stress environments.
- Athletes or workers performing heavy exercise in a hot environment should be advised of the early symptoms of heat injury.
    1. Excessive sweating
    2. Abdominal cramps
    3. Headache
    4. Nausea
    5. Dizziness
    6. Cessation of sweating with piloerection on upper chest and arms
    7. Gradual impairment of consciousness
- Individuals perspiring heavily should be advised to limit water loss to no more than 5% of body weight; a loss of 7% or more is dangerous.
- The following facts pertain to competitive runners:[22]
    1. Prior training in the heat can reduce the risk of heat injury.
    2. Consumption of 500 ml of water immediately prior to the run and 250 ml at every water station during the run can help forestall the development of heat injury.
    3. Competitors should run with a partner and agree to be mutually responsible for the other's well-being during the run.
- Planners of competitive runs should[23]
    1. Plan races to avoid the hottest months of July and August
    2. Choose a course with some shade and provide adequate supplies of fluids every 2 km to 3 km
    3. Alert local hospitals of the event
    4. Provide facilities at the race site to treat heat disorders

References

1. Sprung C: Heatstroke: Modern approach to an ancient disease. Chest 77, No. 4, 462, 1980
2. Ibid
3. Burch G, Knochel J, Murphy R: Stay on guard against heat syndromes. Patient Care, June 30, 1979, p 80

4. Knochel J: Environmental heat illness—an eclectic review. Arch Intern Med 133:849, 1974
5. Maxwell M, Kleeman C (eds): Clinical Disorders of Fluid & Electrolyte Metabolism, 3rd ed, p 1530. New York, McGraw-Hill, 1980
6. Burch G, Knochel J, Murphy R, p 75
7. Wheeler M: Heatstroke in the elderly. Med Clin North Am 60, No. 6:1290, 1976
8. Haraguchi K: Nurses can take the heat off of workers. Occup Health Saf, July–Aug, 1978, p 40
9. Maxwell M, Kleeman C, p 1531
10. Wheeler M, p 1291
11. Burch G, Knochel J, Murphy R, p 69
12. Wheeler M, p 1292
13. Costrini A, Pitt H, Gustafson A, Uddin D: Cardiovascular and metabolic manifestations of heatstroke and severe heat exhaustion. Am J Med 66:301, 1979
14. Sprung C, et al: The metabolic and respiratory alterations of heatstroke. Arch Int Med 140:669, 1980
15. Burch G, Knochel J, Murphy R, p 74
16. Ibid, p 73
17. Levine J: Heatstroke in the aged. Am J Med 47:251, 1969
18. Maxwell M, Kleeman C, op cit, pp 1550–1551
19. Wheeler M, p 1295
20. Burch G, Knochel J, Murphy R, p 78
21. Hodges R: Nutrition in Medical Practice p 181. Philadelphia, WB Saunders, 1980
22. Hughson R: Primary prevention of heatstroke in Canadian long-distance runners. Can Med Assoc J 122:1119, 1980
23. Ibid

chapter 26

Toxemia of Pregnancy

- The syndrome of toxemia of pregnancy has no known origin. Some theorize that toxemia is the result of autoimmunity or allergic reaction caused by the fetal presence. Some feel that malnutrition, especially reduced protein intake, may play a role.[1] Others feel that there is no scientific basis for believing that excesses or deficiencies of any essential nutrients predispose to toxemia. Socioeconomic factors have been implicated, as has slow, disseminated intravascular coagulation (DIC).[2]

- Toxemia can be divided into two types, depending upon the severity: (1) preeclampsia, or toxemia without convulsions; and (2) the extremely serious eclampsia, or toxemia with convulsions.

- Patients with underlying vascular or renal disease are not included under the diagnosis of toxemia of pregnancy; yet, they are often difficult to distinguish from patients with toxemia, especially during the last trimester.

- Five percent of patients with preeclampsia progress to eclampsia; 10 to 15% of those who develop eclampsia die.[3] Perinatal deaths in eclampsia range from 8.6% to 27.8%.[4]

- Although toxemia may occur earlier, it usually begins after the 32nd week of pregnancy. Toxemia can also occur in the postpartum period (usually 24–48 hrs after delivery), with hypertension and convulsions; it has been reported to occur as many as 11 days postpartum.

- Toxemia is primarily a disease of first pregnancy, occurring with higher frequency in the young (teenager) and in the older (over 35 years) primigravida.[5]

365

- When toxemia occurs without proteinuria, particularly in the multiparous woman, it may be an early sign of essential hypertension.
- A twin pregnancy or hydramnios appears to predispose to toxemia, as may preexisting hypertension.
- About one-third of the women who have toxemia will develop the disease in a subsequent pregnancy.[6] The young primigravida who has experienced toxemia has a much more favorable prognosis than the older multigravida with chronic hypertension and superimposed preeclampsia–eclampsia.[7]
- It seems the disease occurs more frequently in lower socioeconomic classes in which obesity and short stature are more common.
- Recent literature refers to toxemia as pregnancy-induced hypertension (PIH): PIH is synonymous with preeclampsia–eclampsia.[8]

Pathophysiology

- While the cause of toxemia remains unknown, generalized vasoconstriction explains many of the signs and symptoms.
- Widespread vasoconstriction of arterioles affects the placental circulation, kidneys, and eye grounds. Generalized vasoconstriction causes increased vascular resistance and thus hypertension; it may also account for the visual problems experienced in severe eclampsia.
- Small degenerative infarcts appear in the placenta, apparently caused by vasoconstriction; the damaged placenta may separate prematurely or may fail to nourish the fetus adequately.
- A slight decrease in renal blood flow and glomerular filtration rate (GFR) occurs in the toxemic patient. (Recall that in normal pregnancy, both the GFR and the renal blood flow increase.)
- In addition to the above changes, the toxemic patient develops fibrinoid deposits in the basement membrane of the glomerular tufts.
- It now appears incorrect that control of weight gain and limitation of sodium intake alone reduce the incidence of preeclampsia; this belief confuses cause and effect. Although preeclamptic patients tend to retain water and sodium, weight gain does not cause preeclampsia.
- Certain endocrine or metabolic disorders, or both, may be implicated in the genesis of the disease.
- Autopsies of patients who die of eclampsia reveal hemorrhagic lesions in the placenta, liver, kidneys, brain, spleen, adrenals, and pancreas. In addition, necrosis, tissue infarction, fibrin deposits, and evidence of DIC may be found. Pallor of the renal cortex may be observed, with little blood in the glomerular capillaries.

Clinical Manifestations

- The signs of onset of toxemia include: *hypertension, edema* (not only of the ankles, but also of the hands and face) and, in some patients, *proteinuria.*
- A significant rise in blood pressure consists of an increase in systolic pressure greater than 15 mm Hg and in the diastolic pressure greater than 10 mm Hg.
- Proteinuria becomes significant at a level of 500 mg/dl or more/24 hr; this level correlates well with a 2+ urinary protein.[9] (Proteinuria in toxemia can range from as much as 10–30 Gm/24 hr to as little as 0.5 Gm/24 hr.)[10]
- A weight gain of more than 5 pounds a week or a blood pressure greater than 140/90 should cause the nurse to consider early toxemia. (However, it is important to remember that almost half of the patients in whom toxemia develops do not display excessive weight gain.)
- Some authorities have established generalized criteria to differentiate between mild and severe toxemia. *Mild toxemia* may be defined as consisting of minimal to moderate hypertension (systolic < 160, diastolic <110), nondependent edema, and proteinuria (< 2 Gm in 24 hr). *Severe toxemia* is characterized by one or more of the following: a blood pressure greater than 160/110, nondependent edema, and proteinuria more than 5 Gm in 24 hours (3+ to 4+), and oliguria.[11] Other advanced symptoms consist of headache, irritability, visual disturbances, epigastric pain, oliguria, and nausea and vomiting—a cluster of symptoms that warn of the approach of eclampsia. Headache and irritability are due in part to cerebral edema; visual impairment may be due to retinal edema, hemorrhage, and even detachment.
- Eclampsia is heralded by generalized tonic–clonic convulsions. While apprehension and hyperreflexia often precede the convulsion, this is not always true. An aura usually does not precede the convulsion. Cerebral vasospasm and cerebral edema are thought to be causes of eclamptic convulsions, but are unproven etiologic reasons.[12]
- Laboratory findings in eclampsia often include hemoconcentration, elevated blood urea nitrogen (BUN), and elevated serum uric acid. Toxemia is also associated with thrombocytopenia.

  Hemoconcentration of toxemia is apparently related to the vasoconstrictive state; administration of a vasodilator allows remobilization of sequestered fluid from the interstitial space back to the vascular space.[13] The serum uric acid level rises in toxemia due to a decrease in urate clearance.

- With eclampsia, death can occur during a convulsion, or it can result from cerebral hemorrhage, which is reported to be the most frequent cause of death in eclampsia.

- The fetus may also be seriously threatened, either by the toxemia or by the eclamptic convulsion.
- Should the pregnancy be terminated either therapeutically or naturally, symptoms of toxemia promptly diminish.

Treatment and Nursing Interventions

- The incidence of severe toxemia can be reduced by proper prenatal care with attention given to monitoring of blood pressure. As stated earlier, a rise in systolic blood pressure greater than 15 mm Hg or diastolic pressure greater than 10 mm Hg is significant.
- The presence of proteinuria is considered by many physicians to be an indicator for need for hospitalization; during hospitalization, the course of the hypertension, proteinuria, and renal function can be closely monitored.
- The patient with mild toxemia (blood pressure no higher than 140/90) may be treated with bed rest, salt restriction to 0.5 to 1.0 Gm/24 hr, sedation with phenobarbital or diazepam, and a diuretic if edema is present.[14] Some authorities avoid the use of diuretics unless pulmonary congestion is present.[15]
- Because there is depletion of effective intravascular volume in toxemia (due to vasoconstriction), many authorities feel that the use of diuretics is generally not recommended. Further depletion of effective intravascular volume by diuretics and severe salt restriction further compromises placental perfusion.
- Administration of plasma volume expanders (dextran or Plasmanate) has been reported to cause improvement in the status of preeclamptic patients.[16]
- In severe toxemia in which the diastolic blood pressure is higher than 110 mm Hg, *parenteral antihypertensive therapy* (such as intravenous hydralazine) is often used.
- Definitive care of eclampsia is fetal delivery.[17] Delivery of the fetus is indicated in severe toxemia after the blood pressure and convulsions have been controlled; convulsions increase both fetal and maternal mortality. (Fetal bradycardia is common immediately after a convulsion, probably due to hypoxia and acidosis induced by heavy muscular activity.)[18]
- All eclamptic women should be typed and crossmatched for blood. As a result of her contracted blood volume, the patient with toxemia cannot tolerate the same degree of blood loss at delivery as can a normal woman.[19] Dangerous hypoperfusion of vital organs is especially likely to follow blood loss in eclamptic patients.

Magnesium Sulfate

- Magnesium sulfate ($MgSO_4$) may be prescribed primarily not only for its anticonvulsant effect, but also for its antihypertensive effect. Magnesium sulfate controls convulsions by blocking neuromuscular transmission; it depresses the central nervous system and produces an initial hypotensive effect because of peripheral vasodilating effect.
- The reported therapeutic maintenance range is a serum magnesium level of 4 to 7 mEq/liter.[20]
  1. Deep-tendon reflexes may be depressed when the serum level exceeds 4 mEq/liter.
  2. The patellar reflex disappears when the serum magnesium level reaches 7 to 10 mEq/liter.
  3. Respiratory depression can occur when the concentration reaches 10 to 15 mEq/liter, hence the importance of frequent checks on the patellar reflex.
  4. Concentrations above 13 mEq/liter can cause ECG changes (prolonged P–R interval and widened QRS complex). Cardiac arrest can occur at 30 mEq/liter.[21]
- Magnesium sulfate may be given by intravenous (IV) infusion or direct IV, or by the intramuscular (IM) route to toxemic patients, because oral doses fail to produce satisfactory blood levels.
  1. The IV administration of $MgSO_4$ to prevent convulsions may consist of adding 10 Gm of $MgSO_4$ to 1000 ml of D_5W and administering the drug at a rate of 1 Gm/hr; if signs of impending convulsions persist, the rate may be increased to 2 Gm/hr.[22]
  2. The presence of convulsions may necessitate the administration of $MgSO_4$ by direct IV. One regimen calls for the administration of 4 Gm of 10% $MgSO_4$ IV over a 4 minute period, followed by a continuous maintenance infusion of 1 Gm of $MgSO_4$/hr in 75 ml of D_5W if maternal depression is not present.[23] (Sometimes the direct IV dose is followed by maintenance IM injections every 4 hr rather than by continued IV infusion of the drug.)

Nursing Considerations

- Intravenous infusion of $MgSO_4$ solutions requires the use of an infusion pump, close monitoring of the patient's response to the infusion, documentation of adequate renal function, and the ability to obtain serum magnesium levels quickly.
- Since IM doses of $MgSO_4$ are painful, they should be given by deep injection in the upper outer quadrant of the gluteus maximus; large doses should be equally divided between both buttocks. (Some

authors report adding Xylocaine to each dose of MgSO$_4$ for analgesia.)[24]

- After the initial dose of MgSO$_4$, *subsequent doses should be given only if*

  1. The patellar reflex is present—a poor to absent patellar reflex may be indicative of hypermagnesemia and should be reported to the physician.

  2. Urine flow has been 100 ml or more in the previous 4 hours—99% of the magnesium administered parenterally is excreted by the kidneys. Obviously, if renal function is impaired, magnesium excretion is decreased.

  3. There is no respiratory depression—as a rule, the physician should be notified if the respiratory rate decreases below 14 to 16 per minute.

- To counteract possible magnesium toxicity, an ampule of 10% calcium gluconate, with a 20-ml syringe, should be kept at the bedside. Calcium is an antidote for magnesium excess because calcium and magnesium are mutually antagonistic. The dosage for respiratory depression is 10 ml of a 10% calcium gluconate solution administered slowly IV.[25]

- Equipment for respiratory resuscitation should be available for emergency use.

- Be aware that MgSO$_4$ should be used cautiously in patients with impaired renal function and in those receiving digitalis; doses of other CNS depressants (such as narcotics and barbiturates) should be reduced when given in conjunction with MgSO$_4$.[26]

- Neonatal hypermagnesemia after maternal MgSO$_4$ administration is common, especially if MgSO$_4$ is given by continuous IV infusion to the mother for longer than 24 hours.[27]

Nursing Considerations in the Care of the Convulsing Patient

- Turn the convulsing patient on her side to avoid aspiration of vomitus and mucus.

- Insert a rubber airway or padded tongue blade between the teeth to prevent biting the tongue and to maintain an airway.

- Have suction apparatus on hand to aspirate fluid from the airway if necessary. Be aware that a chest film may be ordered after the convulsion to check for aspiration of fluid into the lungs.

- Administer oxygen, if necessary, by nasal cannula as soon as the convulsion ceases. Masks and catheters may produce excessive stimulation.[28]

- Use padded siderails to prevent the convulsing or heavily sedated patient from falling or otherwise injuring herself.

- Note the duration and character of each convulsion as well as the depth of coma that follows.

- Check fetal heart tones as often as possible.

- Monitor maternal blood pressure at frequent intervals.
- Observe the convulsing patient for signs of rapid labor; abruptio placentae and excessive bleeding may occur. (Typed and cross-matched blood should be available; remember that the toxemic patient tolerates blood loss poorly.) (See earlier discussion.)
- Be aware that an indwelling urinary catheter should be in place to allow hourly measurement of urine volume.
- Be aware that the patient with toxemia may convulse during or after labor, as well as before. (Half of all convulsions are reported to occur before labor, one-fourth during labor, and most of the remainder within 48 hours postpartum.)[29]
- Provide a quiet room with subdued light, since a stage of agitation is common after the patient regains consciousness.
- Be familiar with the safe use of $MgSO_4$ in the management of the convulsing or convulsion-prone patient. (See earlier discussion.)

References

1. Krupp M, Chatton M: Current Medical Diagnosis and Treatment, p 478. Los Altos, Lange Medical Publications, 1980
2. Rivlin M, Morrison J, Bates G (eds): Manual of Clinical Problems in Obstetrics and Gynecology, p 31. Boston, Little, Brown & Co, 1982
3. Krupp M, Chatton M, p 43
4. Rivlin M, Morrison J, Bates G, p 36
5. Ibid, p 31
6. Maxwell M, Kleeman C (eds): Clinical Disorders of Fluid and Electrolyte Metabolism, 3rd ed, p 1395. New York, McGraw-Hill, 1980
7. Rivlin M, Morrison J, Bates G, p 38
8. Ibid, p 31
9. Ibid
10. Maxwell M, Kleeman C, p 1393
11. Rivlin M, Morrison J, Bates G, p 32
12. Ibid, p 36
13. Maxwell M, Kleeman C, p 1393
14. Ibid
15. Pritchard J, MacDonald P: William's Obstetrics, 16th ed, p 690. New York, Appleton-Century-Crofts, 1980
16. Sehgal N, Hitt J: Plasma volume expansion in the treatment of preeclampsia. Am J Obstet 138:165, 1980
17. Rivlin M, Morrison J, Bates G, p 38
18. Pritchard J, MacDonald P, p 690
19. Rivlin M, Morrison J, Bates G, p 33
20. Ibid
21. Ibid
22. Ibid, p 32
23. Ibid, p 36

24. Ibid, p 32
25. Ibid, p 37
26. Gahart B: Intravenous Medications, 3rd ed, p 129. St Louis, CV Mosby, 1981
27. Maxwell M, Kleeman C, p 1112
28. Krupp M, Chatton M, p 478
29. Rivlin M, Morrison J, Bates G, p 36

chapter 27

Salicylate Intoxication

- Salicylate intoxication is commonly seen in children and accounts for 12% of all toxic ingestions.
  1. Leaving aspirin in the reach of the small child, particularly flavored aspirin, has resulted in salicylate poisoning. The recent use of safety caps, however, has decreased the occurrence of this problem.
  2. Half of the cases of salicylate intoxication in children are now due to parent or practitioner errors; that is, some children are sensitive to theoretically permissible doses.[1]
  3. Occasionally an older child may attempt suicide by aspirin ingestion.
- Of course, salicylate intoxication also occurs in adults exceeding therapeutic doses of aspirin.
- Salicylate intoxication can result from ingesting sodium salicylate tablets or from drinking methyl salicylate (oil of wintergreen).

Pathophysiology

- Salicylate poisoning results in a complex acid–base disturbance caused by the direct effects of salicylates on both respiration and metabolism.
- Initially, salicylates stimulate the medullary respiratory center and cause a marked hyperventilation with primary respiratory alkalosis.
- After 3 to 24 hours, sometimes less, another disturbance occurs as the salicylate disrupts carbohydrate metabolism and depletes liver glycogen. An accumulation of organic acids (ketones with lactic and pyruvic acids) occurs, resulting in primary metabolic acidosis.

- There is an age-related difference in susceptibility to metabolic derangements; respiratory alkalosis frequently dominates in adults, and metabolic acidosis usually occurs early and dominates in children. (Adults and older children can cope with the impaired metabolism of salicylate intoxication better than young children.)
- Hyperglycemia may be present. Hypokalemia often accompanies the respiratory alkalosis and is made more severe by potassium loss in vomitus. Also, a diminished ionized calcium fraction results with the alkalosis.
- Prothrombin and fibrinogen formation in the liver may be disrupted in the presence of salicylate intoxication, predisposing to bleeding.

Clinical Manifestations

- The patient with salicylate intoxication is usually conscious and gives a history of salicylate ingestion; this is helpful in differentiating between salicylate intoxication and diabetic ketoacidosis. (Note that there are marked similarities between the clinical manifestations of these two conditions.)
- Overt symptoms
  1. Hyperpnea
  2. Roaring in the ears
  3. Difficulty hearing
  4. Visual blurring
  5. Nausea and vomiting
  6. Fever
  7. Flushed appearance
  8. Sweating
  9. Numbness and tingling of the face, legs and extremities
  10. Confusion, excitement, disorientation or delirium
  11. Muscle twitching
  12. Lethargy progressing to coma and death
- Laboratory findings
  1. Serum salicylate level greater than 30 mg/100 ml
     (Some individuals may display toxicity at lower blood levels; others may exceed the toxic blood level and not display clinical signs of toxicity.)
  2. Plasma pH depends on which imbalance is most prominent
     - If respiratory alkalosis is more severe than metabolic acidosis, as is frequently the case in adults, the pH is elevated above normal.
     - If metabolic acidosis is more severe than respiratory alkalosis,

as is often the case in small children, the pH is decreased below normal.

- Sometimes the two imbalances neutralize each other and the pH remains normal. (See example below.)

3. Hypernatremia (may be present due to excessive water loss in hyperpnea, sweating, and vomiting).
4. Hypokalemia (associated with alkalosis and loss of K^+ in vomitus).
5. Either hyperglycemia or hypoglycemia.
6. Ketonemia and ketonuria.
7. White blood cell count may be elevated.

Treatment

- Emergency therapy is directed at removing as much of the salicylate from the stomach as possible before it is absorbed. Syrup of ipecac may be used to induce vomiting or the stomach may be lavaged with an isotonic solution of sodium chloride. Oral administration of an aqueous suspension of activated charcoal may also reduce salicylate absorption.
- After the stomach contents have been emptied, attention is given to supplying adequate fluids to promote excretion of salicylates by the kidneys, which account for the excretion of 80% of ingested salicylates.
- The amount of parenteral fluids required depends largely on the length of time elapsed since poisoning occurred. Some patients who have not received prompt therapy suffer severe fluid volume deficit (FVD), which must be repaired.
- Bicarbonate administration must be guided by careful study of acid–base parameters.
  1. During the respiratory alkalosis phase, D_5W solution may be given intravenously to promote renal excretion of salicylate. The free water helps replace the water vapor lost through hyperpnea; the glucose helps prevent excessive lowering of the glucose content of brain cells.[2]
  2. In the presence of metabolic acidosis, particularly when the pH is very low, sodium bicarbonate may be given intravenously to alkalinize the plasma and reduce salicylate toxicity. (Metabolic acidosis is associated with increased movement of salicylate ions out of the blood into the cells.)[3]
  3. When renal function is unimpaired, salicylate excretion can be promoted by alkalinization of the urine. This can be accomplished by the intravenous administration of bicarbonate-con-

taining fluids. However, this must be done cautiously since even a relatively small increase in the serum bicarbonate concentration in the presence of an overriding respiratory alkalosis could cause dangerous alkalosis (metabolic, plus respiratory). An attempt is made to keep the plasma pH no higher than 7.3 when bicarbonate is administered to avoid this problem.

- Potassium is administered to correct hypokalemia.
- Calcium gluconate can be given to relieve symptoms of ionized calcium deficit.
- Vitamin K-1 can be used to prevent excessive bleeding.
- Dialysis is indicated when potentially fatal serum salicylate levels exist (such as 70 mg/100 ml, or higher) or when oliguria or anuria are present (interfering with salicylate elimination).

Example of Mixed Imbalance in Patient With Toxic Salicylate Ingestion

pH = 7.4	(normal range = 7.35–7.45)
$PaCO_2$ = 18 mm Hg	(normal range = 38–42 mm Hg)
HCO_3^- = 11 mEq/liter	(normal range = 22–26 mEq/liter)
Base Excess (BE) = − 10 mEq/liter	(normal range = −2 to +2 mEq/liter)
Na^+ = 140 mEq/liter	(normal range = 135–145 mEq/liter)
Cl^- = 106 mEq/liter	(normal range = 100–106 mEq/liter)

- Note that the simultaneous appearance of respiratory alkalosis and metabolic acidosis has produced a normal pH. (This is not always the case; one imbalance may be stronger than the other and cause an abnormal pH in its favor.)
- If the only problem in this situation were respiratory alkalosis, the HCO_3^- would be expected to be approximately 20 mEq/liter. Note that it is lower than expected (indicating the presence of metabolic acidosis). Also, the negative BE indicates metabolic acidosis.
- If the only problem were metabolic acidosis, the $PaCO_2$ would be expected to be approximately 33 mEq/liter. Note that it is lower than expected (indicating respiratory alkalosis).
- This is a high Anion Gap metabolic acidosis, indicating the presence of excessive organic acids.

$$\text{Anion Gap} = Na^+ - (HCO_3^- + Cl^-) = 12\text{--}15 \text{ mEq/liter}$$
$$140 - (11 + 106) = 23 \text{ mEq/liter}$$

References

1. Kempe C, et al: Current Pediatric Diagnosis and Treatment, 6th ed, p 1000. Los Altos, Lange Medical Publications, 1980
2. Goldberger E: A Primer of Water, Electrolyte and Acid–Base Syndromes, 6th ed, p 262. Philadelphia, Lea & Febiger, 1980
3. Ibid

Glossary

Acidosis Blood pH below normal (less than 7.38 in arterial blood). Acidosis can be either metabolic (bicarbonate deficit) or respiratory (carbonic acid excess).

Aldosterone A hormone secreted from the adrenal cortex; the principal mineralocorticoid. Aldosterone is instrumental in the regulation of body sodium and potassium content.

Alkalosis Blood pH above normal (greater than 7.42 in arterial blood). Alkalosis can be metabolic (bicarbonate excess) or respiratory (carbonic acid deficit).

Anion A negatively charged ion (*e.g.*, Cl^- or HCO_3^-)

Anion Gap The anion gap is made up of unmeasured anions in plasma, such as sulfates and the anions of organic acids (*e.g.*, ketones and lactic acid). It is measured in patients with metabolic acidosis to help determine the cause of the acidosis. For example, a high-anion-gap acidosis accompanies ketoacidosis and lactic acidosis. A normal-anion-gap acidosis occurs with diarrhea.

Antidiuretic hormone (ADH) A hormone secreted from the pituitary mechanism that causes the kidney to conserve water. It is sometimes referred to as the "water-conserving hormone."

"Balanced" electrolyte solutions Parenteral solutions containing proportions of sodium and chloride similar to those of plasma. The solutions also contain bicarbonate (or a precursor, such as lactate or acetate). Some contain small amounts of potassium, calcium, magnesium, and other electrolytes to make the solution more similar to plasma (*e.g.*, lactated Ringer's solution).

Base excess A metabolic (nonrespiratory) body disturbance that may be primary or compensatory. The normal range is -2 to $+2$ mEq/liter. It is always positive in metabolic alkalosis and always negative in metabolic acidosis. The laboratory arrives at this figure by multiplying the deviation of standard bicarbonate from normal by a factor of 1.2, which represents the buffer action of red blood cells.

Carbon dioxide (CO$_2$) content   A laboratory test that measures total bicarbonate (HCO$_3$) and carbonic acid (H$_2$CO$_3$) in plasma

Cation   A positively charged ion (*e.g.*, Na$^+$, K$^+$, Ca^{++} and Mg^{++})

Chvostek's sign   A sign elicited by tapping the facial nerve about 2 cm anterior to the earlobe, just below the zygomatic process. The response is a spasm of the muscles supplied by the facial nerve.

Colloid   A substance that does not dissolve into a true solution and is not capable of passing through a semipermeable membrane (*e.g.*, blood, plasma, albumin, and dextran). The physical opposite of a colloid is a crystalloid.

Crystalloid   A substance that forms a true solution and therefore is capable of passing through a semipermeable membrane (*e.g.*, lactated Ringer's solution and isotonic saline). The physical opposite of a crystalloid is a colloid.

Dehydration   A deficit of body water

Diabetes insipidus   A disease in which there is a deficiency of antidiuretic hormone, leading to severe polyuria. It is sometimes abbreviated as DI, and it may be either neurogenic (central) or nephrogenic.

Electrolyte   A substance that ionizes (develops an electrical charge) when dissolved in water (*e.g.*, sodium and potassium)

Extracellular fluid   Body fluid located outside the cells. It consists of two types: interstitial (tissue fluid) and intravascular (plasma).

Fluid volume deficit   Decreased body fluid volume. The term usually refers to a deficit of extracellular fluid.

Fluid volume excess   Increased body fluid volume. The term usually refers to an excess of extracellular fluid.

Homeostasis   The ability to restore equilibrium under stress

Hypercalcemia   An excess of calcium (Ca) in the blood

Hyperchloremia   An excess of chloride (Cl) in the blood

Hyperkalemia   An excess of potassium (K) in the blood

Hypermagnesemia   An excess of magnesium (Mg) in the blood

Hypernatremia   An excess of sodium (Na) in the blood

Hyperosmolal fluid   A fluid having an osmolality greater than that with which it is compared (usually plasma)

Hyperphosphatemia   An excess of phosphate in the blood

Hypertonic solution   A solution more concentrated than that with which it is compared (*e.g.*, a fluid having a concentration greater than 0.9%, the normal tonicity of body fluids)

Hypocalcemia   A low calcium (Ca) concentration in the blood

Hypochloremia   A low chloride (Cl) concentration in the blood

Hypokalemia   A low potassium (K) concentration in the blood

Hypomagnesemia   A low magnesium (Mg) concentration in the blood

Hyponatremia   A low sodium (Na) concentration in the blood

Hypo-osmolal fluid   A fluid having a lower osmolality than that to which it is compared (usually plasma)

Hypophosphatemia A low phosphate concentration in the blood

Hypotonic solution A solution less concentrated than that with which it is compared (*e.g.*, a fluid having a concentration less than 0.9%, the normal tonicity of body fluids)

Hypovolemia A decrease in the fluid volume in the vascular compartment

Insensible fluid loss Water lost from the lungs and skin (nonvisible perspiration)

Interstitial fluid Fluid between the cells (part of the extracellular fluid)

Intracellular fluid Fluid inside the cells

Intravascular fluid The fluid portion of blood (plasma). It is part of the extracellular fluid.

Ion An electrically charged atom or group of atoms

Iso-osmolal fluid A fluid having the same osmolality as the fluid with which it is compared (usually plasma)

Isotonic fluid A fluid having the same concentration as that with which it is compared

Jejunostomy A surgical opening into the jejunum made through the abdominal wall

Jejunum The second portion of the small intestine, located between the duodenum and ileum

Ketone bodies Substances formed by the liver as a step in the combustion of fats. They are acetone, β-hydroxybutyric acid, and acetoacetic acid

Ketosis A condition in which there is an excessive accumulation of ketone bodies in the body

Kilogram (kg) A metric unit of weight equal to 1000 g and roughly equivalent to 2.2 pounds

Liter A metric unit of volume equal to 1000 ml and roughly equivalent to 1 quart

Milliequivalent (mEq) A unit of chemical activity. One milliequivalent equals the activity of 1 mg of hydrogen

Osmolality A measure of solute concentration. The term refers to the number of osmols per kilogram of water; hence the total volume of a solution is 1 liter of water plus the small volume occupied by the solutes. It is measured in mOsm/kg water

Osmolarity A measure of solute concentration. The term refers to the number of osmols per liter of solution; hence the volume of water is less than 1 liter by the amount equal to the solute volume. It is measured in mOsm/liter

Osmole The standard unit of osmotic pressure

Osmosis Movement of water molecules from an area of lesser solute concentration to an area of higher solute concentration

Osmotic pressure The drawing power for water, determined by the number of particles per unit volume

PaCO$_2$ The partial pressure of carbon dioxide in arterial blood

PaO$_2$ The partial pressure of oxygen in arterial blood *p*H Hydrogen ion (H^+) concentration. Acidity increases as H^+ increases. Because of an inverse relationship between PH and H^+ concentration, the PH value decreases as acidity increases

Plasma The liquid portion of blood in which the blood cells are suspended

Precursor An agent that normally precedes or may develop into another

Sensible perspiration Visible perspiration (as opposed to insensible, or invisible, perspiration through the skin)

Serum Plasma from which fibrinogen has been separated in the process of clotting

SIADH Syndrome of inappropriate antidiuretic hormone (ADH) secretion, a condition in which excessive ADH is secreted, resulting in hyponatremia

Solute The substance that is dissolved in a liquid to form a solution

Specific gravity The weight of a substance divided by the weight of an equal volume of water

Spironolactone An aldosterone-blocking agent that produces diuresis

Standard bicarbonate A bicarbonate concentration in the plasma of blood that has been equilibrated at a $PaCO_2$ of 40 mm Hg, and with O_2, in order to saturate the hemoglobin fully

Tetany Continuous tonic spasm of a muscle

Trousseau's sign A spasm of the hand elicited when the blood supply to the hand is decreased or nerve of the hand is stimulated by pressure. It is elicited within several minutes by applying a blood-pressure cuff inflated above systolic pressure (see Fig. 9-2 in Chap. 9)

Index

An *f* following a page number represents a figure; a *t* indicates tabular material. Numbers in **boldface** represent glossary definitions.

acid–base balance, 141–162
  regulation of, 141–145, 142t, 144f
acid–base imbalance, 145–162. *See also individual types*
  mixed, 156–159
acidosis, **377**
  high anion gap, 146
  metabolic, 145–148, 147t. *See also metabolic acidosis*
  normal anion gap, 146
  respiratory, 150–154, 153t. *See also respiratory acidosis*
acquired immune deficiency syndrome (AIDS), blood transfusion and, 219–220
adrenal gland, homeostasis and, 11–12
adrenal insufficiency, surgery and, 192
aged, water and electrolyte homeostasis in, 13–16
AIDS (acquired immune deficiency syndrome), blood transfusion and, 219–220
air embolism, in intravenous fluid administration, 75–77, 76f
albumin
  in hypovolemic shock, 204
  serum, 51t
    serum calcium and, 116
alcohol, serum, 51t
alcohol solution, parenteral, 71
alcoholism, magnesium deficit in, 128
aldosterone, **377**
  function of, 11–12
  potassium balance and, 103

alkalosis, **377**
  metabolic, 148–150, 149t, 223
  respiratory, 154–156, 156t
allergic reaction
  to blood administration, 215–216
  to drugs, 191
ammonia formation, in cirrhosis of liver, 244–245
ammonium chloride solution, 63t
anabolism, 188
anemia, in acute renal failure, 305
anion, **377**
anion gap, 52t, **377**
anion gap acidosis, 146
antidiuretic hormone (ADH), 11, **377**. *See also* inappropriate secretion of ADH
arterial blood gas samples, 159–162. *See also* blood gas values
arterial puncture, blood sample from, 160–161
ascites, in cirrhosis of liver, 237
aspiration pneumonia, 177–178
assessment
  clinical, 28–45
  nursing, 20–45
azotemia, 175–176

barbiturate therapy, in increased intracranial pressure, 290–291
base excess, 142t, **377–378**
  blood gas values of, 55t

bicarbonate
  blood gas values of, 55t
  replacement, in diabetic ketoacidosis,
      263–264
  standard, **380**
bicarbonate–carbonic acid ratio, 144f
bile, loss of, 27
blood
  citrated, 212–214
    hypocalcemia and, 212–213
    *p*H and, 213–214
  laboratory values for, 47t–52t
blood administration, 209–220
  acquired immune deficiency syn-
      drome (AIDS) and, 219–220
  acute hemolytic reaction and, 218–
      219
  allergic reactions in, 215–216
  bacterial contamination in, 216
  circulatory overload with, 214–215
  citrated blood in, 212–214
  coagulation problems and, 217
  complications of, 211–220
  2,3 diphosphoglycerate in, 214
  febrile reactions to, 215
  hepatitis and, 217
  hyperammonemia in, 214
  hypothermia and, 216
  in hypovolemic shock, 204
  safe, rules for, 209–211
  stored blood and potassium excess
      and, 211
blood gas values, 54t–55t, 142t
  in metabolic acidosis, 147t
  in metabolic alkalosis, 149t
  obtaining samples for, 159–162
  in respiratory acidosis, 153t
  systematic interpretation of, 157–158
blood glucose determination, Ames
      Glucometer for, 268f
blood pressure
  in aged, 15
  assessment of, 39
blood urea nitrogen, 47t
blood vessel, homeostasis and, 10
body fluid. *See also* fluid *entries*
  in aged, 13–16
  amount and composition of, 3–4, 4t
  body weight in relation to age and
      sex and, 4t
  electrolytes as, 7–8, 7t–8t. *See also*
      electrolytes
  homeostatic mechanisms and, 8–13,
      9f
  in infants and children, 16–18
  location of, 4–5, 6f

  regulation of, 3–19
  routes of loss of, 5–7
body surface area
  conversion table for, 59t
  in infant, 17–18
body temperature, assessment of, 37–
      38
body weight
  assessment of, 37
  in diabetic ketoacidosis, 265–266
bone changes, in chronic renal failure,
      310–311
buffer, chemical, 141–143
BUN, 47t
BUN:creatinine ratio, 48t
burns, 331–352
  depth of, 331–333, 332t
  gastrointestinal complications in, 349
  pain control in, 349–350
  parenteral fluid therapy in, 351
  percentage of surface involved in,
      333
  pulmonary edema in, 351
  respiratory, 341–342
  rule of nines in, 333
  severity of, 331–333
  shock in, 340–341
  treatment of, 335–351
    early, 335
    in fluid accumulation phase, 335–
        350
    in fluid remobilization phase, 350–
        351
    initial patient evaluation in, 335,
        340
    intravenous fluid replacement in,
        342–348, 345t
      formulas for, 344, 345t
      kinds of, 343–344
      monitoring, 344–348
    oral electrolyte solutions in, 348–
        349
  urinary output in, 341, 350–351
  water and electrolyte changes in,
      333–335, 336t–339t

calcitonin, calcium imbalance and, 117
calcium
  serum, 49t
  urine, 53t
calcium deficit, 117–122, 120f
  in acute renal failure, 301–302
  citrated blood and, 212–213

calcium excess, 122–125
calcium imbalance, 116–125
  in chronic renal failure, 310–311
calcium solution, nursing consider-
    ations in administration of, 66
carbohydrate solution, parenteral, 67–
    68, 68t
carbon dioxide, serum, 50t
carbon dioxide content, **378**
cardiogenic shock, 208
catabolism, reduction of, 299
cathartics, 229
cation, **378**
cation exchange resins, for potassium
    excretion, 298–299
central venous pressure, assessment of,
    40–44, 44f
cerebral edema, 286–293. *See also* intra-
    cranial pressure, elevated
cervical venous pressure, estimation of,
    39
chemical buffers, 141–143
children, water and electrolyte homeo-
    stasis in, 16–18
chloride
  serum, 50t
  urine, 53t
Chvostek's sign, 44–45, 119, **378**
circulatory overload
  in blood administration, 214–215
  in intravenous fluid administration,
    74–75
cirrhosis of liver, 237–247. *See also* liver,
    cirrhosis of
clean-air center for mixing IV solu-
    tions, 167f
clinical assessment, 28–45
colloid, **378**
colon studies, 230
congestive heart failure, 315–330
  clinical manifestations of, 320t–321t
  decreased left ventricular function in,
    adaptive mechanisms in, 316–
    317
  decreasing myocardial workload in,
    322–323
  decreasing venous return in, 323–
    326
  digitalis in, 326–328
  diuretics in, 325–326
  fluid restriction in, 326
  hemodynamic monitoring in, 318,
    322
  increasing myocardial contractility
    and cardiac output in, 326–
    328
    left-sided, 315–316
    nursing assessment in, 318, 322
    pulmonary edema in, 328–330
    right-sided, 316
    sodium-restricted diets in, 323–325
    treatment of, 322–328
    water and electrolyte disturbances in,
      319t
copper deficiency, in total parenteral
    nutrition, 171
cough, postoperative, assisting patient
    in, 199–200
creatinine, serum, 47
crystalloid, **378**

Davol Uri-Meter drainage bag, 32f
dehydration, **378**
dextran, in hypovolemic shock, 204
diabetes insipidus, **378**
  central, 281–284
    clinical manifestations of, 282
    nursing measures for detection of,
      283–284
    treatment of, 283
    water and electrolyte disturbances
      in, 282–283
diabetes mellitus, surgery and, 192
diabetic ketoacidosis, 253–271
  body weight in, 265–266
  clinical manifestations of, 257t–259t
  electrolytes in, 254–256, 270
  fluid volume deficit in, monitoring
    degree of, 264–266
  hyperglycemia and hyperosmolality
    in, 254
    determining degree of, 266–269,
      267t, 268f
  hypoglycemic coma vs., 257
  ketosis in, 256–257
    degree of, 269–270
  metabolic consequences of, 254–257
  nursing assessment in response to
    therapy in, 264–270
  osmotic diuresis and fluid volume
    deficit in, 254
  patient education in, 271
  plasma osmolality in, 266
  treatment of, 260–264
    electrolyte replacement in, 262–
      264
    fluid replacement in, 260–261
    insulin administration in, 261–262
urine output in, 265
vital signs in, 264–265

diarrhea, 226–228
 fluid imbalance with, 27
 in tube feeding, 178–179
digitalis, in congestive heart failure,
 326–328
disease states, fluid imbalance and,
 22t–23t
diuretic therapy
 in cirrhosis of liver, 242
 in elevated intracranial pressure,
 288–289
drug allergy or idiosyncrasy, 191
duodenal replacement solution, 63t

edema
 assessment of, 35, 36f
 cerebral, 286–293. *See also* intracranial pressure, elevated
 pulmonary, 328–330, 351
elderly. *See* aged
electrolytes, 7–8, 7t–8t, **378**
 in aged, 13–16
 in gastrointestinal fluid, 222t
 in infants and children, 16–18
 plasma, 7, 7t
 serum, 49t–50t
 urine, 52t–53t
electrolyte no. 48, 64t
electrolyte no. 75, 64t
electrolyte solution, 60–67, 61t–64t. *See*
 also individual solutions
 balanced, **377**
 in hypovolemic shock, 203–204
embolism, air, in intravenous fluid administration, 75–77, 76f
endocrine response
 in aged, 16
 to stress, 186–188, 187f
enemas, prolonged use of, 229
extracellular fluid, **378**
 electrolytes in, 7

facial appearance, in fluid volume deficit, 34, 35f
fat emulsion, parenteral, 69–71
febrile reaction, in blood administration, 215
Fenwal dry heat blood warmer, 212f
fistula, intestinal, 228–229
Flexiflo enteral nutrition pump, 184f
Flexiflo gravity gavage set, 183f

Flexitainer enteral nutrition container,
 183f
fluid challenge test, 83
fluid intake, postoperative, 195–196
fluid loss, insensible, **379**
fluid therapy
 in diabetic ketoacidosis, 260–261
 in intestinal obstruction, 233–236
 operative, 193
 preoperative, 193
fluid volume deficit, 81–83, **378**
 in diabetic ketoacidosis, 254, 264–
 266
 in diarrhea, 226
 in infant, 5
 in intestinal obstruction, 232–233
 in vomiting, 223
fluid volume excess, 84–85, **378**
 in acute renal failure, 299–300
 postoperative, 197
fluid volume imbalance, 81–85. *See also*
 body fluid

gastric distension, in tube feeding, 180
gastric juice, loss of, 21
gastric replacement solution, 63t
gastric suction, 224–226
gastritis, hemorrhagic, in burn, 349
gastrointestinal circulation, 221
gastrointestinal fluid
 characteristics of, 221, 222t
 electrolyte composition of, 222t
 *p*H of, 222t
gastrointestinal fluid loss, 27, 221–230
 colon studies and, 230
 in diarrhea, intestinal suction, ileostomy, 226–228
 fistula in, 228–229
 gastrointestinal tract in, 7
 laxatives and enemas in, 229
 postoperative, 200
 in vomiting and gastric suction, 222–
 226
gastrointestinal function, in aged, 15
globulin, serum, 51t
glucose, serum, 49t. *See also* hyperglycemia; hypoglycemia

hand vein, assessment of, 39–40, 41f–
 42f
Hartmann's solution, 62t

head injury, 278–294
    cerebral edema and elevated intracranial pressure in, 286–293
    fluid balance problems in, 278–279
    hypernatremia in, 281–286
    hyponatremia in, 279–281
    nonketotic hyperosmolar coma in, 286
heart disease, surgery and, 192
heart function
    in aged, 14–15
    calcium balance and, 117, 118f, 124
    homeostasis and, 10
    magnesium deficit and, 129
    potassium balance and, 104, 105f
heat cramps, 354
heat disorder, 353–364
heat exhaustion, 354–355
heatstroke, 355–361
    body temperature reduction in, 359–360
    characteristics of, 357–358
    diuretics in, 360
    emergency measures in, 361
    fluid administration in, 360
    prevention of, 361–363
    rest in, 361
    risk factors in, 356–357
    treatment of, 358–361
    types of, 356
hematocrit, 48t
hemolytic blood reaction, 218–219
hepatic flap, 238
hepatitis, blood administration and, 217
history, patient, 20–28, 22t–26t
homeostasis, **378**
    in aged, 13–16
    in infants and children, 16–18
    mechanisms of, 8–13, 9f
hyperammonemia, blood administration and, 214
hypercalcemia, 122–125, **378**
hyperchloremia, **378**
hyperglycemia
    in diabetic ketoacidosis, 254, 266–269, 267t, 268f
    in total parenteral nutrition, 167–170
hyperkalemia, 111–115, **378**. See also potassium excess
hypermagnesemia, 130–132, 303, **378**
hypernatremia, 100–102, **378**. See also sodium excess
hyperosmolal fluid, **378**
hyperosmolality, in diabetic ketoacidosis, 254, 266–269, 267t, 268f

hyperosmolar syndrome, in total parenteral nutrition, 167–170
hyperphosphatemia, 135, **378**
hypertonic solution, **378**
hyperventilation syndrome, 155
hypo-osmolal fluid, **378**
hypocalcemia, 117–122, 120f, **378**. See also also calcium deficit
hypochloremia, **378**
hypoglycemia, post-infusion, in total parenteral nutrition, 170
hypoglycemic coma, diabetic coma vs., 257
hypokalemia, 106–111, 109t, **378**. See also potassium deficit
hypomagnesemia, 128–130, 171, 223–224, **378**
hyponatremia, 90–92, **378**. See also sodium deficit
hypoparathyroidism, calcium deficit and, 117–118
hypophosphatemia, 134–135, **379**
    in total parenteral nutrition, 171
hypothermia, blood administration and, 216
hypotonic solution, **379**
hypovolemia, **379**
hypovolemic shock, 201–205
    blood in, 204
    clinical signs of, 201–202
    dextran in, 204
    electrolyte solutions in, 203–204
    fluid replacement in, 202–204
    oxygen administration in, 205
    plasma and albumin in, 204
    positioning in, 202

iatrogenic fluid imbalance, 24t–26t
ileostomy, 226–228
ileus
    postoperative, 200–201
    reflex, in burn, 349
immobilization, hypercalcemia in, 123
inappropriate secretion of ADH, 92–99, 279–281. See also antidiuretic hormone
    clinical manifestations of, 95–97
    drugs favoring, 95
    in head injured patient, 279–281
    nursing measures to detect, 97
    nursing responsibilities during therapy for, 99
    pathophysiology of, 92–93, 93f

inappropriate secretion of ADH, *(continued)*
  patients at risk for, 93–94
  treatment of, 98
indwelling arterial catheter, blood sample from, 161–162
infant
  extracellular fluid volume of, 5
  water and electrolyte homeostasis in, 16–18
infection
  in acute renal failure, 305–306
  in total parenteral nutrition, 172–173
infiltration, in intravenous fluid administration, 73
insulin therapy, in diabetic ketoacidosis, 261–262
intake–output measurement, 29–30, 29t
  postoperative, 193–194
interstitial fluid, **379**
intestinal obstruction, 231–236
  fluid and electrolyte imbalances in, 232–233
  fluid replacement therapy in, 233–236
  nursing assessment in, 234–236
  simple mechanical, 231–232
  types of, 231
intestinal suction, 226–228
intracellular fluid, **379**
  electrolytes in, 8t
intracranial pressure, elevated, 286–293
  barbiturate therapy in, 290–291
  body position in, 290
  clinical signs of, 287
  controlled ventilation in, 289–290
  fluid therapy in, 288
  nonosmotic diuretics in, 289
  osmotic diuretics in, 288–289
  pathophysiology of, 286–287
  steroid therapy in, 291–292
  treatment and nursing considerations in, 287–293
intravascular fluid, **379**
intravenous fluid replacement, in burn, 342–348, 345t
ion, **379**
iso-osmolal fluid, **379**
Isolyte E solution, 62t
isotonic fluid, **379**

jejunostomy, **379**
jejunum, **379**
jugular vein, assessment of, 39, 40f

ketoacidosis, during illness, patient instruction sheet for, 272f–273f
ketone, serum, 51t
ketone bodies, **379**
ketosis, 269–270, **379**
  in diabetic ketoacidosis, 256–257
kidney
  in acid–base balance, 143
  in aged, 13–14
  fluid loss and, 5
  homeostasis and, 8–10
kilogram (kg), **379**

laboratory data evaluation, 46–56
lactate, serum, 51t
lactated Ringer's solution, 62t
laxatives, prolonged use of, 229
left ventricular function, decreased, 316–317
liter, **379**
lithium
  serum, 51t
  toxicity of, 136–138
liver, cirrhosis of, 237–247
  activity limitation in, 243
  ammonia buildup in, 244–245
  ascites in, 237
  clinical manifestations of, 238
  diuretic therapy in, 242
  medical therapy in, 242–247
  nursing assessment in, 245–247
  paracentesis in, 243–244
  pathophysiology of, 237
  peritoneovenous shunt in, 244
  sodium restriction in, 242
  water and electrolyte disturbances in, 238, 239t–241t
  water restriction in, 243
lung
  in acid–base balance, 143–145
  in aged, 14
  homeostasis and, 10–11
  water loss and, 6

magnesium, serum, 49t
magnesium, deficit, 128–130
  in vomiting and gastric suction, 223–224
magnesium excess, 130–132
  in acute renal failure, 303
magnesium balance, 127–132

magnesium solution, nursing considerations in administration of, 66–67
magnesium sulfate
   intravenous, 130
   in toxemia of pregnancy, 369–370
malignancy, hypercalcemia in, 124–125
medication, postoperative period and, 190–191
metabolic acidosis, 145–148, 147t
   in acute renal failure, 300–301
   in chronic renal failure, 309
   in diarrhea, 226
   in infant, 18
metabolic alkalosis, 148–150, 149t
   in vomiting and gastric suction, 223
milliequivalent (mEq), **379**

nausea, in tube feeding, 180
neck vein, assessment of, 38, 39, 40f
neurogenic shock, 205–206
neurologic change, in aged, 15–16
neuromuscular irritability, assessment of, 44–45
nonketotic hyperosmolar coma, 271–276
   clinical manifestations in, 274–275
   in head injury, 286
   pathophysiology in, 271
   predisposing factors in, 274
   prevention of, 276
   treatment of, 275–276
   in tube feeding, 180
nursing assessment, 20–45
   clinical assessment in, 28–45
   patient history in, 20–28, 22t–26f
nutrition, postoperative period and, 189–190

oral cavity, moisture in, 33
oral electrolyte solution, in burn therapy, 348–349
osmolality, **379**
   vs. osmolarity, 55t
   serum, 48t
   urine, 53t
osmolarity, **379**
   vs. osmolality, 55t
osmole, **379**
osmosis, **379**
osmotic pressure, **379**

osteoporosis, postmenopausal, 121–122
oxygen therapy
   in chronic respiratory acidosis, 153
   in hypovolemic shock, 205
oxytocin, water intoxication with, 98–100

$PaCO_2$, 55t, 142t, **379**
pancreatic juice, loss of, 27
pancreatitis, acute, 247–252
   clinical signs of, 247–248
   fluid and electrolyte problems in, 248–249
   medical management in, 249–250
   nursing considerations in, 250–252
PaO_2, 55t, 142t, **379**
paracentesis, in cirrhosis of liver, 243–244
parathyroid
   calcium imbalance and, 117
   homeostasis and, 12–13
parenteral fluid therapy, 57–78. *See also* total parenteral nutrition
   in burns, 351
   calculation of fluid needs in, 58–60, 59t
   clinical status assessment in, 60
   complications of, 72–77, 76f
   factors affecting fluid orders in, 57–60, 59t
   nutrients in, 67–72
   renal function and, 58
   water and electrolyte solutions for, 60–67, 61t–64t
patient history, 20–28, 22t–26t
peritoneovenous shunt, 244
perspiration
   fluid imbalance and, 27–28
   insensible, 5–6
   sensible, **380**
   visible, 5
*p*H, **380**
   arterial blood, 54t
   bicarbonate–carbonic acid ratio and, 144f
   calcium balance and, 116–117
   citrated blood and, 213–214
   of gastrointestinal fluid, 222t
   potassium balance and, 104–106
   urinary, 54t
phosphate administration, in nonketotic hyperosmolar coma, 276
phosphorus, serum, 50t
phosphorus administration, in diabetic ketoacidosis, 263

phosphorus deficiency, 134–135
  in diabetic ketoacidosis, 255–256,
    270
phosphorus excess, 135
  in acute renal failure, 301–302
phosphorus imbalance, 133–135
  in chronic renal failure, 310–311
pitting edema, 35, 36f
pituitary gland, homeostasis and, 11
plasma, **380**
  electrolytes in, 7t
  in hypovolemic shock, 204
  osmolality of, in diabetic
    ketoacidosis, 266
Plasma-Lyte solution, 62t
pneumonia, aspiration, 177–178
postoperative period, 186–208
  chronic illness and, 191–192
  diminished respiration in, 197–200
  drug allergies or idiosyncrasies in,
    191
  fluid intake in, 195–196
  fluid volume excess in, 107
  gastrointestinal fluid loss in, 200
  ileus in, 200–201
  intake–output measurement in, 193–
    194
  medications and, 190–191
  nutritional status and, 189–190
  potassium replacement in, 195
  prevention of complications in, 189–
    193
  respiratory acidosis in, 197–200
  steroids and, 190–191
  urinary output in, 194–195
  water and electrolyte gains and losses
    in, 193–196
  water and electrolyte problems in,
    196–201
  water intoxication in, 196–197
potassium
  postoperative, 195
  serum, 49t
  urine, 53t
potassium chloride solution, 62t
potassium deficit, 106–111, 109t
  in diabetic ketoacidosis, 255
  in diarrhea, 226–227
  in vomiting and gastric suction, 223
  in total parenteral nutrition, 171
potassium excess, 111–114
  in acute renal failure, 296–299
  cation exchange resins in, 298–299
  reducing catabolism in, 299
  stored blood and, 211
potassium imbalance, 103–115
  in chronic renal failure, 308–309
  in diabetic ketoacidosis, 270

potassium substitutes, 109t
potassium supplement
  in diabetic ketoacidosis, 262–263
  in nonketotic hyperosmolar coma,
    276
  nursing considerations in administra-
    tion of, 64–66, 111
precursor, **380**
pregnancy, toxemia of, 365–372
preoperative period, fluid therapy in,
    193
protein, serum, 50t
protein solution, parenteral, 68–69
pulmonary disease, surgery and, 192
pulmonary edema, 328–330, 351
pulmonary ventilation, nursing actions
    to improve, 154
pulse, assessment of, 38
pyrogenic reaction, in intravenous fluid
    administration, 72–73

renal failure, 295–314
  acute, 295–307
    anemia in, 305
    calcium and phosphorus derange-
      ments in, 301–302
    diuretic phase of, 306–307
    extracellular fluid volume excess
      in, 299–300
    hyperkalemia in, 296–299
    infection in, 305–306
    magnesium excess in, 303
    metabolic acidosis in, 300–301
    oliguric phase of, 296–306
    sodium balance in, 302–303
    uremia in, 303–304
    urine changes in, 306
  chronic, 307–313
    bone changes in, 310–311
    calcium and phosphorus derange-
      ments in, 310–311
    classification of, 308
    fluid and dietary implications of,
      312–313
    metabolic acidosis in, 309
    potassium imbalance in, 308–309
    sodium balance in, 309–310
    surgery and, 192
    uremia in, 311–312
renal function
  in aged, 13–14
  in infants, 17
  parenteral fluid therapy and, 58
respiration
  assessment of, 38

diminished, postoperative, 197–200
respiratory acidosis, 150–154, 153t
  postoperative, 197–200
respiratory alkalosis, 154–156, 156t
respiratory burn, 341–342
rule of nines, 333

salicylate, serum, 51t
salicylate intoxication, 373–376
saline
  3% and 5%, 61t
  half-strength, 61t
  isotonic, 61t
  third-strength, 61t
salt substitutes, 109t
septic shock, 208–209
serum, **380**
serum values, 47t–52t
shock, 201–208
  burn, 340–341
  cardiogenic, 208
  hypovolemic, 201–205
  neurogenic, 205–206
  septic, 206–207
SIADH, **380**. *See also* inappropriate se-
    cretion of ADH
skin, fluid loss and, 5–6
skin change, in aged, 15
skin temperature, 34
skin turgor, 33
sodium
  serum, 49t
  urine, 52t
sodium balance, 90–102
  in acute renal failure, 302–303
  in chronic renal failure, 309–310
  lithium and, 136
sodium bicarbonate solution, 63t
sodium deficit, 90–92
  in diabetic ketoacidosis, 254–255
  in diarrhea, 227
  in SIADH, 279–281
  in tube feeding, 176–177
  in vomiting and gastric suction, 223
sodium excess, 100–102
  causes of, 284–285
  clinical signs of, 285
  in diabetes insipidus, 281–284
  treatment of, 285–286
  in tube feeding, 174–176
sodium lactate solution, 63t
sodium restriction
  in cirrhosis of liver, 242
  in congestive heart failure, 323–325

sodium therapy, in diabetic
    ketoacidosis, 263
solute, **380**
specific gravity, **380**
  urine, 53t–54t
spironolactone, **380**
starvation effect, 189
steroids
  in elevated intracranial pressure,
      291–292
  postoperative period and, 190–191
stress reaction, 186–188, 187f
surgical trauma, 186–189, 187f
  endocrine response (stress reaction)
      to, 186–188, 187f
  fluid therapy and, 193
  starvation effect of, 189
  tissue injury in, 188
sweat, 5
  fluid imbalance and, 27–28

temperature, body, assessment of, 37–
    38
tendon, deep, reflexes of, 45
tetany, **380**
third-space fluid shift, 86–89
  clinical situations in, 86–87
  fluid accumulation in, 87–88
  phases of, 87–89
  shift back to vascular space in, 89
thirst
  in aged, 16
  assessment of, 34
thrombophlebitis, in intravenous fluid
    administration, 74
tissue injury, 188
tongue turgor, 33
total parenteral nutrition, 67–72, 165–
    173. *See also* parenteral fluid
    therapy
  clean-air center for mixing IV solu-
      tions for, 167f
  composition of solutions for, 165–
      166
  electrolyte disturbances in, 170–171
  hyperglycemia and hyperosmolar
      syndrome in, 167–170
  infectious complications of, 172–173
  metabolic complications of, 167–172
  post-infusion hypoglycemia in, 170
  preparation and storage of solutions
      for, 166–167, 167f
toxemia of pregnancy, 365–372
  clinical manifestations of, 367–368
  magnesium sulfate in, 369–370

_mia of pregnancy, *(continued)*
   pathophysiology of, 366
   treatment and nursing implications
      of, 368–371
treatment-induced fluid imbalance,
   24t–26t
Trousseau's sign, 119, 120f, **380**
tube feeding, 174–185
   administration of, 180–182, 181f
   aspiration pneumonia in, 177–178
   commercial, osmolality of, 175t
   complications of, 174–180
   diarrhea in, 178–179
   equipment for, 182–183, 183f–184f
   hypernatremia in, 174–176, 284
   hyponatremia in, 176–177
   nausea and gastric distension in, 180
   nonketotic hyperosmolar syndrome
      in, 180
turgor, skin and tongue, 33

uremia
   in acute renal failure, 303–304
   in chronic renal failure, 311–312
urine
   laboratory values for, 52t–55t

volume and concentration of, 31–32,
   32f
urine output
   in diabetic ketoacidosis, 265
   postoperative, 194–195
urine tests, for sugar content, 267t

ventilation, mechanical, hypernatremia
   in, 284
vitamin, parenteral, 71–72
vomiting, 21, 222–224

water intoxication, postoperative, 196–
   197

zinc, serum, 50t
zinc deficiency, 139
   in total parenteral nutrition, 171
zinc excess, 139–140
zinc imbalance, 138–140